The Definitive Guide to the ARM Cortex-M3

Second Edition

The Definitive Guide to the ARM Cortex-M3

Second Edition

Joseph Yiu

AMSTERDAM • BOSTON • HEIDELBERG • LONDON
NEW YORK • OXFORD • PARIS • SAN DIEGO
SAN FRANCISCO • SINGAPORE • SYDNEY • TOKYO

Newnes is an imprint of Elsevier

ELSEVIER

Newnes

Newnes is an imprint of Elsevier
30 Corporate Drive, Suite 400
Burlington, MA 01803, USA

The Boulevard, Langford Lane
Kidlington, Oxford, OX5 1GB, UK

Notices

Knowledge and best practice in this field are constantly changing. As new research and experience broaden our understanding, changes in research methods, professional practices, or medical treatment may become necessary.

Practitioners and researchers must always rely on their own experience and knowledge in evaluating and using any information, methods, compounds, or experiments described herein. In using such information or methods they should be mindful of their own safety and the safety of others, including parties for whom they have a professional responsibility.

To the fullest extent of the law, neither the Publisher nor the authors, contributors, or editors, assume any liability for any injury and/or damage to persons or property as a matter of products liability, negligence or otherwise, or from any use or operation of any methods, products, instructions, or ideas contained in the material herein.

Library of Congress Cataloging-in-Publication Data
Yiu, Joseph.
    The definitive guide to the ARM Cortex-M3 / Joseph Yiu.
        p.   cm.
    Includes bibliographical references and index.
    ISBN 978-1-85617-963-8 (alk. paper)
1. Embedded computer systems. 2. Microprocessors. I. Title.
    TK7895.E42Y58 2010
    621.39'16—dc22

                                    2009040437

British Library Cataloguing-in-Publication Data
A catalogue record for this book is available from the British Library.

For information on all Academic Press publications
visit our Web site at *www.elsevierdirect.com*

Printed in the United States
09 10 11 12 13   10 9 8 7 6 5 4 3 2 1

Contents

Foreword xvii

Foreword xviii

Preface..xix

Acknowledgments..xix

Conventions..xx

Terms and Abbreviations...xxi

CHAPTER 1   Introduction... 1

  1.1   What Is the ARM Cortex-M3 Processor? 1

  1.2   Background of ARM and ARM Architecture.............................2

      1.2.1 A Brief History...2

      1.2.2 Architecture Versions 3

      1.2.3 Processor Naming 5

  1.3   Instruction Set Development 7

  1.4   The Thumb-2 Technology and Instruction Set Architecture.......8

  1.5   Cortex-M3 Processor Applications 9

  1.6   Organization of This Book 10

  1.7   Further Reading...10

CHAPTER 2   Overview of the Cortex-M3 11

  2.1   Fundamentals 11

  2.2   Registers..12

      2.2.1 R0–R12: General-Purpose Registers...............................12

      2.2.2 R13: Stack Pointers..12

      2.2.3 R14: The Link Register 13

      2.2.4 R15: The Program Counter 13

      2.2.5 Special Registers 14

  2.3   Operation Modes..14

  2.4   The Built-In Nested Vectored Interrupt Controller 15

      2.4.1 Nested Interrupt Support 15

      2.4.2 Vectored Interrupt Support 16

      2.4.3 Dynamic Priority Changes Support 16

      2.4.4 Reduction of Interrupt Latency 16

      2.4.5 Interrupt Masking...16

  2.5   The Memory Map 16

  2.6   The Bus Interface 17

  2.7   The MPU 18

  2.8   The Instruction Set 18

2.9 Interrupts and Exceptions...19
    2.9.1 Low Power and High Energy Efficiency..................................20
2.10 Debugging Support 21
2.11 Characteristics Summary 22
    2.11.1 High Performance 22
    2.11.2 Advanced Interrupt-Handling Features 22
    2.11.3 Low Power Consumption 23
    2.11.4 System Features 23
    2.11.5 Debug Supports...23

CHAPTER 3   **Cortex-M3 Basics** **25**
3.1 Registers..25
    3.1.1 General Purpose Registers R0 through R7............................25
    3.1.2 General Purpose Registers R8 through R12..........................25
    3.1.3 Stack Pointer R13...26
    3.1.4 Link Register R14 28
    3.1.5 Program Counter R15 28
3.2 Special Registers...29
    3.2.1 Program Status Registers 29
    3.2.2 PRIMASK, FAULTMASK, and BASEPRI Registers 30
    3.2.3 The Control Register 31
3.3 Operation Mode 32
3.4 Exceptions and Interrupts..35
3.5 Vector Tables 36
3.6 Stack Memory Operations..36
    3.6.1 Basic Operations of the Stack 37
    3.6.2 Cortex-M3 Stack Implementation..37
    3.6.3 The Two-Stack Model in the Cortex-M3 39
3.7 Reset Sequence..40

CHAPTER 4   **Instruction Sets** **43**
4.1 Assembly Basics 43
    4.1.1 Assembler Language: Basic Syntax 43
    4.1.2 Assembler Language: Use of Suffixes 44
    4.1.3 Assembler Language: Unified Assembler Language 45
4.2 Instruction List 46
    4.2.1 Unsupported Instructions 51
4.3 Instruction Descriptions 52
    4.3.1 Assembler Language: Moving Data.......................................53
    4.3.2 LDR and ADR Pseudo-Instructions 56
    4.3.3 Assembler Language: Processing Data 57

4.3.4 Assembler Language: Call and Unconditional Branch 60

4.3.5 Assembler Language: Decisions and Conditional Branches.........................62

4.3.6 Assembler Language: Combined Compare and Conditional Branch 65

4.3.7 Assembler Language: Instruction Barrier and Memory
Barrier Instructions...67

4.3.8 Assembly Language: Saturation Operations 68

4.4 Several Useful Instructions in the Cortex-M3...70

4.4.1 MSR and MRS 70

4.4.2 More on the IF-THEN Instruction Block..70

4.4.3 SDIV and UDIV 72

4.4.4 REV, REVH, and REVSH 73

4.4.5 Reverse Bit 73

4.4.6 SXTB, SXTH, UXTB, and UXTH 73

4.4.7 Bit Field Clear and Bit Field Insert 74

4.4.8 UBFX and SBFX 74

4.4.9 LDRD and STRD...74

4.4.10 Table Branch Byte and Table Branch Halfword.....................................75

CHAPTER 5   Memory Systems.. 79

5.1 Memory System Features Overview 79

5.2 Memory Maps...79

5.3 Memory Access Attributes 82

5.4 Default Memory Access Permissions...83

5.5 Bit-Band Operations...84

5.5.1 Advantages of Bit-Band Operations...87

5.5.2 Bit-Band Operation of Different Data Sizes 90

5.5.3 Bit-Band Operations in C Programs 90

5.6 Unaligned Transfers 92

5.7 Exclusive Accesses...93

5.8 Endian Mode 95

CHAPTER 6   Cortex-M3 Implementation Overview 99

6.1 The Pipeline 99

6.2 A Detailed Block Diagram..101

6.3 Bus Interfaces on the Cortex-M3 104

6.3.1 The I-Code Bus 104

6.3.2 The D-Code Bus 104

6.3.3 The System Bus...104

6.3.4 The External PPB 104

6.3.5 The DAP Bus..105

6.4 Other Interfaces on the Cortex-M3 105

6.5 The External PPB 105

6.6 Typical Connections 106

6.7 Reset Types and Reset Signals 107

CHAPTER 7   Exceptions... 109

7.1 Exception Types 109

7.2 Definitions of Priority 111

7.3 Vector Tables 117

7.4 Interrupt Inputs and Pending Behavior 118

7.5 Fault Exceptions 120

7.5.1 Bus Faults...121

7.5.2 Memory Management Faults 122

7.5.3 Usage Faults 123

7.5.4 Hard Faults 125

7.5.5 Dealing with Faults 125

7.6 Supervisor Call and Pendable Service Call..126

CHAPTER 8   The Nested Vectored Interrupt Controller and Interrupt Control 131

8.1 Nested Vectored Interrupt Controller Overview..131

8.2 The Basic Interrupt Configuration 132

8.2.1 Interrupt Enable and Clear Enable 132

8.2.2 Interrupt Set Pending and Clear Pending..132

8.2.3 Priority Levels 132

8.2.4 Active Status..134

8.2.5 PRIMASK and FAULTMASK Special Registers 135

8.2.6 The BASEPRI Special Register 136

8.2.7 Configuration Registers for Other Exceptions 137

8.3 Example Procedures in Setting Up an Interrupt...138

8.4 Software Interrupts...140

8.5 The SYSTICK Timer 141

CHAPTER 9   Interrupt Behavior 145

9.1 Interrupt/Exception Sequences..145

9.1.1 Stacking..145

9.1.2 Vector Fetches 147

9.1.3 Register Updates 147

9.2 Exception Exits 147

9.3 Nested Interrupts 148

9.4 Tail-Chaining Interrupts 148

9.5 Late Arrivals 149

9.6 More on the Exception Return Value 149

9.7 Interrupt Latency 152
9.8 Faults Related to Interrupts 152
    9.8.1 Stacking 152
    9.8.2 Unstacking 153
    9.8.3 Vector Fetches 153
    9.8.4 Invalid Returns 153

CHAPTER 10 Cortex-M3 Programming... **155**
10.1 Overview 155
10.2 A Typical Development Flow...155
10.3 Using C..156
    10.3.1 Example of a Simple C Program Using RealView Development Site........157
    10.3.2 Compile the Same Example Using Keil MDK-ARM...........................159
    10.3.3 Accessing Memory-Mapped Registers in C..161
    10.3.4 Intrinsic Functions..163
    10.3.5 Embedded Assembler and Inline Assembler.......................................163
10.4 CMSIS...164
    10.4.1 Background of CMSIS..164
    10.4.2 Areas of Standardization 165
    10.4.3 Organization of CMSIS..166
    10.4.4 Using CMSIS 167
    10.4.5 Benefits of CMSIS 168
10.5 Using Assembly 169
    10.5.1 The Interface between Assembly and C..170
    10.5.2 The First Step in Assembly Programming 170
    10.5.3 Producing Outputs...171
    10.5.4 The "Hello World" Example 172
    10.5.5 Using Data Memory 176
10.6 Using Exclusive Access for Semaphores 177
10.7 Using Bit Band for Semaphores...179
10.8 Working with Bit Field Extract and Table Branch 181

CHAPTER 11 Exception Programming.. **183**
11.1 Using Interrupts...183
    11.1.1 Stack Setup..183
    11.1.2 Vector Table Setup...184
    11.1.3 Interrupt Priority Setup 185
    11.1.4 Enable the Interrupt...186
11.2 Exception/Interrupt Handlers 188
11.3 Software Interrupts..189
11.4 Example of Vector Table Relocation 190

11.5 Using SVC 193
11.6 SVC Example: Use for Text Message Output Functions......................194
11.7 Using SVC with C..197

CHAPTER 12 Advanced Programming Features and System Behavior 201
    12.1 Running a System with Two Separate Stacks 201
    12.2 Double-Word Stack Alignment 204
    12.3 Nonbase Thread Enable 205
    12.4 Performance Considerations 206
    12.5 Lockup Situations..208
        12.5.1 What Happens During Lockup? 208
        12.5.2 Avoiding Lockup 210
    12.6 FAULTMASK 210

CHAPTER 13 The Memory Protection Unit 211
    13.1 Overview 211
    13.2 MPU Registers 212
    13.3 Setting Up the MPU 218
    13.4 Typical Setup...225
        13.4.1 Example Use of the Subregion Disable 225

CHAPTER 14 Other Cortex-M3 Features..229
    14.1 The SYSTICK Timer 229
    14.2 Power Management...232
        14.2.1 Sleep Modes 232
        14.2.2 Sleep-On-Exit Feature..234
        14.2.3 Wakeup Interrupt Controller 234
    14.3 Multiprocessor Communication 236
    14.4 Self-Reset Control..241

CHAPTER 15 Debug Architecture 243
    15.1 Debugging Features Overview 243
    15.2 CoreSight Overview...244
        15.2.1 Processor Debugging Interface 244
        15.2.2 The Debug Host Interface 244
        15.2.3 DP Module, AP Module, and DAP 245
        15.2.4 Trace Interface..246
        15.2.5 CoreSight Characteristics...246
    15.3 Debug Modes 248
    15.4 Debugging Events 250
    15.5 Breakpoint in the Cortex-M3 251

15.6 Accessing Register Content in Debug...253

15.7 Other Core Debugging Features..254

CHAPTER 16   Debugging Components 255

16.1 Introduction...255

16.1.1 The Trace System in the Cortex-M3 255

16.2 Trace Components: DWT 256

16.3 Trace Components: ITM 258

16.3.1 Software Trace with the ITM 259

16.3.2 Hardware Trace with ITM and DWT 260

16.3.3 ITM Timestamp..260

16.4 Trace Components: ETM 260

16.5 Trace Components: TPIU 261

16.6 The Flash Patch and Breakpoint Unit 262

16.6.1 Breakpoint Feature 262

16.6.2 Flash Patch Feature 262

16.6.3 Comparators 263

16.7 The Advanced High-Performance Bus Access Port.......................264

16.8 ROM Table 265

CHAPTER 17   Getting Started with the Cortex-M3 Processor............................ 269

17.1 Choosing a Cortex-M3 Product 269

17.2 Development Tools..270

17.2.1 C Compiler and Debuggers...271

17.2.2 Embedded OS Support..272

17.3 Differences between the Cortex-M3 Revision 0 and Revision 1 272

17.3.1 Revision 1 Change: Moving from JTAG-DP to SWJ-DP.............274

17.4 Differences between the Cortex-M3 Revision 1 and Revision 2 274

17.4.1 Default Configuration of Double Word Stack Alignment............275

17.4.2 Auxiliary Control Register 275

17.4.3 ID Register Values Updates...275

17.4.4 Debug Features...275

17.4.5 Sleep Features 276

17.5 Benefits and Effects of the Revision 2 New Features 277

17.6 Differences between the Cortex-M3 and Cortex-M0.........................278

17.6.1 Programmer's Model...278

17.6.2 Exceptions and NVIC 279

17.6.3 Instruction Set 279

17.6.4 Memory System Features..280

17.6.5 Debug Features...280

17.6.6 Compatibility..280

CHAPTER 18 Porting Applications from the ARM7 to the Cortex-M3.................................. **283**
    18.1 Overview 283
    18.2 System Characteristics 283
        18.2.1 Memory Map..284
        18.2.2 Interrupts 284
        18.2.3 MPU 285
        18.2.4 System Control..285
        18.2.5 Operation Modes...285
    18.3 Assembly Language Files 286
        18.3.1 Thumb State 286
        18.3.2 ARM State 286
    18.4 C Program Files..288
    18.5 Precompiled Object Files 288
    18.6 Optimization...289

CHAPTER 19 Starting Cortex-M3 Development Using the GNU Tool Chain **291**
    19.1 Background 291
    19.2 Getting the GNU Tool Chain 292
    19.3 Development Flow 292
    19.4 Examples 294
        19.4.1 Example 1: The First Program 294
        19.4.2 Example 2: Linking Multiple Files 296
        19.4.3 Example 3: A Simple "Hello World" Program 297
        19.4.4 Example 4: Data in RAM...299
        19.4.5 Example 5: C Program 300
        19.4.6 Example 6: C with Retargeting 302
        19.4.7 Example 7: Implement Your Own Vector Table 304
    19.5 Accessing Special Registers..304
    19.6 Using Unsupported Instructions..305
    19.7 Inline Assembler in the GNU C Compiler 305

**CHAPTER 20 Getting Started with the Keil RealView Microcontroller
                  Development Kit**... **307**
    20.1 Overview 307
    20.2 Getting Started with μVision..308
    20.3 Outputting the "Hello World" Message Via Universal Asynchronous
            Receiver/Transmitter 314
    20.4 Testing the Software..317
    20.5 Using the Debugger...318
    20.6 The Instruction Set Simulator 325

20.7 Modifying the Vector Table 326
20.8 Stopwatch Example with Interrupts with CMSIS 327
20.9 Porting Existing Applications to Use CMSIS 334

CHAPTER 21 **Programming the Cortex-M3 Microcontrollers in NI LabVIEW** **335**
21.1 Overview 335
21.2 What Is LabVIEW 335
    21.2.1 Typical Application Areas 337
    21.2.2 What You Need to Use LabVIEW and ARM 337
21.3 Development Flow 337
21.4 Example of a LabVIEW Project 339
    21.4.1 Create the Project 339
    21.4.2 Define Inputs and Outputs 341
    21.4.3 Create the Program 341
    21.4.4 Build the Design and Test the Application 342
21.5 How It Works 343
21.6 Additional Features in LabVIEW 344
21.7 Porting to Another ARM Processor 345

APPENDIX A **The Cortex-M3 Instruction Set, Reference Material** **349**
A.1 Instruction Set Summary 349
A.2 About the Instruction Descriptions 353
    A.2.1 Operands 353
    A.2.2 Restrictions When Using PC or SP 353
    A.2.3 Flexible Second Operand 353
    A.2.4 Shift Operations 354
    A.2.5 Address Alignment 357
    A.2.6 PC-Relative Expressions 358
    A.2.7 Conditional Execution 358
    A.2.8 Instruction Width Selection 360
A.3 Memory Access Instructions 361
    A.3.1 ADR 362
    A.3.2 LDR and STR, Immediate Offset 362
    A.3.3 LDR and STR, Register Offset 365
    A.3.4 LDR and STR, Unprivileged 366
    A.3.5 LDR, PC-Relative 367
    A.3.6 LDM and STM 368
    A.3.7 PUSH and POP 370
    A.3.8 LDREX and STREX 371
    A.3.9 CLREX 372

A.4 General Data-Processing Instructions 373

A.4.1 ADD, ADC, SUB, SBC, and RSB...374

A.4.2 AND, ORR, EOR, BIC, and ORN 376

A.4.3 ASR, LSL, LSR, ROR, and RRX 377

A.4.4 CLZ 378

A.4.5 CMP and CMN 378

A.4.6 MOV and MVN 379

A.4.7 MOVT 381

A.4.8 REV, REV16, REVSH, and RBIT 381

A.4.9 TST and TEQ 382

A.5 Multiply and Divide Instructions 383

A.5.1 MUL, MLA, and MLS 383

A.5.2 UMULL, UMLAL, SMULL, and SMLAL.................................385

A.5.3 SDIV and UDIV 386

A.6 Saturating Instructions 386

A.6.1 SSAT and USAT 386

A.7 Bitfield Instructions..388

A.7.1 BFC and BFI 388

A.7.2 SBFX and UBFX 389

A.7.3 SXT and UXT 390

A.8 Branch and Control Instructions 391

A.8.1 B, BL, BX, and BLX..391

A.8.2 CBZ and CBNZ..393

A.8.3 IT 393

A.8.4 TBB and TBH 395

A.9 Miscellaneous Instructions..397

A.9.1 BKPT..397

A.9.2 CPS...398

A.9.3 DMB..398

A.9.4 DSB...399

A.9.5 ISB...399

A.9.6 MRS 400

A.9.7 MSR 400

A.9.8 NOP...401

A.9.9 SEV 402

A.9.10 SVC 402

A.9.11 WFE...403

A.9.12 WFI..403

APPENDIX B   **The 16-Bit Thumb Instructions and Architecture Versions** **405**

APPENDIX C **Cortex-M3 Exceptions Quick Reference** **407**
    C.1 Exception Types and Enables 407
    C.2 Stack Contents After Exception Stacking 408

APPENDIX D **Nested Vectored Interrupt Controller and System Control Block Registers Quick Reference** **409**

APPENDIX E **Cortex-M3 Troubleshooting Guide** **421**
    E.1 Overview 421
    E.2 Developing Fault Handlers 422
        E.2.1 Report Fault Status Registers 423
        E.2.2 Report Stacked PC and Other Stacked Registers 423
        E.2.3 Read Fault Address Register 425
        E.2.4 Clear Fault Status Bits 425
        E.2.5 Others 426
    E.3 Understanding the Cause of the Fault 426
    E.4 Other Possible Problems 429

APPENDIX F **Example Linker Script for CodeSourcery G++** **433**
    F.1 Example Linker Script for Cortex-M3 433

APPENDIX G **CMSIS Core Access Functions Reference** **439**
    G.1 Exception and Interrupt Numbers 439
    G.2 NVIC Access Functions 440
    G.3 System and SYSTICK Functions 443
    G.4 Core Registers Access Functions 443
    G.5 CMSIS Intrinsic Functions 444
    G.6 Debug Message Output Function 445

APPENDIX H **Connectors for Debug and Tracers** **447**
    H.1 Overview 447
    H.2 The 20-Pin Cortex Debug + ETM Connector 447
    H.3 The 10-Pin Cortex Debug Connector 449
    H.4 Legacy 20-Pin IDC Connector 449
    H.5 Legacy 38-Pin Mictor Connector 449

REFERENCES **451**

INDEX **453**

Foreword

Progress in the ARM microcontroller community since the publication of the first edition of this book has been impressive, significantly exceeding our expectations and it is no exaggeration to say that it is revolutionizing the world of Microcontroller Units (MCUs). There are many thousands of end users of ARM-powered MCUs, making it the fastest growing MCU technology on the market. As such, the second edition of Joseph's book is very timely and provides a good opportunity to present updated information on MCU technology.

As a community, progress has been made in many important areas including the number of companies building Cortex™-M3 processor-based devices (now over 30), development of the Cortex Microcontroller Software Interface Standard (CMSIS) enabling simpler code portability between Cortex processors and silicon vendors, improved versions of development tool chains, and the release of the Cortex-M0 processor to take ARM MCUs into even the lowest cost designs.

With such a rate of change it is certainly an exciting time to be developing embedded solutions based on the Cortex-M3 processor!

—Richard York
Director of Product Marketing, ARM

Foreword

Microcontroller programmers, by nature, are truly resourceful beings. From a fixed design, they create fantastic new products by using the microcontroller in a unique way. Constantly, they demand highly efficient computing from the most frugal of system designs. The primary ingredient used to perform this alchemy is the tool chain environment, and it is for this reason that engineers from ARM's own tool chain division joined forces with CPU designers to form a team that would rationalize, simplify, and improve the ARM7TDMI processor design.

The result of this combination, the ARM Cortex™-M3, represents an exciting development to the original ARM architecture. The device blends the best features from the 32-bit ARM architecture with the highly successful Thumb-2 instruction set design while adding several new capabilities. Despite these changes, the Cortex-M3 retains a simplified programmer's model that will be easily recognizable to all existing ARM aficionados.

—Wayne Lyons
Director of Embedded Solutions, ARM

Preface

This book is for both hardware and software engineers who are interested in the ARM Cortex™-M3 processor. The *Cortex-M3 Technical Reference Manual* (TRM) and the *ARMv7-M Architecture Application Level Reference Manual* already provide lots of information on this processor, but they are very detailed and can be challenging for novice readers.

This book is intended to be a lighter read for programmers, embedded product designers, system-on-chip (SoC) engineers, electronics enthusiasts, academic researchers, and others who are investigating the Cortex-M3 processor, with some experience of microcontrollers or microprocessors. The text includes an introduction to the architecture, an instruction set summary, examples of some instructions, information on hardware features, and an overview of the processor's advanced debug system. It also provides application examples, including basic steps in software development for the Cortex-M3 processor using ARM tools as well as the Gnu's Not Unix tool chain. This book is also suitable for engineers who are migrating their software from ARM7TDMI to the Cortex-M3 processor because it covers the differences between the two processors, and the porting of application software from the ARM7TDMI to the Cortex-M3.

ACKNOWLEDGMENTS

I would like to thank the following people for providing me with help, advice, and feedback to the first or the second edition of this book:

Richard York, Andrew Frame, Reinhard Keil, Nick Sampays, Dev Banerjee, Robert Boys, Dominic Pajak, Alan Tringham, Stephen Theobald, Dan Brook, David Brash, Haydn Povey, Gary Campbell, Kevin McDermott, Richard Earnshaw, Shyam Sadasivan, Simon Craske, Simon Axford, Takashi Ugajin, Wayne Lyons, Samin Ishtiaq, and Simon Smith.

I would like to thank Ian Bell and Jamie Brettle at National Instruments for their help in reviewing the materials covering NI LabVIEW and for their support. I would also like to express my gratitude to Carlos O'Donell, Brian Barrera, and Daniel Jacobowitz from CodeSourcery for their support and help in reviewing the materials covering software development with the CodeSourcery tool chain. And, of course, thanks to the staff at Elsevier for their professional work toward the publication of this book.

Finally, a special thank-you to Peter Cole and Ivan Yardley for their continuous support and advice during this project.

Conventions

Various typographical conventions have been used in this book, as follows:

- Normal assembly program codes:
  ```
  MOV   R0, R1;  Move data from Register R1 to Register R0
  ```
- Assembly code in generalized syntax; items inside < > must be replaced by real register names:
  ```
  MRS   <reg>, <special_reg>
  ```
- C program codes:
  ```
  for (i=0;i<3;i++) { func1(); }
  ```
- Pseudocode:
  ```
  if (a > b) { ...
  ```
- Values:
  1. 4'hC, 0x123 are both hexadecimal values
  2. *#3* indicates item number 3 (e.g., IRQ #3 means IRQ number 3)
  3. *#immed_12* refers to 12-bit immediate data
- Register bits:
  Typically used to illustrate a part of a value based on bit position; for example, bit[15:12] means bit number 15 down to 12.
- Register access types are as follows:
  1. R is Read only
  2. W is Write only
  3. R/W is Read or Write accessible
  4. R/Wc is Readable and clear by a Write access

Terms and Abbreviations

Abbreviation	Meaning
ADK	AMBA Design Kit
AHB	Advanced High-Performance Bus
AHB-AP	AHB Access Port
AMBA	Advanced Microcontroller Bus Architecture
APB	Advanced Peripheral Bus
ARM ARM	ARM Architecture Reference Manual
ASIC	Application-specific integrated circuit
ATB	Advanced Trace Bus
BE8	Byte-invariant big endian mode
CMSIS	Cortex Microcontroller Software Interface Standard
CPI	Cycles per instruction
CPU	Central processing unit
CS3	CodeSourcery Common Start-up Code Sequence
DAP	Debug Access Port
DSP	Digital Signal Processor/Digital Signal Processing
DWT	Data Watchpoint and Trace unit
EABI/ABI	Embedded application binary interface
ETM	Embedded Trace Macrocell
FPB	Flash Patch and Breakpoint unit
FPGA	Field Programmable Gate Array
FSR	Fault status register
HTM	CoreSight AHB Trace Macrocell
ICE	In-circuit emulator
IDE	Integrated Development Environment
IRQ	Interrupt Request (normally refers to external interrupts)
ISA	Instruction set architecture
ISR	Interrupt Service Routine
ITM	Instrumentation Trace Macrocell
JTAG	Joint Test Action Group (a standard of test/debug interfaces)
JTAG-DP	JTAG Debug Port
LR	Link register
LSB	Least Significant Bit
LSU	Load/store unit
MCU	Microcontroller Unit
MDK-ARM	Keil Microcontroller Development Kit for ARM
MMU	Memory management unit
MPU	Memory Protection Unit
MSB	Most Significant Bit
MSP	Main Stack Pointer
NMI	Nonmaskable interrupt

NVIC	Nested Vectored Interrupt Controller
OS	Operating system
PC	Program counter
PMU	Power management unit
PSP	Process Stack Pointer
PPB	Private Peripheral Bus
PSR	Program Status Register
SCB	System control block
SCS	System control space
SIMD	Single Instruction, Multiple Data
SoC	System-on-Chip
SP	Stack pointer
SRPG	State retention power gating
SW	Serial-Wire
SW-DP	Serial-Wire Debug Port
SWJ-DP	Serial-Wire JTAG Debug Port
SWV	Serial-Wire Viewer (an operation mode of TPIU)
TCM	Tightly coupled memory (Cortex-M1 feature)
TPA	Trace Port Analyzer
TPIU	Trace Port Interface Unit
TRM	Technical Reference Manual
UAL	Unified Assembly Language
UART	Universal Asynchronous Receiver Transmitter
WIC	Wakeup Interrupt Controller

Introduction

IN THIS CHAPTER

What Is the ARM Cortex-M3 Processor?.. 1
Background of ARM and ARM Architecture 2
Instruction Set Development.. 7
The Thumb-2 Technology and Instruction Set Architecture.. 8
Cortex-M3 Processor Applications... 9
Organization of This Book 10
Further Reading 10

1.1 WHAT IS THE ARM CORTEX-M3 PROCESSOR?

The microcontroller market is vast, with more than 20 billion devices per year estimated to be shipped in 2010. A bewildering array of vendors, devices, and architectures is competing in this market. The requirement for higher performance microcontrollers has been driven globally by the industry's changing needs; for example, microcontrollers are required to handle more work without increasing a product's frequency or power. In addition, microcontrollers are becoming increasingly connected, whether by Universal Serial Bus (USB), Ethernet, or wireless radio, and hence, the processing needed to support these communication channels and advanced peripherals are growing. Similarly, general application complexity is on the increase, driven by more sophisticated user interfaces, multimedia requirements, system speed, and convergence of functionalities.

The ARM Cortex™-M3 processor, the first of the Cortex generation of processors released by ARM in 2006, was primarily designed to target the 32-bit microcontroller market. The Cortex-M3 processor provides excellent performance at low gate count and comes with many new features previously available only in high-end processors. The Cortex-M3 addresses the requirements for the 32-bit embedded processor market in the following ways:

- *Greater performance efficiency*: allowing more work to be done without increasing the frequency or power requirements
- *Low power consumption*: enabling longer battery life, especially critical in portable products including wireless networking applications

- *Enhanced determinism*: guaranteeing that critical tasks and interrupts are serviced as quickly as possible and in a known number of cycles
- *Improved code density*: ensuring that code fits in even the smallest memory footprints
- *Ease of use*: providing easier programmability and debugging for the growing number of 8-bit and 16-bit users migrating to 32 bits
- *Lower cost solutions*: reducing 32-bit-based system costs close to those of legacy 8-bit and 16-bit devices and enabling low-end, 32-bit microcontrollers to be priced at less than US$1 for the first time
- *Wide choice of development tools*: from low-cost or free compilers to full-featured development suites from many development tool vendors

Microcontrollers based on the Cortex-M3 processor already compete head-on with devices based on a wide variety of other architectures. Designers are increasingly looking at reducing the system cost, as opposed to the traditional device cost. As such, organizations are implementing device aggregation, whereby a single, more powerful device can potentially replace three or four traditional 8-bit devices.

Other cost savings can be achieved by improving the amount of code reuse across all systems. Because Cortex-M3 processor-based microcontrollers can be easily programmed using the C language and are based on a well-established architecture, application code can be ported and reused easily, reducing development time and testing costs.

It is worthwhile highlighting that the Cortex-M3 processor is not the first ARM processor to be used to create generic microcontrollers. The venerable ARM7 processor has been very successful in this market, with partners such as NXP (Philips), Texas Instruments, Atmel, OKI, and many other vendors delivering robust 32-bit Microcontroller Units (MCUs). The ARM7 is the most widely used 32-bit embedded processor in history, with over 1 billion processors produced each year in a huge variety of electronic products, from mobile phones to cars.

The Cortex-M3 processor builds on the success of the ARM7 processor to deliver devices that are significantly easier to program and debug and yet deliver a higher processing capability. Additionally, the Cortex-M3 processor introduces a number of features and technologies that meet the specific requirements of the microcontroller applications, such as nonmaskable interrupts for critical tasks, highly deterministic nested vector interrupts, atomic bit manipulation, and an optional Memory Protection Unit (MPU). These factors make the Cortex-M3 processor attractive to existing ARM processor users as well as many new users considering use of 32-bit MCUs in their products.

1.2 BACKGROUND OF ARM AND ARM ARCHITECTURE

1.2.1 A Brief History

To help you understand the variations of ARM processors and architecture versions, let's look at a little bit of ARM history.

ARM was formed in 1990 as Advanced RISC Machines Ltd., a joint venture of Apple Computer, Acorn Computer Group, and VLSI Technology. In 1991, ARM introduced the ARM6 processor family, and VLSI became the initial licensee. Subsequently, additional companies, including Texas Instruments, NEC, Sharp, and ST Microelectronics, licensed the ARM processor designs, extending the applications of ARM processors into mobile phones, computer hard disks, personal digital assistants (PDAs), home entertainment systems, and many other consumer products.

THE CORTEX-M3 PROCESSOR VERSUS CORTEX-M3-BASED MCUs

The Cortex-M3 processor is the central processing unit (CPU) of a microcontroller chip. In addition, a number of other components are required for the whole Cortex-M3 processor-based microcontroller. After chip manufacturers license the Cortex-M3 processor, they can put the Cortex-M3 processor in their silicon designs, adding memory, peripherals, input/output (I/O), and other features. Cortex-M3 processor-based chips from different manufacturers will have different memory sizes, types, peripherals, and features. This book focuses on the architecture of the processor core. For details about the rest of the chip, readers are advised to check the particular chip manufacturer's documentation.

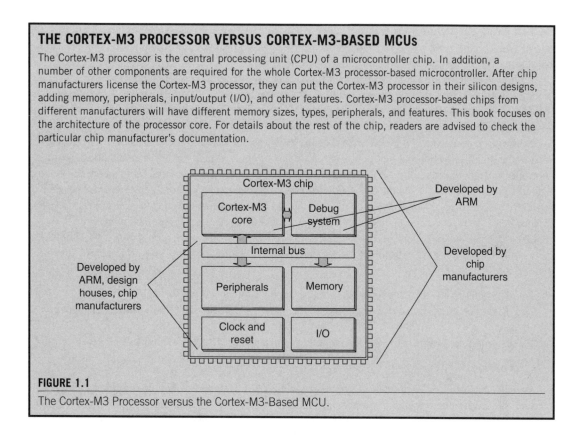

FIGURE 1.1

The Cortex-M3 Processor versus the Cortex-M3-Based MCU.

Nowadays, ARM partners ship in excess of 2 billion ARM processors each year. Unlike many semiconductor companies, ARM does not manufacture processors or sell the chips directly. Instead, ARM licenses the processor designs to business partners, including a majority of the world's leading semiconductor companies. Based on the ARM low-cost and power-efficient processor designs, these partners create their processors, microcontrollers, and system-on-chip solutions. This business model is commonly called intellectual property (IP) licensing.

In addition to processor designs, ARM also licenses systems-level IP and various software IPs. To support these products, ARM has developed a strong base of development tools, hardware, and software products to enable partners to develop their own products.

1.2.2 Architecture Versions

Over the years, ARM has continued to develop new processors and system blocks. These include the popular ARM7TDMI processor and, more recently, the ARM1176TZ(F)-S processor, which is used in high-end applications such as smart phones. The evolution of features and enhancements to the processors over time has led to successive versions of the ARM architecture. Note that architecture version numbers are independent from processor names. For example, the ARM7TDMI processor is based on the ARMv4T architecture (the *T* is for *Thumb*® instruction mode support).

The ARMv5E architecture was introduced with the ARM9E processor families, including the ARM926E-S and ARM946E-S processors. This architecture added "Enhanced" Digital Signal Processing (DSP) instructions for multimedia applications.

With the arrival of the ARM11 processor family, the architecture was extended to the ARMv6. New features in this architecture included memory system features and Single Instruction–Multiple Data (SIMD) instructions. Processors based on the ARMv6 architecture include the ARM1136J(F)-S, the ARM1156T2(F)-S, and the ARM1176JZ(F)-S.

Following the introduction of the ARM11 family, it was decided that many of the new technologies, such as the optimized Thumb-2 instruction set, were just as applicable to the lower cost markets of microcontroller and automotive components. It was also decided that although the architecture needed to be consistent from the lowest MCU to the highest performance application processor, there was a need to deliver processor architectures that best fit applications, enabling very deterministic and low gate count processors for cost-sensitive markets and feature-rich and high-performance ones for high-end applications.

Over the past several years, ARM extended its product portfolio by diversifying its CPU development, which resulted in the architecture version 7 or v7. In this version, the architecture design is divided into three profiles:

- The *A profile* is designed for high-performance open application platforms.
- The *R profile* is designed for high-end embedded systems in which real-time performance is needed.
- The *M profile* is designed for deeply embedded microcontroller-type systems.

Let's look at these profiles in a bit more detail:

- *A Profile (ARMv7-A)*: Application processors which are designed to handle complex applications such as high-end embedded operating systems (OSs) (e.g., Symbian, Linux, and Windows Embedded). These processors requiring the highest processing power, virtual memory system support with memory management units (MMUs), and, optionally, enhanced Java support and a secure program execution environment. Example products include high-end mobile phones and electronic wallets for financial transactions.

- *R Profile (ARMv7-R)*: Real-time, high-performance processors targeted primarily at the higher end of the real-time[1] market—those applications, such as high-end breaking systems and hard drive controllers, in which high processing power and high reliability are essential and for which low latency is important.

- *M Profile (ARMv7-M)*: Processors targeting low-cost applications in which processing efficiency is important and cost, power consumption, low interrupt latency, and ease of use are critical, as well as industrial control applications, including real-time control systems.

The Cortex processor families are the first products developed on architecture v7, and the Cortex-M3 processor is based on one profile of the v7 architecture, called ARM v7-M, an architecture specification for microcontroller products.

---

[1] There is always great debate as to whether we can have a "real-time" system using general processors. By definition, "real time" means that the system can get a response within a guaranteed period. In any processor-based system, you may or may not be able to get this response due to choice of OS, interrupt latency, or memory latency, as well as if the CPU is running a higher priority interrupt.

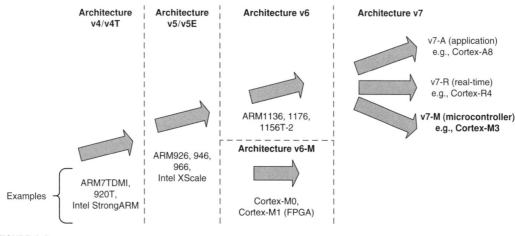

FIGURE 1.2

The Evolution of ARM Processor Architecture.

This book focuses on the Cortex-M3 processor, but it is only one of the Cortex product families that use the ARMv7 architecture. Other Cortex family processors include the Cortex-A8 (application processor), which is based on the ARMv7-A profile, and the Cortex-R4 (real-time processor), which is based on the ARMv7-R profile (see Figure 1.2).

The details of the ARMv7-M architecture are documented in *The ARMv7-M Architecture Application Level Reference Manual* [Ref. 2]. This document can be obtained via the ARM web site through a simple registration process. The ARMv7-M architecture contains the following key areas:

- Programmer's model
- Instruction set
- Memory model
- Debug architecture

Processor-specific information, such as interface details and timing, is documented in the *Cortex-M3 Technical Reference Manual (TRM)* [Ref. 1]. This manual can be accessed freely on the ARM web site. The Cortex-M3 TRM also covers a number of implementation details not covered by the architecture specifications, such as the list of supported instructions, because some of the instructions covered in the ARMv7-M architecture specification are optional on ARMv7-M devices.

1.2.3 Processor Naming

Traditionally, ARM used a numbering scheme to name processors. In the early days (the 1990s), suffixes were also used to indicate features on the processors. For example, with the ARM7TDMI processor, the *T* indicates Thumb instruction support, *D* indicates JTAG debugging, *M* indicates fast multiplier, and *I* indicates an embedded ICE module. Subsequently, it was decided that these features should become standard features of future ARM processors; therefore, these suffixes are no longer added to the new processor family names. Instead, variations on memory interface, cache, and tightly coupled memory (TCM) have created a new scheme for processor naming.

For example, ARM processors with cache and MMUs are now given the suffix "26" or "36," whereas processors with MPUs are given the suffix "46" (e.g., ARM946E-S). In addition, other suffixes are added to indicate synthesizable[2] (*S*) and Jazelle (*J*) technology. Table 1.1 presents a summary of processor names.

With version 7 of the architecture, ARM has migrated away from these complex numbering schemes that needed to be decoded, moving to a consistent naming for families of processors, with Cortex its initial brand. In addition to illustrating the compatibility across processors, this system removes confusion between architectural version and processor family number; for example, the ARM7TDMI is not a v7 processor but was based on the v4T architecture.

Table 1.1 ARM Processor Names

Processor Name	Architecture Version	Memory Management Features	Other Features
ARM7TDMI	ARMv4T		
ARM7TDMI-S	ARMv4T		
ARM7EJ-S	ARMv5E		DSP, Jazelle
ARM920T	ARMv4T	MMU	
ARM922T	ARMv4T	MMU	
ARM926EJ-S	ARMv5E	MMU	DSP, Jazelle
ARM946E-S	ARMv5E	MPU	DSP
ARM966E-S	ARMv5E	DSP	
ARM968E-S	ARMv5E		DMA, DSP
ARM966HS	ARMv5E	MPU (optional)	DSP
ARM1020E	ARMv5E	MMU	DSP
ARM1022E	ARMv5E	MMU	DSP
ARM1026EJ-S	ARMv5E	MMU or MPU	DSP, Jazelle
ARM1136J(F)-S	ARMv6	MMU	DSP, Jazelle
ARM1176JZ(F)-S	ARMv6	MMU + TrustZone	DSP, Jazelle
ARM11 MPCore	ARMv6	MMU + multiprocessor cache support	DSP, Jazelle
ARM1156T2(F)-S	ARMv6	MPU	DSP
Cortex-M0	ARMv6-M		NVIC
Cortex-M1	ARMv6-M	FPGA TCM interface	NVIC
Cortex-M3	ARMv7-M	MPU (optional)	NVIC

---

[2]A synthesizable core design is available in the form of a hardware description language (HDL) such as Verilog or VHDL and can be converted into a design netlist using synthesis software.

Table 1.1 ARM Processor Names *Continued*

Processor Name	Architecture Version	Memory Management Features	Other Features
Cortex-R4	ARMv7-R	MPU	DSP
Cortex-R4F	ARMv7-R	MPU	DSP + Floating point
Cortex-A8	ARMv7-A	MMU + TrustZone	DSP, Jazelle, NEON + floating point
Cortex-A9	ARMv7-A	MMU + TrustZone + multiprocessor	DSP, Jazelle, NEON + floating point

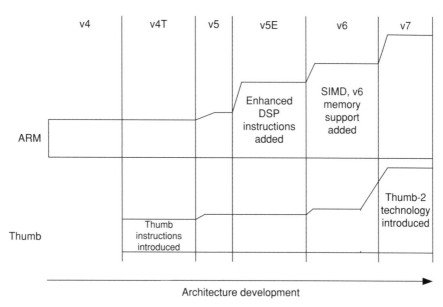

FIGURE 1.3

Instruction Set Enhancement.

1.3 INSTRUCTION SET DEVELOPMENT

Enhancement and extension of instruction sets used by the ARM processors has been one of the key driving forces of the architecture's evolution (see Figure 1.3).

Historically (since ARM7TDMI), two different instruction sets are supported on the ARM processor: the ARM instructions that are 32 bits and Thumb instructions that are 16 bits. During program execution, the processor can be dynamically switched between the ARM state and the Thumb state to use either one of the instruction sets. The Thumb instruction set provides only a subset of the ARM instructions, but it can provide higher code density. It is useful for products with tight memory requirements.

As the architecture version has been updated, extra instructions have been added to both ARM instructions and Thumb instructions. Appendix B provides some information on the change of Thumb instructions during the architecture enhancements. In 2003, ARM announced the Thumb-2 instruction set, which is a new superset of Thumb instructions that contains both 16-bit and 32-bit instructions.

The details of the instruction set are provided in a document called *The ARM Architecture Reference Manual* (also known as the ARM ARM). This manual has been updated for the ARMv5 architecture, the ARMv6 architecture, and the ARMv7 architecture. For the ARMv7 architecture, due to its growth into different profiles, the specification is also split into different documents. For the Cortex-M3 instruction set, the complete details are specified in the *ARM v7-M Architecture Application Level Reference Manual* [Ref. 2]. Appendix A of this book also covers information regarding instruction sets required for software development.

1.4 THE THUMB-2 TECHNOLOGY AND INSTRUCTION SET ARCHITECTURE

The Thumb-2[3] technology extended the Thumb Instruction Set Architecture (ISA) into a highly efficient and powerful instruction set that delivers significant benefits in terms of ease of use, code size, and performance (see Figure 1.4). The extended instruction set in Thumb-2 is a superset of the previous 16-bit Thumb instruction set, with additional 16-bit instructions alongside 32-bit instructions. It allows more complex operations to be carried out in the Thumb state, thus allowing higher efficiency by reducing the number of states switching between ARM state and Thumb state.

Focused on small memory system devices such as microcontrollers and reducing the size of the processor, the Cortex-M3 supports only the Thumb-2 (and traditional Thumb) instruction set. Instead of using ARM instructions for some operations, as in traditional ARM processors, it uses the Thumb-2 instruction set for all operations. As a result, the Cortex-M3 processor is not backward compatible with traditional

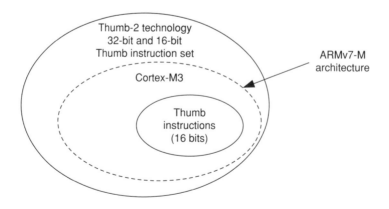

FIGURE 1.4

The Relationship between the Thumb Instruction Set in Thumb-2 Technology and the Traditional Thumb.

[3] Thumb and Thumb-2 are registered trademarks of ARM.

ARM processors. That is, you cannot run a binary image for ARM7 processors on the Cortex-M3 processor. Nevertheless, the Cortex-M3 processor can execute almost all the 16-bit Thumb instructions, including all 16-bit Thumb instructions supported on ARM7 family processors, making application porting easy.

With support for both 16-bit and 32-bit instructions in the Thumb-2 instruction set, there is no need to switch the processor between Thumb state (16-bit instructions) and ARM state (32-bit instructions). For example, in ARM7 or ARM9 family processors, you might need to switch to ARM state if you want to carry out complex calculations or a large number of conditional operations and good performance is needed, whereas in the Cortex-M3 processor, you can mix 32-bit instructions with 16-bit instructions without switching state, getting high code density and high performance with no extra complexity.

The Thumb-2 instruction set is a very important feature of the ARMv7 architecture. Compared with the instructions supported on ARM7 family processors (ARMv4T architecture), the Cortex-M3 processor instruction set has a large number of new features. For the first time, hardware divide instruction is available on an ARM processor, and a number of multiply instructions are also available on the Cortex-M3 processor to improve data-crunching performance. The Cortex-M3 processor also supports unaligned data accesses, a feature previously available only in high-end processors.

1.5 CORTEX-M3 PROCESSOR APPLICATIONS

With its high performance and high code density and small silicon footprint, the Cortex-M3 processor is ideal for a wide variety of applications:

- *Low-cost microcontrollers*: The Cortex-M3 processor is ideally suited for low-cost microcontrollers, which are commonly used in consumer products, from toys to electrical appliances. It is a highly competitive market due to the many well-known 8-bit and 16-bit microcontroller products on the market. Its lower power, high performance, and ease-of-use advantages enable embedded developers to migrate to 32-bit systems and develop products with the ARM architecture.

- *Automotive*: Another ideal application for the Cortex-M3 processor is in the automotive industry. The Cortex-M3 processor has very high-performance efficiency and low interrupt latency, allowing it to be used in real-time systems. The Cortex-M3 processor supports up to 240 external vectored interrupts, with a built-in interrupt controller with nested interrupt supports and an optional MPU, making it ideal for highly integrated and cost-sensitive automotive applications.

- *Data communications*: The processor's low power and high efficiency, coupled with instructions in Thumb-2 for bit-field manipulation, make the Cortex-M3 ideal for many communications applications, such as Bluetooth and ZigBee.

- *Industrial control*: In industrial control applications, simplicity, fast response, and reliability are key factors. Again, the Cortex-M3 processor's interrupt feature, low interrupt latency, and enhanced fault-handling features make it a strong candidate in this area.

- *Consumer products*: In many consumer products, a high-performance microprocessor (or several of them) is used. The Cortex-M3 processor, being a small processor, is highly efficient and low in power and supports an MPU enabling complex software to execute while providing robust memory protection.

There are already many Cortex-M3 processor-based products on the market, including low-end products priced as low as US$1, making the cost of ARM microcontrollers comparable to or lower than that of many 8-bit microcontrollers.

1.6 ORGANIZATION OF THIS BOOK

This book contains a general overview of the Cortex-M3 processor, with the rest of the contents divided into a number of sections:

- Chapters 1 and 2, Introduction and Overview of the Cortex-M3
- Chapters 3 through 6, Cortex-M3 Basics
- Chapters 7 through 9, Exceptions and Interrupts
- Chapters 10 and 11, Cortex-M3 Programming
- Chapters 12 through 14, Cortex-M3 Hardware Features
- Chapters 15 and 16, Debug Supports in Cortex-M3
- Chapters 17 through 21, Application Development with Cortex-M3
- Appendices

1.7 FURTHER READING

This book does not contain all the technical details on the Cortex-M3 processor. It is intended to be a starter guide for people who are new to the Cortex-M3 processor and a supplemental reference for people using Cortex-M3 processor-based microcontrollers. To get further detail on the Cortex-M3 processor, the following documents, available from ARM (*www.arm.com*) and ARM partner web sites, should cover most necessary details:

- *The Cortex-M3 Technical Reference Manual* (*TRM*) [Ref. 1] provides detailed information about the processor, including programmer's model, memory map, and instruction timing.
- *The ARMv7-M Architecture Application Level Reference Manual* [Ref. 2] contains detailed information about the instruction set and the memory model.
- Refer to datasheets for the Cortex-M3 processor-based microcontroller products; visit the manufacturer web site for the datasheets on the Cortex-M3 processor-based product you plan to use.
- *Cortex-M3 User Guides* are available from MCU vendors. In some cases, this user guide is available as a part of a complete microcontroller product manual. This document contains a programmer's model for the ARM Cortex-M3 processor, and instruction set details, and is customized by each MCU vendors to match their microcontroller implementations.
- Refer to *AMBA Specification 2.0* [Ref. 4] for more detail regarding internal AMBA interface bus protocol details.
- C programming tips for Cortex-M3 can be found in the *ARM Application Note 179: Cortex-M3 Embedded Software Development* [Ref. 7].

This book assumes that you already have some knowledge of and experience with embedded programming, preferably using ARM processors. If you are a manager or a student who wants to learn the basics without spending too much time reading the whole book or the *TRM*, Chapter 2 of this book is a good one to read because it provides a summary on the Cortex-M3 processor.

Overview of the Cortex-M3

IN THIS CHAPTER

Fundamentals 11
Registers...12
Operation Modes 14
The Built-In Nested Vectored Interrupt Controller 15
The Memory Map...16
The Bus Interface 17
The MPU 18
The Instruction Set 18
Interrupts and Exceptions...19
Debugging Support 21
Characteristics Summary 22

2.1 FUNDAMENTALS

The Cortex™-M3 is a 32-bit microprocessor. It has a 32-bit data path, a 32-bit register bank, and 32-bit memory interfaces (see Figure 2.1). The processor has a Harvard architecture, which means that it has a separate instruction bus and data bus. This allows instructions and data accesses to take place at the same time, and as a result of this, the performance of the processor increases because data accesses do not affect the instruction pipeline. This feature results in multiple bus interfaces on Cortex-M3, each with optimized usage and the ability to be used simultaneously. However, the instruction and data buses share the same memory space (a unified memory system). In other words, you cannot get 8 GB of memory space just because you have separate bus interfaces.

For complex applications that require more memory system features, the Cortex-M3 processor has an optional Memory Protection Unit (MPU), and it is possible to use an external cache if it's required. Both little endian and big endian memory systems are supported.

The Cortex-M3 processor includes a number of fixed internal debugging components. These components provide debugging operation supports and features, such as breakpoints and watchpoints.

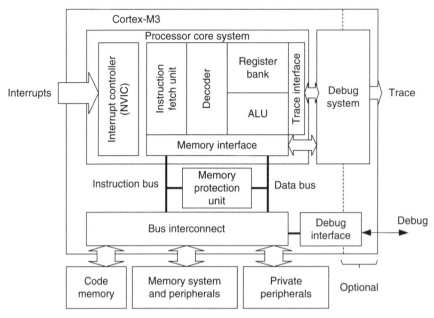

FIGURE 2.1

A Simplified View of the Cortex-M3.

In addition, optional components provide debugging features, such as instruction trace, and various types of debugging interfaces.

2.2 REGISTERS

The Cortex-M3 processor has registers R0 through R15 (see Figure 2.2). R13 (the stack pointer) is banked, with only one copy of the R13 visible at a time.

2.2.1 RO–R12: General-Purpose Registers

R0–R12 are 32-bit general-purpose registers for data operations. Some 16-bit Thumb® instructions can only access a subset of these registers (low registers, R0–R7).

2.2.2 R13: Stack Pointers

The Cortex-M3 contains two stack pointers (R13). They are banked so that only one is visible at a time. The two stack pointers are as follows:

- *Main Stack Pointer (MSP)*: The default stack pointer, used by the operating system (OS) kernel and exception handlers
- *Process Stack Pointer (PSP)*: Used by user application code

The lowest 2 bits of the stack pointers are always 0, which means they are always word aligned.

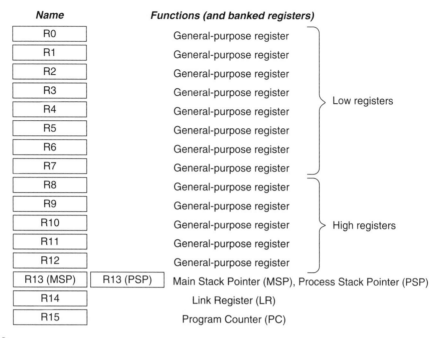

FIGURE 2.2

Registers in the Cortex-M3.

2.2.3 R14: The Link Register

When a subroutine is called, the return address is stored in the link register.

2.2.4 R15: The Program Counter

The program counter is the current program address. This register can be written to control the program flow.

2.2.5 Special Registers

The Cortex-M3 processor also has a number of special registers (see Figure 2.3). They are as follows:

- Program Status registers (PSRs)
- Interrupt Mask registers (PRIMASK, FAULTMASK, and BASEPRI)
- Control register (CONTROL)

These registers have special functions and can be accessed only by special instructions. They cannot be used for normal data processing (see Table 2.1).

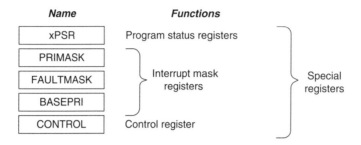

FIGURE 2.3

Special Registers in the Cortex-M3.

Table 2.1 Special Registers and Their Functions	
Register	**Function**
xPSR	Provide arithmetic and logic processing flags (zero flag and carry flag), execution status, and current executing interrupt number
PRIMASK	Disable all interrupts except the nonmaskable interrupt (NMI) and hard fault
FAULTMASK	Disable all interrupts except the NMI
BASEPRI	Disable all interrupts of specific priority level or lower priority level
CONTROL	Define privileged status and stack pointer selection
For more information on these registers, see Chapter 3.	

2.3 OPERATION MODES

The Cortex-M3 processor has two modes and two privilege levels. The operation modes (thread mode and handler mode) determine whether the processor is running a normal program or running an exception handler like an interrupt handler or system exception handler (see Figure 2.4). The privilege levels (privileged level and user level) provide a mechanism for safeguarding memory accesses to critical regions as well as providing a basic security model.

When the processor is running a main program (thread mode), it can be either in a privileged state or a user state, but exception handlers can only be in a privileged state. When the processor exits reset, it is in thread mode, with privileged access rights. In the privileged state, a program has access to all memory ranges (except when prohibited by MPU settings) and can use all supported instructions.

Software in the privileged access level can switch the program into the user access level using the control register. When an exception takes place, the processor will always switch back to the privileged state and return to the previous state when exiting the exception handler. A user program cannot change back to the privileged state by writing to the control register (see Figure 2.5). It has to go through an exception handler that programs the control register to switch the processor back into the privileged access level when returning to thread mode.

The separation of privilege and user levels improves system reliability by preventing system configuration registers from being accessed or changed by some untrusted programs. If an MPU is available,

	Privileged	User
When running an exception handler	Handler mode	
When not running an exception handler (e.g., main program)	Thread mode	Thread mode

FIGURE 2.4

Operation Modes and Privilege Levels in Cortex-M3.

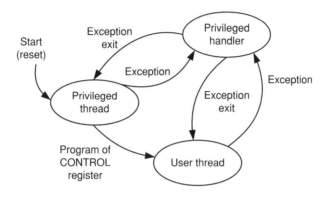

FIGURE 2.5

Allowed Operation Mode Transitions.

it can be used in conjunction with privilege levels to protect critical memory locations, such as programs and data for OSs.

For example, with privileged accesses, usually used by the OS kernel, all memory locations can be accessed (unless prohibited by MPU setup). When the OS launches a user application, it is likely to be executed in the user access level to protect the system from failing due to a crash of untrusted user programs.

2.4 THE BUILT-IN NESTED VECTORED INTERRUPT CONTROLLER

The Cortex-M3 processor includes an interrupt controller called the Nested Vectored Interrupt Controller (NVIC). It is closely coupled to the processor core and provides a number of features as follows:

- Nested interrupt support
- Vectored interrupt support
- Dynamic priority changes support
- Reduction of interrupt latency
- Interrupt masking

2.4.1 Nested Interrupt Support

The NVIC provides nested interrupt support. All the external interrupts and most of the system exceptions can be programmed to different priority levels. When an interrupt occurs, the NVIC compares the priority of this interrupt to the current running priority level. If the priority of the new interrupt is higher than the current level, the interrupt handler of the new interrupt will override the current running task.

2.4.2 Vectored Interrupt Support

The Cortex-M3 processor has vectored interrupt support. When an interrupt is accepted, the starting address of the interrupt service routine (ISR) is located from a vector table in memory. There is no need to use software to determine and branch to the starting address of the ISR. Thus, it takes less time to process the interrupt request.

2.4.3 Dynamic Priority Changes Support

Priority levels of interrupts can be changed by software during run time. Interrupts that are being serviced are blocked from further activation until the ISR is completed, so their priority can be changed without risk of accidental reentry.

2.4.4 Reduction of Interrupt Latency

The Cortex-M3 processor also includes a number of advanced features to lower the interrupt latency. These include automatic saving and restoring some register contents, reducing delay in switching from one ISR to another, and handling of late arrival interrupts. Details of these optimization features are covered in Chapter 9.

2.4.5 Interrupt Masking

Interrupts and system exceptions can be masked based on their priority level or masked completely using the interrupt masking registers BASEPRI, PRIMASK, and FAULTMASK. They can be used to ensure that time-critical tasks can be finished on time without being interrupted.

2.5 THE MEMORY MAP

The Cortex-M3 has a predefined memory map. This allows the built-in peripherals, such as the interrupt controller and the debug components, to be accessed by simple memory access instructions. Thus, most system features are accessible in C program code. The predefined memory map also allows the Cortex-M3 processor to be highly optimized for speed and ease of integration in system-on-a-chip (SoC) designs.

Overall, the 4 GB memory space can be divided into ranges as shown in Figure 2.6.

The Cortex-M3 design has an internal bus infrastructure optimized for this memory usage. In addition, the design allows these regions to be used differently. For example, data memory can still be put into the CODE region, and program code can be executed from an external Random Access Memory (RAM) region.

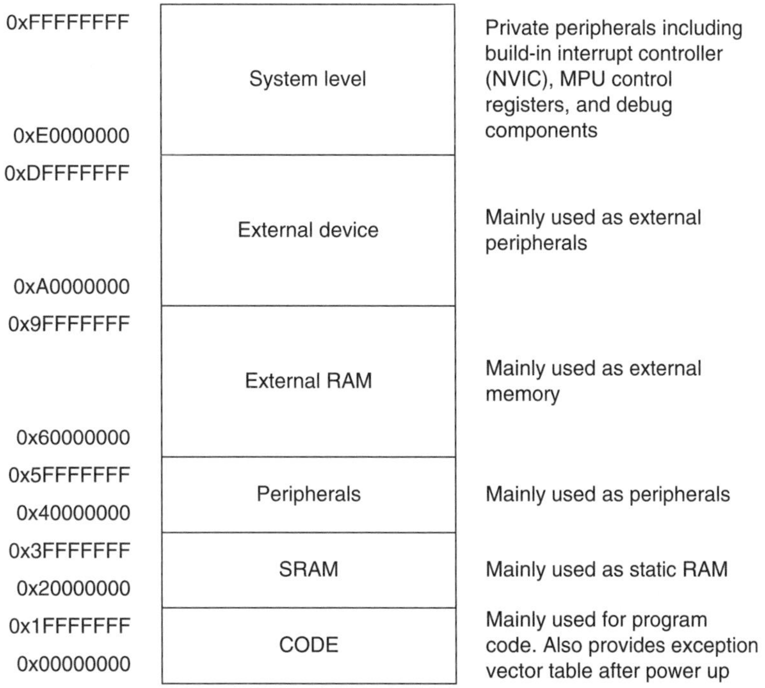

FIGURE 2.6

The Cortex-M3 Memory Map.

The system-level memory region contains the interrupt controller and the debug components. These devices have fixed addresses, detailed in Chapter 5. By having fixed addresses for these peripherals, you can port applications between different Cortex-M3 products much more easily.

2.6 THE BUS INTERFACE

There are several bus interfaces on the Cortex-M3 processor. They allow the Cortex-M3 to carry instruction fetches and data accesses at the same time. The main bus interfaces are as follows:

- Code memory buses
- System bus
- Private peripheral bus

The code memory region access is carried out on the code memory buses, which physically consist of two buses, one called I-Code and other called D-Code. These are optimized for instruction fetches for best instruction execution speed.

The system bus is used to access memory and peripherals. This provides access to the Static Random Access Memory (SRAM), peripherals, external RAM, external devices, and part of the system-level memory regions.

The private peripheral bus provides access to a part of the system-level memory dedicated to private peripherals, such as debugging components.

2.7 THE MPU

The Cortex-M3 has an optional MPU. This unit allows access rules to be set up for privileged access and user program access. When an access rule is violated, a fault exception is generated, and the fault exception handler will be able to analyze the problem and correct it, if possible.

The MPU can be used in various ways. In common scenarios, the OS can set up the MPU to protect data use by the OS kernel and other privileged processes to be protected from untrusted user programs. The MPU can also be used to make memory regions read-only, to prevent accidental erasing of data or to isolate memory regions between different tasks in a multitasking system. Overall, it can help make embedded systems more robust and reliable.

The MPU feature is optional and is determined during the implementation stage of the microcontroller or SoC design. For more information on the MPU, refer to Chapter 13.

2.8 THE INSTRUCTION SET

The Cortex-M3 supports the Thumb-2 instruction set. This is one of the most important features of the Cortex-M3 processor because it allows 32-bit instructions and 16-bit instructions to be used together for high code density and high efficiency. It is flexible and powerful yet easy to use.

In previous ARM processors, the central processing unit (CPU) had two operation states: a 32-bit ARM state and a 16-bit Thumb state. In the ARM state, the instructions are 32 bits and can execute all supported instructions with very high performance. In the Thumb state, the instructions are 16 bits, so there is a much higher instruction code density, but the Thumb state does not have all the functionality of ARM instructions and may require more instructions to complete certain types of operations.

To get the best of both worlds, many applications have mixed ARM and Thumb codes. However, the mixed-code arrangement does not always work best. There is overhead (in terms of both execution time and instruction space, see Figure 2.7) to switch between the states, and ARM and Thumb codes might need to be compiled separately in different files. This increases the complexity of software development and reduces maximum efficiency of the CPU core.

With the introduction of the Thumb-2 instruction set, it is now possible to handle all processing requirements in one operation state. There is no need to switch between the two. In fact, the Cortex-M3 does not support the ARM code. Even interrupts are now handled with the Thumb state. (Previously, the ARM core entered interrupt handlers in the ARM state.) Since there is no need to switch between states, the Cortex-M3 processor has a number of advantages over traditional ARM processors, such as:

- No state switching overhead, saving both execution time and instruction space
- No need to separate ARM code and Thumb code source files, making software development and maintenance easier
- It's easier to get the best efficiency and performance, in turn making it easier to write software, because there is no need to worry about switching code between ARM and Thumb to try to get the best density/performance

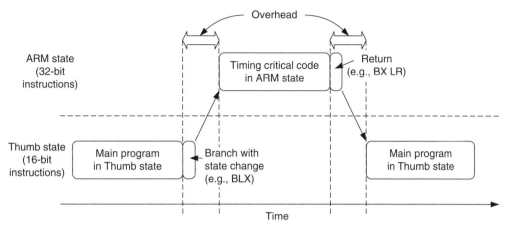

FIGURE 2.7

Switching between ARM Code and Thumb Code in Traditional ARM Processors Such as the ARM7.

The Cortex-M3 processor has a number of interesting and powerful instructions. Here are a few examples:

- *UFBX, BFI, and BFC*: Bit field extract, insert, and clear instructions
- *UDIV and SDIV*: Unsigned and signed divide instructions
- *WFE, WFI, and SEV*: Wait-For-Event, Wait-For-Interrupts, and Send-Event; these allow the processor to enter sleep mode and to handle task synchronization on multiprocessor systems
- *MSR and MRS*: Move to special register from general-purpose register and move special register to general-purpose register; for access to the special registers

Since the Cortex-M3 processor supports the Thumb-2 instruction set only, existing program code for ARM needs to be ported to the new architecture. Most C applications simply need to be recompiled using new compilers that support the Cortex-M3. Some assembler codes need modification and porting to use the new architecture and the new unified assembler framework.

Note that not all the instructions in the Thumb-2 instruction set are implemented on the Cortex-M3. The *ARMv7-M Architecture Application Level Reference Manual* [Ref. 2] only requires a subset of the Thumb-2 instructions to be implemented. For example, coprocessor instructions are not supported on the Cortex-M3 (external data processing engines can be added), and Single Instruction–Multiple Data (SIMD) is not implemented on the Cortex-M3. In addition, a few Thumb instructions are not supported, such as Branch with Link and Exchange (BLX) with immediate (used to switch processor state from Thumb to ARM), a couple of change process state (CPS) instructions, and the SETEND (Set Endian) instructions, which were introduced in architecture v6. For a complete list of supported instructions, refer to Appendix A.

2.9 INTERRUPTS AND EXCEPTIONS

The Cortex-M3 processor implements a new exception model, introduced in the ARMv7-M architecture. This exception model differs from the traditional ARM exception model, enabling very efficient exception handling. It has a number of system exceptions plus a number of external Interrupt Request (IRQs) (external interrupt inputs). There is no fast interrupt (FIQ) (fast interrupt in ARM7/ARM9/ARM10/ARM11) in the Cortex-M3; however, interrupt priority handling and nested interrupt support are now included in the interrupt architecture. Therefore, it is easy to set up a system that supports nested interrupts (a higher-priority interrupt can override or preempt a lower-priority interrupt handler) and that behaves just like the FIQ in traditional ARM processors.

The interrupt features in the Cortex-M3 are implemented in the NVIC. Aside from supporting external interrupts, the Cortex-M3 also supports a number of internal exception sources, such as system fault handling. As a result, the Cortex-M3 has a number of predefined exception types, as shown in Table 2.2.

2.9.1 Low Power and High Energy Efficiency

The Cortex-M3 processor is designed with various features to allow designers to develop low power and high energy efficient products. First, it has sleep mode and deep sleep mode supports, which can work with various system-design methodologies to reduce power consumption during idle period.

Table 2.2 Cortex-M3 Exception Types

Exception Number	Exception Type	Priority (Default to 0 if Programmable)	Description
0	NA	NA	No exception running
1	Reset	−3 (Highest)	Reset
2	NMI	−2	NMI (external NMI input)
3	Hard fault	−1	All fault conditions, if the corresponding fault handler is not enabled
4	MemManage fault	Programmable	Memory management fault; MPU violation or access to illegal locations
5	Bus fault	Programmable	Bus error (prefetch abort or data abort)
6	Usage fault	Programmable	Program error
7–10	Reserved	NA	Reserved
11	SVCall	Programmable	Supervisor call
12	Debug monitor	Programmable	Debug monitor (break points, watchpoints, or external debug request)
13	Reserved	NA	Reserved
14	PendSV	Programmable	Pendable request for system service
15	SYSTICK	Programmable	System tick timer
16	IRQ #0	Programmable	External interrupt #0
17	IRQ #1	Programmable	External interrupt #1
...	...	...	...
255	IRQ #239	Programmable	External interrupt #239

The number of external interrupt inputs is defined by chip manufacturers. A maximum of 240 external interrupt inputs can be supported. In addition, the Cortex-M3 also has an NMI interrupt input. When it is asserted, the NMI-ISR is executed unconditionally.

Second, its low gate count and design techniques reduce circuit activities in the processor to allow active power to be reduced. In addition, since Cortex-M3 has high code density, it has lowered the program size requirement. At the same time, it allows processing tasks to be completed in a short time, so that the processor can return to sleep modes as soon as possible to cut down energy use. As a result, the energy efficiency of Cortex-M3 is better than many 8-bit or 16-bit microcontrollers.

Starting from Cortex-M3 revision 2, a new feature called Wakeup Interrupt Controller (WIC) is available. This feature allows the whole processor core to be powered down, while processor states are retained and the processor can be returned to active state almost immediately when an interrupt takes place. This makes the Cortex-M3 even more suitable for many ultra-low power applications that previously could only be implemented with 8-bit or 16-bit microcontrollers.

2.10 DEBUGGING SUPPORT

The Cortex-M3 processor includes a number of debugging features, such as program execution controls, including halting and stepping, instruction breakpoints, data watchpoints, registers and memory accesses, profiling, and traces.

The debugging hardware of the Cortex-M3 processor is based on the CoreSight™ architecture. Unlike traditional ARM processors, the CPU core itself does not have a Joint Test Action Group (JTAG) interface. Instead, a debug interface module is decoupled from the core, and a bus interface called the Debug Access Port (DAP) is provided at the core level. Through this bus interface, external debuggers can access control registers to debug hardware as well as system memory, even when the processor is running. The control of this bus interface is carried out by a Debug Port (DP) device. The DPs currently available are the Serial-Wire JTAG Debug Port (SWJ-DP) (supports the traditional JTAG protocol as well as the Serial-Wire protocol) or the SW-DP (supports the Serial-Wire protocol only). A JTAG-DP module from the ARM CoreSight product family can also be used. Chip manufacturers can choose to attach one of these DP modules to provide the debug interface.

Chip manufacturers can also include an Embedded Trace Macrocell (ETM) to allow instruction trace. Trace information is output via the Trace Port Interface Unit (TPIU), and the debug host (usually a Personal Computer [PC]) can then collect the executed instruction information via external trace-capturing hardware.

Within the Cortex-M3 processor, a number of events can be used to trigger debug actions. Debug events can be breakpoints, watchpoints, fault conditions, or external debugging request input signals. When a debug event takes place, the Cortex-M3 processor can either enter halt mode or execute the debug monitor exception handler.

The data watchpoint function is provided by a Data Watchpoint and Trace (DWT) unit in the Cortex-M3 processor. This can be used to stop the processor (or trigger the debug monitor exception routine) or to generate data trace information. When data trace is used, the traced data can be output via the TPIU. (In the CoreSight architecture, multiple trace devices can share one single trace port.)

In addition to these basic debugging features, the Cortex-M3 processor also provides a Flash Patch and Breakpoint (FPB) unit that can provide a simple breakpoint function or remap an instruction access from Flash to a different location in SRAM.

An Instrumentation Trace Macrocell (ITM) provides a new way for developers to output data to a debugger. By writing data to register memory in the ITM, a debugger can collect the data via a trace interface and display or process them. This method is easy to use and faster than JTAG output.

All these debugging components are controlled via the DAP interface bus on the Cortex-M3 or by a program running on the processor core, and all trace information is accessible from the TPIU.

2.11 CHARACTERISTICS SUMMARY

Why is the Cortex-M3 processor such a revolutionary product? What are the advantages of using the Cortex-M3? The benefits and advantages are summarized in this section.

2.11.1 High Performance

The Cortex-M3 processor delivers high performance in microcontroller products:

- Many instructions, including multiply, are single cycle. Therefore, the Cortex-M3 processor outperforms most microcontroller products.
- Separate data and instruction buses allow simultaneous data and instruction accesses to be performed.
- The Thumb-2 instruction set makes state switching overhead history. There's no need to spend time switching between the ARM state (32 bits) and the Thumb state (16 bits), so instruction cycles and program size are reduced. This feature has also simplified software development, allowing faster time to market, and easier code maintenance.
- The Thumb-2 instruction set provides extra flexibility in programming. Many data operations can now be simplified using shorter code. This also means that the Cortex-M3 has higher code density and reduced memory requirements.
- Instruction fetches are 32 bits. Up to two instructions can be fetched in one cycle. As a result, there's more available bandwidth for data transfer.
- The Cortex-M3 design allows microcontroller products to operate at high clock frequency (over 100 MHz in modern semiconductor manufacturing processes). Even running at the same frequency as most other microcontroller products, the Cortex-M3 has a better clock per instruction (CPI) ratio. This allows more work per MHz or designs can run at lower clock frequency for lower power consumption.

2.11.2 Advanced Interrupt-Handling Features

The interrupt features on the Cortex-M3 processor are easy to use, very flexible, and provide high interrupt processing throughput:

- The built-in NVIC supports up to 240 external interrupt inputs. The vectored interrupt feature considerably reduces interrupt latency because there is no need to use software to determine which IRQ handler to serve. In addition, there is no need to have software code to set up nested interrupt support.

- The Cortex-M3 processor automatically pushes registers R0–R3, R12, Link register (LR), PSR, and PC in the stack at interrupt entry and pops them back at interrupt exit. This reduces the IRQ handling latency and allows interrupt handlers to be normal C functions (as explained later in Chapter 8).
- Interrupt arrangement is extremely flexible because the NVIC has programmable interrupt priority control for each interrupt. A minimum of eight levels of priority are supported, and the priority can be changed dynamically.
- Interrupt latency is reduced by special optimization, including late arrival interrupt acceptance and tail-chain interrupt entry.
- Some of the multicycle operations, including Load-Multiple (LDM), Store-Multiple (STM), PUSH, and POP, are now interruptible.
- On receipt of an NMI request, immediate execution of the NMI handler is guaranteed unless the system is completely locked up. NMI is very important for many safety-critical applications.

2.11.3 Low Power Consumption

The Cortex-M3 processor is suitable for various low-power applications:

- The Cortex-M3 processor is suitable for low-power designs because of the low gate count.
- It has power-saving mode support (SLEEPING and SLEEPDEEP). The processor can enter sleep mode using WFI or WFE instructions. The design has separated clocks for essential blocks, so clocking circuits for most parts of the processor can be stopped during sleep.
- The fully static, synchronous, synthesizable design makes the processor easy to be manufactured using any low power or standard semiconductor process technology.

2.11.4 System Features

The Cortex-M3 processor provides various system features making it suitable for a large number of applications:

- The system provides bit-band operation, byte-invariant big endian mode, and unaligned data access support.
- Advanced fault-handling features include various exception types and fault status registers, making it easier to locate problems.
- With the shadowed stack pointer, stack memory of kernel and user processes can be isolated. With the optional MPU, the processor is more than sufficient to develop robust software and reliable products.

2.11.5 Debug Supports

The Cortex-M3 processor includes comprehensive debug features to help software developers design their products:

- Supports JTAG or Serial-Wire debug interfaces
- Based on the CoreSight debugging solution, processor status or memory contents can be accessed even when the core is running

- Built-in support for six breakpoints and four watchpoints
- Optional ETM for instruction trace and data trace using DWT
- New debugging features, including fault status registers, new fault exceptions, and Flash Patch operations, make debugging much easier
- ITM provides an easy-to-use method to output debug information from test code
- PC sampler and counters inside the DWT provide code-profiling information

Cortex-M3 Basics

IN THIS CHAPTER

Registers..25
Special Registers 29
Operation Mode 32
Exceptions and Interrupts...35
Vector Tables 36
Stack Memory Operations 36
Reset Sequence...40

3.1 REGISTERS

As we've seen, the Cortex™-M3 processor has registers R0 through R15 and a number of special registers. R0 through R12 are general purpose, but some of the 16-bit Thumb® instructions can only access R0 through R7 (low registers), whereas 32-bit Thumb-2 instructions can access all these registers. Special registers have predefined functions and can only be accessed by special register access instructions.

3.1.1 General Purpose Registers R0 through R7

The R0 through R7 general purpose registers are also called *low registers*. They can be accessed by all 16-bit Thumb instructions and all 32-bit Thumb-2 instructions. They are all 32 bits; the reset value is unpredictable.

3.1.2 General Purpose Registers R8 through R12

The R8 through R12 registers are also called *high registers*. They are accessible by all Thumb-2 instructions but not by all 16-bit Thumb instructions. These registers are all 32 bits; the reset value is unpredictable (see Figure 3.1).

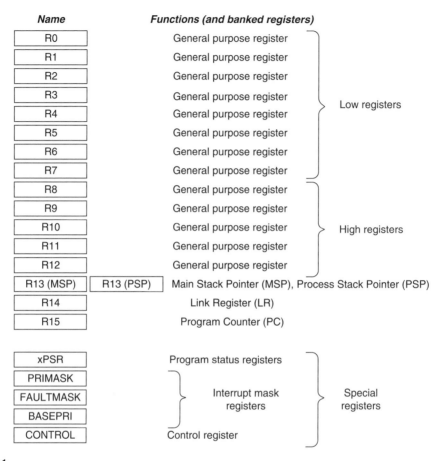

FIGURE 3.1

Registers in the Cortex-M3.

3.1.3 Stack Pointer R13

R13 is the stack pointer (SP). In the Cortex-M3 processor, there are two SPs. This duality allows two separate stack memories to be set up. When using the register name R13, you can only access the current SP; the other one is inaccessible unless you use special instructions to move to special register from general-purpose register (MSR) and move special register to general-purpose register (MRS). The two SPs are as follows:

- *Main Stack Pointer (MSP) or SP_main in ARM documentation*: This is the default SP; it is used by the operating system (OS) kernel, exception handlers, and all application codes that require privileged access.
- *Process Stack Pointer (PSP) or SP_process in ARM documentation*: This is used by the base-level application code (when not running an exception handler).

STACK PUSH AND POP

Stack is a memory usage model. It is simply part of the system memory, and a pointer register (inside the processor) is used to make it work as a first-in/last-out buffer. The common use of a stack is to save register contents before some data processing and then restore those contents from the stack after the processing task is done.

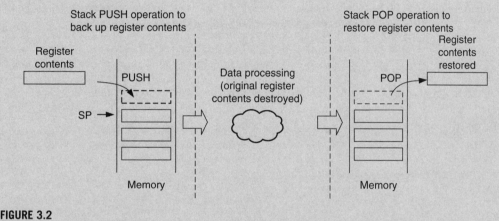

FIGURE 3.2

Basic Concept of Stack Memory.

When doing PUSH and POP operations, the pointer register, commonly called stack pointer, is adjusted automatically to prevent next stack operations from corrupting previous stacked data. More details on stack operations are provided on later part of this chapter.

It is not necessary to use both SPs. Simple applications can rely purely on the MSP. The SPs are used for accessing stack memory processes such as PUSH and POP.

In the Cortex-M3, the instructions for accessing stack memory are PUSH and POP. The assembly language syntax is as follows (text after each semicolon [;] is a comment):

```
PUSH {R0} ; R13=R13-4, then Memory[R13] = R0
POP  {R0} ; R0 = Memory[R13], then R13 = R13 + 4
```

The Cortex-M3 uses a full-descending stack arrangement. (More detail on this subject can be found in the "Stack Memory Operations" section of this chapter.) Therefore, the SP decrements when new data is stored in the stack. PUSH and POP are usually used to save register contents to stack memory at the start of a subroutine and then restore the registers from stack at the end of the subroutine. You can PUSH or POP multiple registers in one instruction:

```
subroutine_1
    PUSH   {R0-R7, R12, R14} ; Save registers
    ...                      ; Do your processing
    POP    {R0-R7, R12, R14} ; Restore registers
    BX     R14               ; Return to calling function
```

Instead of using *R13,* you can use *SP* (for SP) in your program codes. It means the same thing. Inside program code, both the MSP and the PSP can be called *R13/SP*. However, you can access a particular one using special register access instructions (MRS/MSR).

The MSP, also called *SP_main* in ARM documentation, is the default SP after power-up; it is used by kernel code and exception handlers. The PSP, or *SP_process* in ARM documentation, is typically used by thread processes in system with embedded OS running.

Because register PUSH and POP operations are always word aligned (their addresses must be 0x0, 0x4, 0x8, ...), the SP/R13 bit 0 and bit 1 are hardwired to 0 and always read as zero (RAZ).

3.1.4 Link Register R14

R14 is the link register (LR). Inside an assembly program, you can write it as either *R14* or *LR*. LR is used to store the return program counter (PC) when a subroutine or function is called—for example, when you're using the branch and link (BL) instruction:

```
main ; Main program
    ...
    BL function1 ; Call function1 using Branch with Link instruction.
                 ; PC = function1 and
                 ; LR = the next instruction in main
    ...
function1
    ...          ; Program code for function 1
    BX LR        ; Return
```

Despite the fact that bit 0 of the PC is always 0 (because instructions are word aligned or half word aligned), the LR bit 0 is readable and writable. This is because in the Thumb instruction set, bit 0 is often used to indicate ARM/Thumb states. To allow the Thumb-2 program for the Cortex-M3 to work with other ARM processors that support the Thumb-2 technology, this least significant bit (LSB) is writable and readable.

3.1.5 Program Counter R15

R15 is the PC. You can access it in assembler code by either R15 or PC. Because of the pipelined nature of the Cortex-M3 processor, when you read this register, you will find that the value is different than the location of the executing instruction, normally by 4. For example:

```
0x1000 : MOV R0, PC ; R0 = 0x1004
```

In other instructions like literal load (reading of a memory location related to current PC value), the effective value of PC might not be instruction address plus 4 due to alignment in address calculation. But the PC value is still at least 2 bytes ahead of the instruction address during execution.

Writing to the PC will cause a branch (but LRs do not get updated). Because an instruction address must be half word aligned, the LSB (bit 0) of the PC read value is always 0. However, in branching, either by writing to PC or using branch instructions, the LSB of the target address should be set to 1 because it is used to indicate the Thumb state operations. If it is 0, it can imply trying to switch to the ARM state and will result in a fault exception in the Cortex-M3.

3.2 SPECIAL REGISTERS

The special registers in the Cortex-M3 processor include the following (see Figures 3.3 and 3.4):

- Program Status registers (PSRs)
- Interrupt Mask registers (PRIMASK, FAULTMASK, and BASEPRI)
- Control register (CONTROL)

Special registers can only be accessed via MSR and MRS instructions; they do not have memory addresses:

```
MRS <reg>, <special_reg>; Read special register
MSR <special_reg>, <reg>; write to special register
```

3.2.1 Program Status Registers

The PSRs are subdivided into three status registers:

- Application Program Status register (APSR)
- Interrupt Program Status register (IPSR)
- Execution Program Status register (EPSR)

The three PSRs can be accessed together or separately using the special register access instructions MSR and MRS. When they are accessed as a collective item, the name *xPSR* is used.

You can read the PSRs using the MRS instruction. You can also change the APSR using the MSR instruction, but EPSR and IPSR are read-only. For example:

```
MRS     r0, APSR    ; Read Flag state into R0
MRS     r0, IPSR    ; Read Exception/Interrupt state
MRS     r0, EPSR    ; Read Execution state
MSR     APSR, r0    ; Write Flag state
```

	31	30	29	28	27	26:25	24	23:20	19:16	15:10	9	8	7	6	5	4:0
APSR	N	Z	C	V	Q											
IPSR													Exception number			
EPSR					ICI/IT	T			ICI/IT							

FIGURE 3.3

Program Status Registers (PSRs) in the Cortex-M3.

	31	30	29	28	27	26:25	24	23:20	19:16	15:10	9	8	7	6	5	4:0
xPSR	N	Z	C	V	Q	ICI/IT	T			ICI/IT			Exception number			

FIGURE 3.4

Combined Program Status Registers (xPSR) in the Cortex-M3.

Table 3.1 Bit Fields in Cortex-M3 Program Status Registers

Bit	Description
N	Negative
Z	Zero
C	Carry/borrow
V	Overflow
Q	Sticky saturation flag
ICI/IT	Interrupt-Continuable Instruction (ICI) bits, IF-THEN instruction status bit
T	Thumb state, always 1; trying to clear this bit will cause a fault exception
Exception number	Indicates which exception the processor is handling

	31	30	29	28	27	26:25	24	23:20	19:16	15:10	9	8	7	6	5	4:0
ARM (general)	N	Z	C	V	Q	IT	J	Reserved	GE[3:0]	IT	E	A	I	F	T	M[4:0]
ARM7 TDMI	N	Z	C	V	Reserved								I	F	T	M[4:0]

FIGURE 3.5

Current Program Status Registers in Traditional ARM Processors.

In ARM assembler, when accessing xPSR (all three PSRs as one), the symbol *PSR* is used:

```
MRS     r0, PSR    ; Read the combined program status word
MSR     PSR, r0    ; Write combined program state word
```

The descriptions for the bit fields in PSR are shown in Table 3.1.

If you compare this with the Current Program Status register (CPSR) in ARM7, you might find that some bit fields that were used in ARM7 are gone. The Mode (M) bit field is gone because the Cortex-M3 does not have the operation mode as defined in ARM7. Thumb-bit (T) is moved to bit 24. Interrupt status (I and F) bits are replaced by the new interrupt mask registers (PRIMASKs), which are separated from PSR. For comparison, the CPSR in traditional ARM processors is shown in Figure 3.5.

3.2.2 PRIMASK, FAULTMASK, and BASEPRI Registers

The PRIMASK, FAULTMASK, and BASEPRI registers are used to disable exceptions (see Table 3.2).

The PRIMASK and BASEPRI registers are useful for temporarily disabling interrupts in timing-critical tasks. An OS could use FAULTMASK to temporarily disable fault handling when a task has crashed. In this scenario, a number of different faults might be taking place when a task crashes. Once the core starts cleaning up, it might not want to be interrupted by other faults caused by the crashed process. Therefore, the FAULTMASK gives the OS kernel time to deal with fault conditions.

Table 3.2 Cortex-M3 Interrupt Mask Registers

Register Name	Description
PRIMASK	A 1-bit register, when this is set, it allows nonmaskable interrupt (NMI) and the hard fault exception; all other interrupts and exceptions are masked. The default value is 0, which means that no masking is set.
FAULTMASK	A 1-bit register, when this is set, it allows only the NMI, and all interrupts and fault handling exceptions are disabled. The default value is 0, which means that no masking is set.
BASEPRI	A register of up to 8 bits (depending on the bit width implemented for priority level). It defines the masking priority level. When this is set, it disables all interrupts of the same or lower level (larger priority value). Higher priority interrupts can still be allowed. If this is set to 0, the masking function is disabled (this is the default).

To access the PRIMASK, FAULTMASK, and BASEPRI registers, a number of functions are available in the device driver libraries provided by the microcontroller vendors. For example, the following:

```
x = __get_BASEPRI();  // Read BASEPRI register
x = __get_PRIMARK();  // Read PRIMASK register
x = __get_FAULTMASK();  // Read FAULTMASK register
__set_BASEPRI(x);  // Set new value for BASEPRI
__set_PRIMASK(x);  // Set new value for PRIMASK
__set_FAULTMASK(x);  // Set new value for FAULTMASK
__disable_irq();  // Clear PRIMASK, enable IRQ
__enable_irq();  // Set PRIMASK, disable IRQ
```

Details of these core register access functions are covered in Appendix G. A detailed introduction of Cortex Microcontroller Software Interface Standard (CMSIS) can be found in Chapter 10.

In assembly language, the MRS and MSR instructions are used. For example:

```
MRS    r0, BASEPRI    ; Read BASEPRI register into R0
MRS    r0, PRIMASK    ; Read PRIMASK register into R0
MRS    r0, FAULTMASK  ; Read FAULTMASK register into R0
MSR    BASEPRI, r0    ; Write R0 into BASEPRI register
MSR    PRIMASK, r0    ; Write R0 into PRIMASK register
MSR    FAULTMASK, r0  ; Write R0 into FAULTMASK register
```

The PRIMASK, FAULTMASK, and BASEPRI registers cannot be set in the user access level.

3.2.3 The Control Register

The control register is used to define the privilege level and the SP selection. This register has 2 bits, as shown in Table 3.3.

CONTROL[1]

In the Cortex-M3, the CONTROL[1] bit is always 0 in handler mode. However, in the thread or base level, it can be either 0 or 1.

Table 3.3 Cortex-M3 Control Register

Bit	Function
CONTROL[1]	Stack status: 1 = Alternate stack is used 0 = Default stack (MSP) is used If it is in the thread or base level, the alternate stack is the PSP. There is no alternate stack for handler mode, so this bit must be 0 when the processor is in handler mode.
CONTROL[0]	0 = Privileged in thread mode 1 = User state in thread mode If in handler mode (not thread mode), the processor operates in privileged mode.

This bit is writable only when the core is in thread mode and privileged. In the user state or handler mode, writing to this bit is not allowed. Aside from writing to this register, another way to change this bit is to change bit 2 of the LR when in exception return. This subject is discussed in Chapter 8, where details on exceptions are described.

CONTROL[0]

The CONTROL[0] bit is writable only in a privileged state. Once it enters the user state, the only way to switch back to privileged is to trigger an interrupt and change this in the exception handler.

To access the control register in C, the following CMSIS functions are available in CMSIS compliant device driver libraries:

```
x = __get_CONTROL(); // Read the current value of CONTROL
__set_CONTROL(x); // Set the CONTROL value to x
```

To access the control register in assembly, the MRS and MSR instructions are used:

```
MRS    r0, CONTROL ; Read CONTROL register into R0
MSR    CONTROL, r0 ; Write R0 into CONTROL register
```

3.3 OPERATION MODE

The Cortex-M3 processor supports two modes and two privilege levels (see Figure 3.6).

When the processor is running in thread mode, it can be in either the privileged or user level, but handlers can only be in the privileged level. When the processor exits reset, it is in thread mode, with privileged access rights.

In the user access level (thread mode), access to the system control space (SCS)—a part of the memory region for configuration registers and debugging components—is blocked. Furthermore, instructions that access special registers (such as MSR, except when accessing APSR) cannot be used. If a program running at the user access level tries to access SCS or special registers, a fault exception will occur.

Software in a privileged access level can switch the program into the user access level using the control register. When an exception takes place, the processor will always switch to a privileged state and

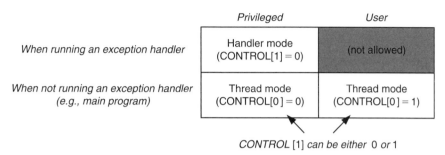

FIGURE 3.6

Operation Modes and Privilege Levels in Cortex-M3.

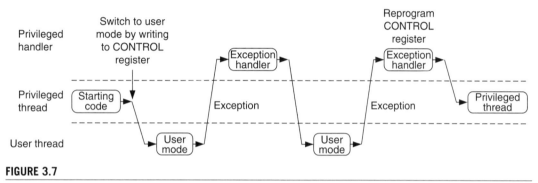

FIGURE 3.7

Switching of Operation Mode by Programming the Control Register or by Exceptions.

return to the previous state when exiting the exception handler. A user program cannot change back to the privileged state directly by writing to the control register. It has to go through an exception handler that programs the control register to switch the processor back into privileged access level when returning to thread mode. (See Figures 3.7).

The support of privileged and user access levels provides a more secure and robust architecture. For example, when a user program goes wrong, it will not be able to corrupt control registers in the Nested Vectored Interrupt Controller (NVIC). In addition, if the Memory Protection Unit (MPU) is present, it is possible to block user programs from accessing memory regions used by privileged processes.

In simple applications, there is no need to separate the privileged and user access levels. In these cases, there is no need to use user access level and no need to program the control register.

You can separate the user application stack from the kernel stack memory to avoid the possibility of crashing a system caused by stack operation errors in user programs. With this arrangement, the user program (running in thread mode) uses the PSP, and the exception handlers use the MSP. The switching of SPs is automatic upon entering or leaving the exception handlers (see section 3.6.3). This topic is discussed in more detail in Chapter 8.

The mode and access level of the processor are defined by the control register. When the control register bit 0 is 0, the processor mode changes when an exception takes place (see Figures 3.8 and 3.9).

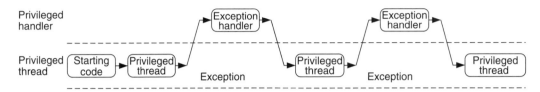

FIGURE 3.8

Simple Applications Do Not Require User Access Level in Thread Mode.

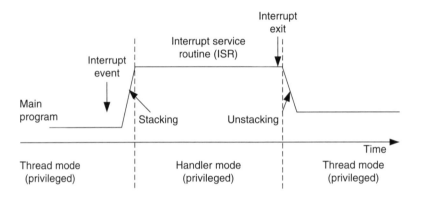

FIGURE 3.9

Switching Processor Mode at Interrupt.

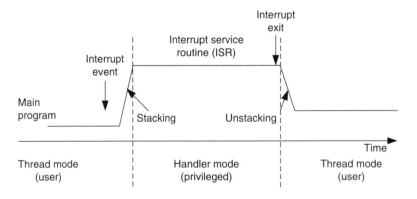

FIGURE 3.10

Switching Processor Mode and Privilege Level at Interrupt.

When control register bit 0 is 1 (thread running user application), both processor mode and access level change when an exception takes place (see Figure 3.10).

Control register bit 0 is programmable only in the privileged level (see Figure 2.5). For a user-level program to switch to privileged state, it has to raise an interrupt (for example, supervisor call [SVC]) and write to CONTROL[0] within the handler.

3.4 EXCEPTIONS AND INTERRUPTS

The Cortex-M3 supports a number of exceptions, including a fixed number of system exceptions and a number of interrupts, commonly called *IRQ*. The number of interrupt inputs on a Cortex-M3 microcontroller depends on the individual design. Interrupts generated by peripherals, except System Tick Timer, are also connected to the interrupt input signals. The typical number of interrupt inputs is 16 or 32. However, you might find some microcontroller designs with more (or fewer) interrupt inputs.

Besides the interrupt inputs, there is also a nonmaskable interrupt (NMI) input signal. The actual use of NMI depends on the design of the microcontroller or system-on-chip (SoC) product you use. In most cases, the NMI could be connected to a watchdog timer or a voltage-monitoring block that warns the processor when the voltage drops below a certain level. The NMI exception can be activated any time, even right after the core exits reset.

The list of exceptions found in the Cortex-M3 is shown in Table 3.4. A number of the system exceptions are fault-handling exceptions that can be triggered by various error conditions. The NVIC also provides a number of fault status registers so that error handlers can determine the cause of the exceptions.

More details on exception operations in the Cortex-M3 processor are discussed in Chapters 7 to 9.

Table 3.4 Exception Types in Cortex-M3

Exception Number	Exception Type	Priority	Function
1	Reset	−3 (Highest)	Reset
2	NMI	−2	Nonmaskable interrupt
3	Hard fault	−1	All classes of fault, when the corresponding fault handler cannot be activated because it is currently disabled or masked by exception masking
4	MemManage	Settable	Memory management fault; caused by MPU violation or invalid accesses (such as an instruction fetch from a nonexecutable region)
5	Bus fault	Settable	Error response received from the bus system; caused by an instruction prefetch abort or data access error
6	Usage fault	Settable	Usage fault; typical causes are invalid instructions or invalid state transition attempts (such as trying to switch to ARM state in the Cortex-M3)
7–10	—	—	Reserved
11	SVC	Settable	Supervisor call via SVC instruction
12	Debug monitor	Settable	Debug monitor
13	—	—	Reserved
14	PendSV	Settable	Pendable request for system service
15	SYSTICK	Settable	System tick timer
16–255	IRQ	Settable	IRQ input #0–239

Table 3.5 Vector Table Definition after Reset

Exception Type	Address Offset	Exception Vector
18–255	0x48–0x3FF	IRQ #2–239
17	0x44	IRQ #1
16	0x40	IRQ #0
15	0x3C	SYSTICK
14	0x38	PendSV
13	0x34	Reserved
12	0x30	Debug monitor
11	0x2C	SVC
7–10	0x1C–0x28	Reserved
6	0x18	Usage fault
5	0x14	Bus fault
4	0x10	MemManage fault
3	0x0C	Hard fault
2	0x08	NMI
1	0x04	Reset
0	0x00	Starting value of the MSP

3.5 VECTOR TABLES

When an exception event takes place on the Cortex-M3 and is accepted by the processor core, the corresponding exception handler is executed. To determine the starting address of the exception handler, a vector table mechanism is used. The *vector table* is an array of word data inside the system memory, each representing the starting address of one exception type. The vector table is relocatable, and the relocation is controlled by a relocation register in the NVIC (see Table 3.5). After reset, this relocation control register is reset to 0; therefore, the vector table is located in address 0x0 after reset.

For example, if the reset is exception type 1, the address of the reset vector is 1 times 4 (each word is 4 bytes), which equals 0x00000004, and NMI vector (type 2) is located in $2 \times 4 = 0x00000008$. The address 0x00000000 is used to store the starting value for the MSP.

The LSB of each exception vector indicates whether the exception is to be executed in the Thumb state. Because the Cortex-M3 can support only Thumb instructions, the LSB of all the exception vectors should be set to 1.

3.6 STACK MEMORY OPERATIONS

In the Cortex-M3, besides normal software-controlled stack PUSH and POP, the stack PUSH and POP operations are also carried out automatically when entering or exiting an exception/interrupt handler. In this section, we examine the software stack operations. (Stack operations during exception handling are covered in Chapter 9.)

3.6.1 Basic Operations of the Stack

In general, stack operations are memory write or read operations, with the address specified by an SP. Data in registers is saved into stack memory by a PUSH operation and can be restored to registers later by a POP operation. The SP is adjusted automatically in PUSH and POP so that multiple data PUSH will not cause old stacked data to be erased.

The function of the stack is to store register contents in memory so that they can be restored later, after a processing task is completed. For normal uses, for each store (PUSH), there must be a corresponding read (POP), and the address of the POP operation should match that of the PUSH operation (see Figure 3.11). When PUSH/POP instructions are used, the SP is incremented/decremented automatically.

When program control returns to the main program, the R0–R2 contents are the same as before. Notice the order of PUSH and POP: The POP order must be the reverse of PUSH.

These operations can be simplified, thanks to PUSH and POP instructions allowing multiple load and store. In this case, the ordering of a register POP is automatically reversed by the processor (see Figure 3.12).

You can also combine RETURN with a POP operation. This is done by pushing the LR to the stack and popping it back to PC at the end of the subroutine (see Figure 3.13).

3.6.2 Cortex-M3 Stack Implementation

The Cortex-M3 uses a full-descending stack operation model. The SP points to the last data pushed to the stack memory, and the SP decrements before a new PUSH operation. See Figure 3.14 for an example showing execution of the instruction PUSH {R0}.

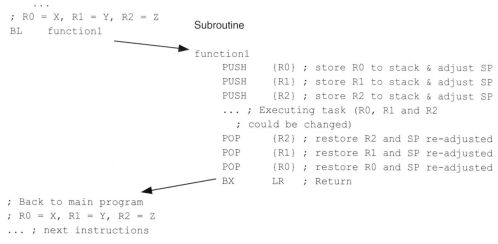

FIGURE 3.11

Stack Operation Basics: One Register in Each Stack Operation.

Main program

```
        . . .
    ; R0 = X, R1 = Y, R2 = Z          Subroutine
    BL    function 1

                                      function 1
                                          PUSH    {R0-R2} ; Store R0, R1, R2 to stack
                                          ... ; Executing task (R0, R1 and R2
                                            ; could be changed)
                                          POP     {R0-R2} ; restore R0, R1, R2
                                          BX      LR    ; Return


    ; Back to main program
    ; R0 = X, R1 = Y, R2 = Z
    ... ; next instructions
```

FIGURE 3.12

Stack Operation Basics: Multiple Register Stack Operation.

Main program

```
        . . .
    ; R0 = X, R1 = Y, R2 = Z          Subroutine
    BL    function 1

                                      function 1
                                          PUSH    {R0-R2, LR} ; Save registers
                                                    ; including link register
                                          ... ; Executing task (R0, R1 and R2
                                            ; could be changed)
                                          POP     {R0-R2, PC} ; Restore registers and
                                                    ; return

    ; Back to main program
    ; R0 = X, R1 = Y, R2 = Z
    ... ; next instructions
```

FIGURE 3.13

Stack Operation Basics: Combining Stack POP and RETURN.

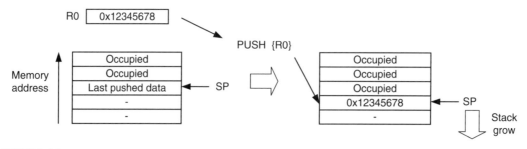

FIGURE 3.14

Cortex-M3 Stack PUSH Implementation.

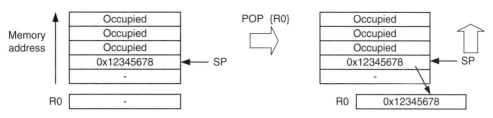

FIGURE 3.15

Cortex-M3 Stack POP Implementation.

For POP operations, the data is read from the memory location pointer by SP, and then, the SP is incremented. The contents in the memory location are unchanged but will be overwritten when the next PUSH operation takes place (see Figure 3.15).

Because each PUSH/POP operation transfers 4 bytes of data (each register contains 1 word, or 4 bytes), the SP decrements/increments by 4 at a time or a multiple of 4 if more than 1 register is pushed or popped.

In the Cortex-M3, R13 is defined as the SP. When an interrupt takes place, a number of registers will be pushed automatically, and R13 will be used as the SP for this stacking process. Similarly, the pushed registers will be restored/popped automatically when exiting an interrupt handler, and the SP will also be adjusted.

3.6.3 The Two-Stack Model in the Cortex-M3

As mentioned before, the Cortex-M3 has two SPs: the MSPS and the PSP. The SP register to be used is controlled by the control register bit 1 (CONTROL[1] in the following text).

When CONTROL[1] is 0, the MSP is used for both thread mode and handler mode (see Figure 3.16). In this arrangement, the main program and the exception handlers share the same stack memory region. This is the default setting after power-up.

When the CONTROL[1] is 1, the PSP is used in thread mode (see Figure 3.17). In this arrangement, the main program and the exception handler can have separate stack memory regions. This can prevent

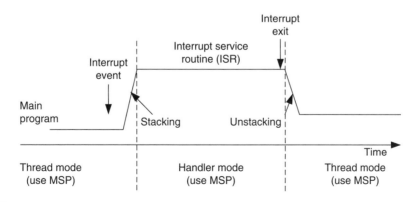

FIGURE 3.16

CONTROL[1]=0: Both Thread Level and Handler Use Main Stack.

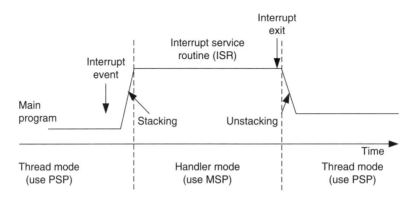

FIGURE 3.17

CONTROL[1]=1: Thread Level Uses Process Stack and Handler Uses Main Stack.

a stack error in a user application from damaging the stack used by the OS (assuming that the user application runs only in thread mode and the OS kernel executes in handler mode).

Note that in this situation, the automatic stacking and unstacking mechanism will use PSP, whereas stack operations inside the handler will use MSP.

It is possible to perform read/write operations directly to the MSP and PSP, without any confusion of which R13 you are referring to. Provided that you are in privileged level, you can access MSP and PSP values:

```
x = __get_MSP(); // Read the value of MSP
__set_MSP(x); // Set the value of MSP
x = __get_PSP(); // Read the value of PSP
__set_PSP(x); // Set the value of PSP
```

In general, it is not recommended to change current selected SP values in a C function, as the stack memory could be used for storing local variables. To access the SPs in assembly, you can use the MRS and MSR instructions:

```
MRS R0, MSP   ; Read Main Stack Pointer to R0
MSR MSP, R0   ; Write R0 to Main Stack Pointer
MRS R0, PSP   ; Read Process Stack Pointer to R0
MSR PSP, R0   ; Write R0 to Process Stack Pointer
```

By reading the PSP value using an MRS instruction, the OS can read data stacked by the user application (such as register contents before SVC). In addition, the OS can change the PSP pointer value—for example, during context switching in multitasking systems.

3.7 RESET SEQUENCE

After the processor exits reset, it will read two words from memory (see Figure 3.18):

- Address 0x00000000: Starting value of R13 (the SP)
- Address 0x00000004: Reset vector (the starting address of program execution; LSB should be set to 1 to indicate Thumb state)

This differs from traditional ARM processor behavior. Previous ARM processors executed program code starting from address 0x0. Furthermore, the vector table in previous ARM devices was instructions (you have to put a branch instruction there so that your exception handler can be put in another location).

In the Cortex-M3, the initial value for the MSP is put at the beginning of the memory map, followed by the vector table, which contains vector address values. (The vector table can be relocated to another location later, during program execution.) In addition, the contents of the vector table are address values

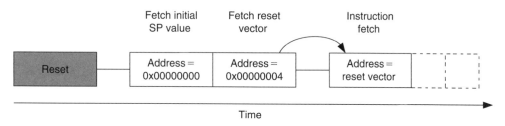

FIGURE 3.18

Reset Sequence.

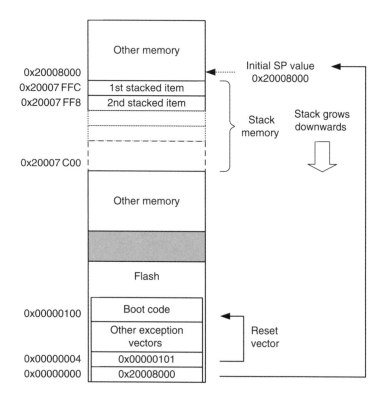

FIGURE 3.19

Initial Stack Pointer Value and Initial Program Counter Value Example.

not branch instructions. The first vector in the vector table (exception type 1) is the reset vector, which is the second piece of data fetched by the processor after reset.

Because the stack operation in the Cortex-M3 is a full descending stack (SP decrement before store), the initial SP value should be set to the first memory after the top of the stack region. For example, if you have a stack memory range from 0x20007C00 to 0x20007FFF (1 KB), the initial stack value should be set to 0x20008000.

The vector table starts after the initial SP value. The first vector is the reset vector. Notice that in the Cortex-M3, vector addresses in the vector table should have their LSB set to 1 to indicate that they are Thumb code. For that reason, the previous example has 0x101 in the reset vector, whereas the boot code starts at address 0x100 (see Figure 3.19). After the reset vector is fetched, the Cortex-M3 can then start to execute the program from the reset vector address and begin normal operations. It is necessary to have the SP initialized, because some of the exceptions (such as NMI) can happen right after reset, and the stack memory could be required for the handler of those exceptions.

Various software development tools might have different ways to specify the starting SP value and reset vector. If you need more information on this topic, it's best to look at project examples provided with the development tools. Simple examples are provided in Chapters 10 and 20 for ARM tools and in Chapter 19 for the GNU tool chain.

Instruction Sets

IN THIS CHAPTER

Assembly Basics 43
Instruction List 46
Instruction Descriptions 52
Several Useful Instructions in the Cortex-M3 70

This chapter provides some insight into the instruction set in the Cortex™-M3 and examples for a number of instructions. You'll also find more information on the instruction set in Appendix A of this book. For complete details of each instruction, refer to the *ARM v7-M Architecture Application Level Reference Manual* [Ref. 2] or user guides from microcontroller vendors.

4.1 ASSEMBLY BASICS

Here, we introduce some basic syntax of ARM assembly to make it easier to understand the rest of the code examples in this book. Most of the assembly code examples in this book are based on the ARM assembler tools, with the exception of those in Chapter 19, which focus on the Gnu's Not Unix tool chain.

4.1.1 Assembler Language: Basic Syntax

In assembler code, the following instruction formatting is commonly used:

```
label
        opcode operand1, operand2, ...; Comments
```

The *label* is optional. Some of the instructions might have a label in front of them so that the address of the instructions can be determined using the label. Then, you will find the opcode (the instruction) followed by a number of operands. Normally, the first operand is the destination of the operation. The number of operands in an instruction depends on the type of instruction, and the syntax format of the operand can also be different. For example, immediate data are usually in the form *#number*, as shown here:

```
MOV R0, #0x12 ; Set R0 = 0x12 (hexadecimal)
MOV R1, #'A'  ; Set R1 = ASCII character A
```

The text after each semicolon (;) is a comment. These comments do not affect the program operation, but they can make programs easier for humans to understand.

You can define constants using EQU, and then use them inside your program code. For example,

```
NVIC_IRQ_SETEN0 EQU 0xE000E100
NVIC_IRQ0_ENABLE EQU 0x1
    ...
    LDR R0,=NVIC_IRQ_SETEN0; ; LDR here is a pseudo-instruction that
                            ; convert to a PC relative load by
                            ; assembler.
    MOV R1,#NVIC_IRQ0_ENABLE ; Move immediate data to register
    STR R1,[R0]              ; Enable IRQ 0 by writing R1 to address
                            ; in R0
```

A number of data definition directives are available for insertion of constants inside assembly code. For example, DCI (Define Constant Instruction) can be used to code an instruction if your assembler cannot generate the exact instruction that you want and if you know the binary code for the instruction.

```
DCI 0xBE00 ; Breakpoint (BKPT 0), a 16-bit instruction
```

We can use DCB (Define Constant Byte) for byte size constant values, such as characters, and Define Constant Data (DCD) for word size constant values to define binary data in your code.

```
        LDR R3,=MY_NUMBER  ; Get the memory address value of MY_NUMBER
        LDR R4,[R3]        ; Get the value code 0x12345678 in R4
        ...
        LDR R0,=HELLO_TXT  ; Get the starting memory address of
                           ; HELLO_TXT
        BL PrintText       ; Call a function called PrintText to
                           ; display string
        ...
MY_NUMBER
        DCD 0x12345678
HELLO_TXT
        DCB "Hello\n",0    ; null terminated string
```

Note that the assembler syntax depends on which assembler tool you are using. Here, the ARM assembler tools syntax is introduced. For syntax of other assemblers, it is best to start from the code examples provided with the tools.

4.1.2 Assembler Language: Use of Suffixes

In assembler for ARM processors, instructions can be followed by suffixes, as shown in Table 4.1.

For the Cortex-M3, the conditional execution suffixes are usually used for branch instructions. However, other instructions can also be used with the conditional execution suffixes if they are inside an IF-THEN instruction block. (This concept is introduced in a later part of this chapter.) In those

Table 4.1 Suffixes in Instructions

Suffix	Description
S	Update Application Program Status register (APSR) (flags); for example: `ADDS RO, R1 ; this will update APSR`
EQ, NE, LT, GT, and so on	Conditional execution; EQ = Equal, NE = Not Equal, LT = Less Than, GT = Greater Than, and so forth. For example: `BEQ <Label> ; Branch if equal`

cases, the *S* suffix and the conditional execution suffixes can be used at the same time. Fifteen condition choices are available, as described later in this chapter.

4.1.3 Assembler Language: Unified Assembler Language

To support and get the best out of the Thumb®-2 instruction set, the Unified Assembler Language (UAL) was developed to allow selection of 16-bit and 32-bit instructions and to make it easier to port applications between ARM code and Thumb code by using the same syntax for both. (With UAL, the syntax of Thumb instructions is now the same as for ARM instructions.)

```
ADD RO, R1       ; RO = RO + R1, using Traditional Thumb syntax
ADD RO, RO, R1 ; Equivalent instruction using UAL syntax
```

The traditional Thumb syntax can still be used. The choice between whether the instructions are interpreted as traditional Thumb code or the new UAL syntax is normally defined by the directive in the assembly file. For example, with ARM assembler tool, a program code header with "CODE16" directive implies the code is in the traditional Thumb syntax, and "THUMB" directive implies the code is in the new UAL syntax.

One thing you need to be careful with reusing traditional Thumb is that some instructions change the flags in APSR, even if the *S* suffix is not used. However, when the UAL syntax is used, whether the instruction changes the flag depends on the *S* suffix. For example,

```
AND  RO, R1     ; Traditional Thumb syntax
ANDS RO, RO, R1 ; Equivalent UAL syntax (S suffix is added)
```

With the new instructions in Thumb-2 technology, some of the operations can be handled by either a Thumb instruction or a Thumb-2 instruction. For example, R0 = R0 + 1 can be implemented as a 16-bit Thumb instruction or a 32-bit Thumb-2 instruction. With UAL, you can specify which instruction you want by adding suffixes:

```
ADDS    RO, #1 ; Use 16-bit Thumb instruction by default
               ; for smaller size
ADDS.N RO, #1 ; Use 16-bit Thumb instruction (N=Narrow)
ADDS.W RO, #1 ; Use 32-bit Thumb-2 instruction (W=wide)
```

The .W (wide) suffix specifies a 32-bit instruction. If no suffix is given, the assembler tool can choose either instruction but usually defaults to 16-bit Thumb code to get a smaller size. Depending on tool support, you may also use the .N (narrow) suffix to specify a 16-bit Thumb instruction.

Again, this syntax is for ARM assembler tools. Other assemblers might have slightly different syntax. If no suffix is given, the assembler might choose the instruction for you, with the minimum code size.

In most cases, applications will be coded in C, and the C compilers will use 16-bit instructions if possible due to smaller code size. However, when the immediate data exceed a certain range or when the operation can be better handled with a 32-bit Thumb-2 instruction, the 32-bit instruction will be used.

The 32-bit Thumb-2 instructions can be half word aligned. For example, you can have a 32-bit instruction located in a half word location.

```
0x1000 : LDR r0,[r1] ;a 16-bit instructions (occupy 0x1000-0x1001)
0x1002 : RBIT.W r0   ;a 32-bit Thumb-2 instruction (occupy
                     ;  0x1002-0x1005)
```

Most of the 16-bit instructions can only access registers R0–R7; 32-bit Thumb-2 instructions do not have this limitation. However, use of PC (R15) might not be allowed in some of the instructions. Refer to the *ARM v7-M Architecture Application Level Reference Manual* [Ref. 2] (section A4.6) if you need to find out more detail in this area.

4.2 INSTRUCTION LIST

The supported instructions are listed in Tables 4.2 through 4.9. The complete details of each instruction are available in the *ARM v7-M Architecture Application Level Reference Manual* [Ref. 2]. There is also information of the supported instruction sets in Appendix A.

Table 4.2 16-Bit Data Processing Instructions

Instruction	Function
ADC	Add with carry
ADD	Add
ADR	Add PC and an immediate value and put the result in a register
AND	Logical AND
ASR	Arithmetic shift right
BIC	Bit clear (Logical AND one value with the logic inversion of another value)
CMN	Compare negative (compare one data with two's complement of another data and update flags)
CMP	Compare (compare two data and update flags)
CPY	Copy (available from architecture v6; move a value from one high or low register to another high or low register); synonym of MOV instruction
EOR	Exclusive OR
LSL	Logical shift left
LSR	Logical shift right
MOV	Move (can be used for register-to-register transfers or loading immediate data)
MUL	Multiply
MVN	Move NOT (obtain logical inverted value)
NEG	Negate (obtain two's complement value), equivalent to RSB

Table 4.2 16-Bit Data Processing Instructions *Continued*

Instruction	Function
ORR	Logical OR
RSB	Reverse subtract
ROR	Rotate right
SBC	Subtract with carry
SUB	Subtract
TST	Test (use as logical AND; Z flag is updated but AND result is not stored)
REV	Reverse the byte order in a 32-bit register (available from architecture v6)
REV16	Reverse the byte order in each 16-bit half word of a 32-bit register (available from architecture v6)
REVSH	Reverse the byte order in the lower 16-bit half word of a 32-bit register and sign extends the result to 32 bits (available from architecture v6)
SXTB	Signed extend byte (available from architecture v6)
SXTH	Signed extend half word (available from architecture v6)
UXTB	Unsigned extend byte (available from architecture v6)
UXTH	Unsigned extend half word (available from architecture v6)

Table 4.3 16-Bit Branch Instructions

Instruction	Function
B	Branch
B<cond>	Conditional branch
BL	Branch with link; call a subroutine and store the return address in LR (this is actually a 32-bit instruction, but it is also available in Thumb in traditional ARM processors)
BLX	Branch with link and change state (BLX <reg> only)[1]
BX <reg>	Branch with exchange state
CBZ	Compare and branch if zero (architecture v7)
CBNZ	Compare and branch if nonzero (architecture v7)
IT	IF-THEN (architecture v7)

Table 4.4 16-Bit Load and Store Instructions

Instruction	Function
LDR	Load word from memory to register
LDRH	Load half word from memory to register
LDRB	Load byte from memory to register

Continued

---

[1]BLX with immediate is not supported because it will always try to change to the ARM state, which is not supported in the Cortex-M3. Attempts to use BLX <reg> to change to the ARM state will also result in a fault exception.

Table 4.4 16-Bit Load and Store Instructions *Continued*

Instruction	Function
LDRSH	Load half word from memory, sign extend it, and put it in register
LDRSB	Load byte from memory, sign extend it, and put it in register
STR	Store word from register to memory
STRH	Store half word from register to memory
STRB	Store byte from register to memory
LDM/LDMIA	Load multiple/Load multiple increment after
STM/STMIA	Store multiple/Store multiple increment after
PUSH	Push multiple registers
POP	Pop multiple registers

Table 4.5 Other 16-Bit Instructions

Instruction	Function
SVC	Supervisor call
SEV	Send event
WFE	Sleep and wait for event
WFI	Sleep and wait for interrupt
BKPT	Breakpoint; if debug is enabled, it will enter debug mode (halted), or if debug monitor exception is enabled, it will invoke the debug exception; otherwise, it will invoke a fault exception
NOP	No operation
CPSIE	Enable PRIMASK (CPSIE i)/FAULTMASK (CPSIE f) register (set the register to 0)
CPSID	Disable PRIMASK (CPSID i)/ FAULTMASK (CPSID f) register (set the register to 1)

Table 4.6 32-Bit Data Processing Instructions

Instruction	Function
ADC	Add with carry
ADD	Add
ADDW	Add wide (#immed_12)
ADR	Add PC and an immediate value and put the result in a register
AND	Logical AND
ASR	Arithmetic shift right
BIC	Bit clear (logical AND one value with the logic inversion of another value)
BFC	Bit field clear
BFI	Bit field insert
CMN	Compare negative (compare one data with two's complement of another data and update flags)

Table 4.6 32-Bit Data Processing Instructions *Continued*

Instruction	Function
CMP	Compare (compare two data and update flags)
CLZ	Count leading zero
EOR	Exclusive OR
LSL	Logical shift left
LSR	Logical shift right
MLA	Multiply accumulate
MLS	Multiply and subtract
MOV	Move
MOVW	Move wide (write a 16-bit immediate value to register)
MOVT	Move top (write an immediate value to the top half word of destination reg)
MVN	Move negative
MUL	Multiply
ORR	Logical OR
ORN	Logical OR NOT
RBIT	Reverse bit
REV	Byte reverse word
REV16	Byte reverse packed half word
REVSH	Byte reverse signed half word
ROR	Rotate right
RSB	Reverse subtract
RRX	Rotate right extended
SBC	Subtract with carry
SBFX	Signed bit field extract
SDIV	Signed divide
SMLAL	Signed multiply accumulate long
SMULL	Signed multiply long
SSAT	Signed saturate
SBC	Subtract with carry
SUB	Subtract
SUBW	Subtract wide (#immed_12)
SXTB	Sign extend byte
SXTH	Sign extend half word
TEQ	Test equivalent (use as logical exclusive OR; flags are updated but result is not stored)
TST	Test (use as logical AND; Z flag is updated but AND result is not stored)
UBFX	Unsigned bit field extract
UDIV	Unsigned divide
UMLAL	Unsigned multiply accumulate long
UMULL	Unsigned multiply long
USAT	Unsigned saturate

Continued

Table 4.6 32-Bit Data Processing Instructions *Continued*

Instruction	Function
UXTB	Unsigned extend byte
UXTH	Unsigned extend half word

Table 4.7 32-Bit Load and Store Instructions

Instruction	Function
LDR	Load word data from memory to register
LDRT	Load word data from memory to register with unprivileged access
LDRB	Load byte data from memory to register
LDRBT	Load byte data from memory to register with unprivileged access
LDRH	Load half word data from memory to register
LDRHT	Load half word data from memory to register with unprivileged access
LDRSB	Load byte data from memory, sign extend it, and put it to register
LDRSBT	Load byte data from memory with unprivileged access, sign extend it, and put it to register
LDRSH	Load half word data from memory, sign extend it, and put it to register
LDRSHT	Load half word data from memory with unprivileged access, sign extend it, and put it to register
LDM/LDMIA	Load multiple data from memory to registers
LDMDB	Load multiple decrement before
LDRD	Load double word data from memory to registers
STR	Store word to memory
STRT	Store word to memory with unprivileged access
STRB	Store byte data to memory
STRBT	Store byte data to memory with unprivileged access
STRH	Store half word data to memory
STRHT	Store half word data to memory with unprivileged access
STM/STMIA	Store multiple words from registers to memory
STMDB	Store multiple decrement before
STRD	Store double word data from registers to memory
PUSH	Push multiple registers
POP	Pop multiple registers

Table 4.8 32-Bit Branch Instructions

Instruction	Function
B	Branch
B<cond>	Conditional branch
BL	Branch and link
TBB	Table branch byte; forward branch using a table of single byte offset
TBH	Table branch half word; forward branch using a table of half word offset

Table 4.9 Other 32-Bit Instructions

Instruction	Function
LDREX	Exclusive load word
LDREXH	Exclusive load half word
LDREXB	Exclusive load byte
STREX	Exclusive store word
STREXH	Exclusive store half word
STREXB	Exclusive store byte
CLREX	Clear the local exclusive access record of local processor
MRS	Move special register to general-purpose register
MSR	Move to special register from general-purpose register
NOP	No operation
SEV	Send event
WFE	Sleep and wait for event
WFI	Sleep and wait for interrupt
ISB	Instruction synchronization barrier
DSB	Data synchronization barrier
DMB	Data memory barrier

Table 4.10 Unsupported Thumb Instructions for Traditional ARM Processors

Unsupported Instruction	Function
BLX label	This is branch with link and exchange state. In a format with immediate data, BLX always changes to ARM state. Because the Cortex-M3 does not support the ARM state, instructions like this one that attempt to switch to the ARM state will result in a fault exception called *usage fault*.
SETEND	This Thumb instruction, introduced in architecture v6, switches the endian configuration during run time. Since the Cortex-M3 does not support dynamic endian, using the SETEND instruction will result in a fault exception.

4.2.1 **Unsupported Instructions**

A number of Thumb instructions are not supported in the Cortex-M3; they are presented in Table 4.10.

A number of instructions listed in the *ARM v7-M Architecture Application Level Reference Manual* are not supported in the Cortex-M3. ARM v7-M architecture allows Thumb-2 coprocessor instructions, but the Cortex-M3 processor does not have any coprocessor support. Therefore, executing the coprocessor instructions shown in Table 4.11 will result in a fault exception (Usage Fault with No-Coprocessor "NOCP" bit in Usage Fault Status Register in NVIC set to 1).

Some of the change process state (CPS) instructions are also not supported in the Cortex-M3 (see Table 4.12). This is because the Program Status register (PSR) definition has changed, so some bits defined in the ARM architecture v6 are not available in the Cortex-M3.

Table 4.11 Unsupported Coprocessor Instructions

Unsupported Instruction	Function
MCR	Move to coprocessor from ARM processor
MCR2	Move to coprocessor from ARM processor
MCRR	Move to coprocessor from two ARM register
MRC	Move to ARM register from coprocessor
MRC2	Move to ARM register from coprocessor
MRRC	Move to two ARM registers from coprocessor
LDC	Load coprocessor; load memory data from a sequence of consecutive memory addresses to a coprocessor
STC	Store coprocessor; stores data from a coprocessor to a sequence of consecutive memory addresses

Table 4.12 Unsupported Change Process State Instructions

Unsupported Instruction	Function
CPS<IEIID>.W A	There is no A bit in the Cortex-M3
CPS.W #mode	There is no mode bit in the Cortex-M3 PSR

Table 4.13 Unsupported Hint Instructions

Unsupported Instruction	Function
DBG	A hint instruction to debug and trace system
PLD	Preload data; this is a hint instruction for cache memory, however, since there is no cache in the Cortex-M3 processor, this instruction behaves as NOP
PLI	Preload instruction; this is a hint instruction for cache memory, however, since there is no cache in the Cortex-M3 processor, this instruction behaves as NOP
YIELD	A hint instruction to allow multithreading software to indicate to hardware that it is doing a task that can be swapped out to improve overall system performance.

In addition, the hint instructions shown in Table 4.13 will behave as NOP in the Cortex-M3.

All other undefined instructions, when executed, will cause the usage fault exception to take place.

4.3 INSTRUCTION DESCRIPTIONS

Here, we introduce some of the commonly used syntax for ARM assembly code. Some of the instructions have various options such as barrel shifter; these will not be fully covered in this chapter.

4.3.1 Assembler Language: Moving Data

One of the most basic functions in a processor is transfer of data. In the Cortex-M3, data transfers can be of one of the following types:

- Moving data between register and register
- Moving data between memory and register
- Moving data between special register and register
- Moving an immediate data value into a register

The command to move data between registers is MOV (move). For example, moving data from register R3 to register R8 looks like this:

```
MOV R8, R3
```

Another instruction can generate the negative value of the original data; it is called MVN (move negative).

The basic instructions for accessing memory are Load and Store. Load (LDR) transfers data from memory to registers, and Store transfers data from registers to memory. The transfers can be in different data sizes (byte, half word, word, and double word), as outlined in Table 4.14.

Multiple Load and Store operations can be combined into single instructions called LDM (Load Multiple) and STM (Store Multiple), as outlined in Table 4.15.

The exclamation mark (!) in the instruction specifies whether the register *Rd* should be updated after the instruction is completed. For example, if R8 equals 0x8000:

```
STMIA.W R8!, {R0-R3} ; R8 changed to 0x8010 after store
                     ; (increment by 4 words)
STMIA.W R8 , {R0-R3} ; R8 unchanged after store
```

ARM processors also support memory accesses with preindexing and postindexing. For preindexing, the register holding the memory address is adjusted. The memory transfer then takes place with the updated address. For example,

```
LDR.W R0,[R1, #offset]! ; Read memory[R1+offset], with R1
                        ; update to R1+offset
```

Table 4.14 Commonly Used Memory Access Instructions

Example	Description
LDRB Rd, [Rn, #offset]	Read byte from memory location Rn + offset
LDRH Rd, [Rn, #offset]	Read half word from memory location Rn + offset
LDR Rd, [Rn, #offset]	Read word from memory location Rn + offset
LDRD Rd1,Rd2, [Rn, #offset]	Read double word from memory location Rn + offset
STRB Rd, [Rn, #offset]	Store byte to memory location Rn + offset
STRH Rd, [Rn, #offset]	Store half word to memory location Rn + offset
STR Rd, [Rn, #offset]	Store word to memory location Rn + offset
STRD Rd1,Rd2, [Rn, #offset]	Store double word to memory location Rn + offset

Table 4.15 Multiple Memory Access Instructions

Example	Description
LDMIA Rd!,<reg list>	Read multiple words from memory location specified by *Rd*; address increment after (IA) each transfer (16-bit Thumb instruction)
STMIA Rd!,<reg list>	Store multiple words to memory location specified by *Rd*; address increment after (IA) each transfer (16-bit Thumb instruction)
LDMIA.W Rd(!),<reg list>	Read multiple words from memory location specified by *Rd*; address increment after each read (.W specified it is a 32-bit Thumb-2 instruction)
LDMDB.W Rd(!),<reg list>	Read multiple words from memory location specified by *Rd*; address Decrement Before (DB) each read (.W specified it is a 32-bit Thumb-2 instruction)
STMIA.W Rd(!),<reg list>	Write multiple words to memory location specified by *Rd*; address increment after each read (.W specified it is a 32-bit Thumb-2 instruction)
STMDB.W Rd(!),<reg list>	Write multiple words to memory location specified by *Rd*; address DB each read (.W specified it is a 32-bit Thumb-2 instruction)

Table 4.16 Examples of Preindexing Memory Access Instructions

Example	Description
LDR.W Rd, [Rn, #offset]! LDRB.W Rd, [Rn, #offset]! LDRH.W Rd, [Rn, #offset]! LDRD.W Rd1, Rd2,[Rn, #offset]!	Preindexing load instructions for various sizes (word, byte, half word, and double word)
LDRSB.W Rd, [Rn, #offset]! LDRSH.W Rd, [Rn, #offset]!	Preindexing load instructions for various sizes with sign extend (byte, half word)
STR.W Rd, [Rn, #offset]! STRB.W Rd, [Rn, #offset]! STRH.W Rd, [Rn, #offset]! STRD.W Rd1, Rd2,[Rn, #offset]!	Preindexing store instructions for various sizes (word, byte, half word, and double word)

The use of the "!" indicates the update of base register R1. The "!" is optional; without it, the instruction would be just a normal memory transfer with offset from a base address. The preindexing memory access instructions include load and store instructions of various transfer sizes (see Table 4.16).

Postindexing memory access instructions carry out the memory transfer using the base address specified by the register and then update the address register afterward. For example,

```
LDR.W R0,[R1], #offset ; Read memory[R1], with R1
                       ; updated to R1+offset
```

When a postindexing instruction is used, there is no need to use the "!" sign, because all postindexing instructions update the base address register, whereas in preindexing you might choose whether to update the base address register or not.

Similarly to preindexing, postindexing memory access instructions are available for different transfer sizes (see Table 4.17).

Table 4.17 Examples of Postindexing Memory Access Instructions

Example	Description

```LDR.W   Rd, [Rn], #offset```
```LDRB.W   Rd, [Rn], #offset```
```LDRH.W   Rd, [Rn], #offset```
```LDRD.W   Rd1, Rd2,[Rn], #offset```
	Postindexing load instructions for various sizes (word, byte, half word, and double word)

```LDRSB.W Rd, [Rn], #offset``` ```LDRSH.W Rd, [Rn], #offset```	Postindexing load instructions for various sizes with sign extend (byte, half word)

```STR.W   Rd, [Rn], #offset```
```STRB.W   Rd, [Rn], #offset```
```STRH.W   Rd, [Rn], #offset```
```STRD.W   Rd1, Rd2,[Rn], #offset```
	Postindexing store instructions for various sizes (word, byte, half word, and double word)

Two other types of memory operation are stack PUSH and stack POP. For example,

```
PUSH {R0, R4-R7, R9} ; Push R0, R4, R5, R6, R7, R9 into
                     ; stack memory
POP {R2,R3}          ; Pop R2 and R3 from stack
```

Usually a PUSH instruction will have a corresponding POP with the same register list, but this is not always necessary. For example, a common exception is when POP is used as a function return:

```
PUSH {R0-R3, LR} ; Save register contents at beginning of
                 ; subroutine
....             ; Processing
POP {R0-R3, PC}  ; restore registers and return
```

In this case, instead of popping the LR register back and then branching to the address in LR, we POP the address value directly in the program counter.

As mentioned in Chapter 3, the Cortex-M3 has a number of special registers. To access these registers, we use the instructions MRS and MSR. For example,

```
MRS R0, PSR      ; Read Processor status word into R0
MSR CONTROL, R1  ; Write value of R1 into control register
```

Unless you're accessing the APSR, you can use MSR or MRS to access other special registers only in privileged mode.

Moving immediate data into a register is a common thing to do. For example, you might want to access a peripheral register, so you need to put the address value into a register beforehand. For small values (8 bits or less), you can use MOVS (move). For example,

```
MOVS R0, #0x12 ; Set R0 to 0x12
```

For a larger value (over 8 bits), you might need to use a Thumb-2 move instruction. For example,

```
MOVW.W R0, #0x789A ; Set R0 to 0x789A
```

Or if the value is 32-bit, you can use two instructions to set the upper and lower halves:

```
MOVW.W R0,#0x789A ; Set R0 lower half to 0x789A
MOVT.W R0,#0x3456 ; Set R0 upper half to 0x3456. Now
                  ; R0=0x3456789A
```

Alternatively, you can also use LDR (a pseudo-instruction provided in ARM assembler). For example,

```
LDR R0, =0x3456789A
```

This is not a real assembler command, but the ARM assembler will convert it into a PC relative load instruction to produce the required data. To generate 32-bit immediate data, using LDR is recommended rather than the MOVW.W and MOVT.W combination because it gives better readability and the assembler might be able to reduce the memory being used if the same immediate data are reused in several places of the same program.

4.3.2 LDR and ADR Pseudo-Instructions

Both LDR and ADR pseudo-instructions can be used to set registers to a program address value. They have different syntaxes and behaviors. For LDR, if the address is a program address value, the assembler will automatically set the LSB to 1. For example,

```
      LDR R0, =address1 ; R0 set to 0x4001
      ...
address1          ; address here is 0x4000
      MOV R0, R1 ; address1 contains program code
      ...
```

You will find that the LDR instruction will put 0x4001 into R1; the LSB is set to 1 to indicate that it is Thumb code. If *address1* is a data address, LSB will not be changed. For example,

```
      LDR R0, =address1 ; R0 set to 0x4000
      ...
address1      ; address here is 0x4000
      DCD 0x0 ; address1 contains data
      ...
```

For ADR, you can load the address value of a program code into a register without setting the LSB automatically. For example,

```
      ADR R0, address1
      ...
address1          ; (address here is 0x4000)
      MOV R0, R1 ; address1 contains program code
      ...
```

You will get 0x4000 in the ADR instruction. Note that there is no equal sign (=) in the ADR statement.

LDR obtains the immediate data by putting the data in the program code and uses a PC relative load to get the data into the register. ADR tries to generate the immediate value by adding or subtracting instructions (for example, based on the current PC value). As a result, it is not possible to create all immediate values using ADR, and the target address label must be in a close range. However, using ADR can generate smaller code sizes compared with LDR.

The 16-bit version of ADR requires that the target address must be word aligned (address value is a multiple of 4). If the target address is not word aligned, you can use the 32-bit version of ADR instruction "ADR.W." If the target address is more than ±4095 bytes of current PC, you can use "ADRL" pseudo-instruction, which gives ±1 MB range.

4.3.3 Assembler Language: Processing Data

The Cortex-M3 provides many different instructions for data processing. A few basic ones are introduced here. Many data operation instructions can have multiple instruction formats. For example, an ADD instruction can operate between two registers or between one register and an immediate data value:

```
ADD   R0, R0, R1   ; R0 = R0 + R1
ADDS  R0, R0, #0x12 ; R0 = R0 + 0x12
ADD.W R0, R1, R2    ; R0 = R1 + R2
```

These are all ADD instructions, but they have different syntaxes and binary coding.

With the traditional Thumb instruction syntax, when 16-bit Thumb code is used, an ADD instruction can change the flags in the PSR. However, 32-bit Thumb-2 code can either change a flag or keep it unchanged. To separate the two different operations, the *S* suffix should be used if the following operation depends on the flags:

```
ADD.W  R0, R1, R2 ; Flag unchanged
ADDS.W R0, R1, R2 ; Flag change
```

Aside from ADD instructions, the arithmetic functions that the Cortex-M3 supports include subtract (SUB), multiply (MUL), and unsigned and signed divide (UDIV/SDIV). Table 4.18 shows some of the most commonly used arithmetic instructions.

Table 4.18 Examples of Arithmetic Instructions

Instruction		Operation
`ADD  Rd, Rn, Rm`	`; Rd = Rn + Rm`	ADD operation
`ADD  Rd, Rd, Rm`	`; Rd = Rd + Rm`	
`ADD  Rd, #immed`	`; Rd = Rd + #immed`	
`ADD  Rd, Rn, # immed`	`; Rd = Rn + #immed`	
`ADC  Rd, Rn, Rm`	`; Rd = Rn + Rm + carry`	ADD with carry
`ADC  Rd, Rd, Rm`	`; Rd = Rd + Rm + carry`	
`ADC  Rd, #immed`	`; Rd = Rd + #immed + carry`	
`ADDW Rd, Rn,#immed`	`; Rd = Rn + #immed`	ADD register with 12-bit immediate value
`SUB  Rd, Rn, Rm`	`; Rd = Rn − Rm`	SUBTRACT
`SUB  Rd, #immed`	`; Rd = Rd − #immed`	
`SUB  Rd, Rn,#immed`	`; Rd = Rn − #immed`	
`SBC   Rd, Rm`	`; Rd = Rd − Rm − borrow`	SUBTRACT with borrow (not carry)
`SBC.W Rd, Rn, #immed`	`; Rd = Rn − #immed − borrow`	
`SBC.W Rd, Rn, Rm`	`; Rd = Rn − Rm − borrow`	
`RSB.W Rd, Rn, #immed`	`; Rd = #immed −Rn`	Reverse subtract
`RSB.W Rd, Rn, Rm`	`; Rd = Rm − Rn`	
`MUL   Rd, Rm`	`; Rd = Rd * Rm`	Multiply
`MUL.W Rd, Rn, Rm`	`; Rd = Rn * Rm`	
`UDIV Rd, Rn, Rm`	`; Rd = Rn/Rm`	Unsigned and signed divide
`SDIV Rd, Rn, Rm`	`; Rd = Rn/Rm`	

These instructions can be used with or without the "S" suffix to determine if the APSR should be updated. In most cases, if UAL syntax is selected and if "S" suffix is not used, the 32-bit version of the instructions would be selected as most of the 16-bit Thumb instructions update APSR.

The Cortex-M3 also supports 32-bit multiply instructions and multiply accumulate instructions that give 64-bit results. These instructions support signed or unsigned values (see Table 4.19).

Another group of data processing instructions are the logical operations instructions and logical operations such as AND, ORR (or), and shift and rotate functions. Table 4.20 shows some of the most commonly used logical instructions. These instructions can be used with or without the "S" suffix to determine if the APSR should be updated. If UAL syntax is used and if "S" suffix is not used, the 32-bit version of the instructions would be selected as all of the 16-bit logic operation instructions update APSR.

The Cortex-M3 provides rotate and shift instructions. In some cases, the rotate operation can be combined with other operations (for example, in memory address offset calculation for load/store instructions). For standalone rotate/shift operations, the instructions shown in Table 4.21 are provided. Again, a 32-bit version of the instruction is used if "S" suffix is not used and if UAL syntax is used.

Table 4.19 32-Bit Multiply Instructions

Instruction	Operation
SMULL RdLo, RdHi, Rn, Rm ; {RdHi,RdLo} = Rn * Rm SMLAL RdLo, RdHi, Rn, Rm ; {RdHi,RdLo} += Rn * Rm	32-bit multiply instructions for signed values
UMULL RdLo, RdHi, Rn, Rm ; {RdHi,RdLo} = Rn * Rm UMLAL RdLo, RdHi, Rn, Rm ; {RdHi,RdLo} += Rn * Rm	32-bit multiply instructions for unsigned values

Table 4.20 Logic Operation Instructions

Instruction	Operation
AND Rd, Rn ; Rd = Rd & Rn AND.W Rd, Rn,#immed ; Rd = Rn & #immed AND.W Rd, Rn, Rm ; Rd = Rn & Rd	Bitwise AND
ORRRd, Rn ; Rd = Rd | Rn ORR.W Rd, Rn,#immed ; Rd = Rn | #immed ORR.W Rd, Rn, Rm ; Rd = Rn | Rd	Bitwise OR
BIC Rd, Rn ; Rd = Rd & (~Rn) BIC.W Rd, Rn,#immed ; Rd = Rn &(~#immed) BIC.W Rd, Rn, Rm ; Rd = Rn &(~Rd)	Bit clear
ORN.W Rd, Rn,#immed ; Rd = Rn | (~#immed) ORN.W Rd, Rn, Rm ; Rd = Rn | (~Rd)	Bitwise OR NOT
EOR Rd, Rn ; Rd = Rd ^ Rn EOR.W Rd, Rn,#immed ; Rd = Rn | #immed EOR.W Rd, Rn, Rm ; Rd = Rn | Rd	Bitwise Exclusive OR

Table 4.21 Shift and Rotate Instructions

Instruction	Operation

`ASR    Rd, Rn,#immed ; Rd = Rn » immed`
`ASRRd, Rn          ; Rd = Rd » Rn`
`ASR.W  Rd, Rn, Rm   ; Rd = Rn » Rm`
	Arithmetic shift right

`LSLRd, Rn,#immed    ; Rd = Rn « immed`
`LSLRd, Rn           ; Rd = Rd « Rn`
`LSL.W  Rd, Rn, Rm    ; Rd = Rn « Rm`
	Logical shift left

`LSRRd, Rn,#immed    ; Rd = Rn » immed`
`LSRRd, Rn           ; Rd = Rd » Rn`
`LSR.W  Rd, Rn, Rm    ; Rd = Rn » Rm`
	Logical shift right

`ROR    Rd, Rn       ; Rd rot by Rn`
`ROR.W  Rd, Rn,#immed ; Rd = Rn rot by immed`
`ROR.W  Rd, Rn, Rm    ; Rd = Rn rot by Rm`
	Rotate right

`RRX.W  Rd, Rn       ; {C, Rd} = {Rn, C}`	Rotate right extended

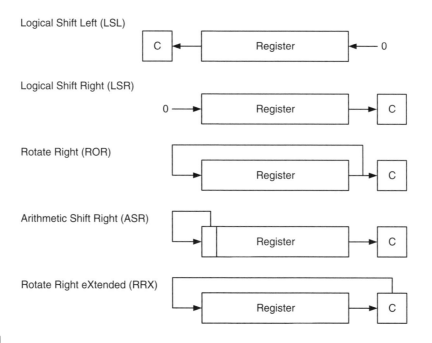

FIGURE 4.1

Shift and Rotate Instructions.

In UAL syntax, the rotate and shift operations can also update the carry flag if the *S* suffix is used (and always update the carry flag if the 16-bit Thumb code is used). See Figure 4.1.

If the shift or rotate operation shifts the register position by multiple bits, the value of the carry flag *C* will be the last bit that shifts out of the register.

WHY IS THERE ROTATE RIGHT BUT NO ROTATE LEFT?

The rotate left operation can be replaced by a rotate right operation with a different rotate offset. For example, a rotate left by 4-bit operation can be written as a rotate right by 28-bit instruction, which gives the same result and takes the same amount of time to execute.

Table 4.22 Sign Extend Instructions

Instruction	Operation
SXTB Rd, Rm ; Rd = signext(Rm[7:0])	Sign extend byte data into word
SXTH Rd, Rm ; Rd = signext(Rm[15:0])	Sign extend half word data into word

Table 4.23 Data Reverse Ordering Instructions

Instruction	Operation
REV Rd, Rn ; Rd = rev(Rn)	Reverse bytes in word
REV16 Rd, Rn ; Rd = rev16(Rn)	Reverse bytes in each half word
REVSH Rd, Rn ; Rd = revsh(Rn)	Reverse bytes in bottom half word and sign extend the result

For conversion of signed data from byte or half word to word, the Cortex-M3 provides the two instructions shown in Table 4.22. Both 16-bit and 32-bit versions are available. The 16-bit version can only access low registers.

Another group of data processing instructions is used for reversing data bytes in a register (see Table 4.23). These instructions are usually used for conversion between little endian and big endian data. See Figure 4.2. Both 16-bit and 32-bit versions are available. The 16-bit version can only access low registers.

The last group of data processing instructions is for bit field processing. They include the instructions shown in Table 4.24. Examples of these instructions are provided in a later part of this chapter.

4.3.4 Assembler Language: Call and Unconditional Branch

The most basic branch instructions are as follows:

```
B label ; Branch to a labeled address
BX reg  ; Branch to an address specified by a register
```

In BX instructions, the LSB of the value contained in the register determines the next state (Thumb/ARM) of the processor. In the Cortex-M3, because it is always in Thumb state, this bit should be set to 1. If it is zero, the program will cause a usage fault exception because it is trying to switch the processor into ARM state (See Figure 4.2.).

To call a function, the branch and link instructions should be used.

```
BL label ; Branch to a labeled address and save return
         ; address in LR
```

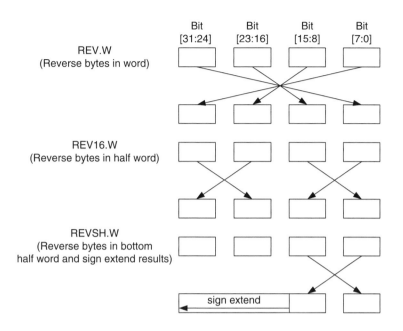

FIGURE 4.2

Operation of Reverse instructions.

Table 4.24 Bit Field Processing and Manipulation Instructions	
Instruction	**Operation**
BFC.W Rd, Rn, #<width>	Clear bit field within a register
BFI.W Rd, Rn, #<lsb>, #<width>	Insert bit field to a register
CLZ.W Rd, Rn	Count leading zero
RBIT.W Rd, Rn	Reverse bit order in register
SBFX.W Rd, Rn, #<lsb>, #<width>	Copy bit field from source and sign extend it
UBFX.W Rd, Rn, #<lsb>, #<width>	Copy bit field from source register

```
BLX reg   ; Branch to an address specified by a register and
          ; save return
          ; address in LR.
```

With these instructions, the return address will be stored in the link register (LR) and the function can be terminated using BX LR, which causes program control to return to the calling process. However, when using BLX, make sure that the LSB of the register is 1. Otherwise the processor will produce a fault exception because it is an attempt to switch to the ARM state.

You can also carry out a branch operation using MOV instructions and LDR instructions. For example,

```
MOV R15, R0   ; Branch to an address inside R0
LDR R15, [R0] ; Branch to an address in memory location
              ; specified by R0
```

```
POP {R15}          ; Do a stack pop operation, and change the
                   ; program counter value
                   ; to the result value.
```

When using these methods to carry out branches, you also need to make sure that the LSB of the new program counter value is 0x1. Otherwise, a usage fault exception will be generated because it will try to switch the processor to ARM mode, which is not allowed in the Cortex-M3 redundancy.

SAVE THE LR IF YOU NEED TO CALL A SUBROUTINE

The BL instruction will destroy the current content of your LR. So, if your program code needs the LR later, you should save your LR before you use BL. The common method is to push the LR to stack in the beginning of your subroutine. For example,

```
main
        ...
        BL functionA
        ...
functionA
        PUSH {LR} ; Save LR content to stack
        ...
        BL functionB
        ...
        POP {PC} ; Use stacked LR content to return to main
functionB
        PUSH {LR}
        ...
        POP {PC} ; Use stacked LR content to return to functionA
```

In addition, if the subroutine you call is a C function, you might also need to save the contents in R0–R3 and R12 if these values will be needed at a later stage. According to *AAPCS* [Ref. 5], the contents in these registers could be changed by a C function.

4.3.5 Assembler Language: Decisions and Conditional Branches

Most conditional branches in ARM processors use flags in the APSR to determine whether a branch should be carried out. In the APSR, there are five flag bits; four of them are used for branch decisions (see Table 4.25).

There is another flag bit at bit[27], called the *Q flag*. It is for saturation math operations and is not used for conditional branches.

Table 4.25 Flag Bits in APSR that Can Be Used for Conditional Branches

Flag	PSR Bit	Description
N	31	Negative flag (last operation result is a negative value)
Z	30	Zero (last operation result returns a zero value)
C	29	Carry (last operation returns a carry out or borrow)
V	28	Overflow (last operation results in an overflow)

> **FLAGS IN ARM PROCESSORS**
>
> Often, data processing instructions change the flags in the PSR. The flags might be used for branch decisions, or they can be used as part of the input for the next instruction. The ARM processor normally contains at least the Z, N, C, and V flags, which are updated by execution of data processing instructions.
>
> - Z (Zero) flag: This flag is set when the result of an instruction has a zero value or when a comparison of two data returns an equal result.
> - N (Negative) flag: This flag is set when the result of an instruction has a negative value (bit 31 is 1).
> - C (Carry) flag: This flag is for unsigned data processing—for example, in add (ADD) it is set when an overflow occurs; in subtract (SUB) it is set when a borrow did not occur (borrow is the invert of carry).
> - V (Overflow) flag: This flag is for signed data processing; for example, in an add (ADD), when two positive values added together produce a negative value, or when two negative values added together produce a positive value.
>
> These flags can also have special results when used with shift and rotate instructions. Refer to the *ARM v7-M Architecture Application Level Reference Manual* [Ref. 2] for details.

With combinations of the four flags (N, Z, C, and V), 15 branch conditions are defined (see Table 4.26). Using these conditions, branch instructions can be written as, for example,

```
BEQ label ; Branch to address 'label' if Z flag is set
```

You can also use the Thumb-2 version if your branch target is further away. For example,

```
BEQ.W label ; Branch to address 'label' if Z flag is set
```

Table 4.26 Conditions for Branches or Other Conditional Operations

Symbol	Condition	Flag
EQ	Equal	Z set
NE	Not equal	Z clear
CS/HS	Carry set/unsigned higher or same	C set
CC/LO	Carry clear/unsigned lower	C clear
MI	Minus/negative	N set
PL	Plus/positive or zero	N clear
VS	Overflow	V set
VC	No overflow	V clear
HI	Unsigned higher	C set and Z clear
LS	Unsigned lower or same	C clear or Z set
GE	Signed greater than or equal	N set and V set, or N clear and V clear (N == V)
LT	Signed less than	N set and V clear, or N clear and V set (N != V)
GT	Signed greater than	Z clear, and either N set and V set, or N clear and V clear (Z == 0, N == V)
LE	Signed less than or equal	Z set, or N set and V clear, or N clear and V set (Z == 1 or N != V)
AL	Always (unconditional)	—

The defined branch conditions can also be used in IF-THEN-ELSE structures. For example,

```
CMP R0, R1     ; Compare R0 and R1
ITTEE GT       ; If R0 > R1 Then
               ; if true, first 2 statements execute,
               ; if false, other 2 statements execute
MOVGT R2, R0 ;     R2 = R0
MOVGT R3, R1 ;     R3 = R1
MOVLE R2, R0 ; Else R2 = R1
MOVLE R3, R1 ;     R3 = R0
```

APSR flags can be affected by the following:

- Most of the 16-bit ALU instructions
- 32-bit (Thumb-2) ALU instructions with the S suffix; for example, ADDS.W
- Compare (e.g., CMP) and Test (e.g., TST, TEQ)
- Write to APSR/xPSR directly

Most of the 16-bit Thumb arithmetic instructions affect the N, Z, C, and V flags. With 32-bit Thumb-2 instructions, the ALU operation can either change flags or not change flags. For example,

```
ADDS.W R0, R1, R2 ; This 32-bit Thumb instruction updates flag
ADD.W  R0, R1, R2 ; This 32-bit Thumb instruction does not
                  ; update flag
```

Be careful when reusing program code from old projects. If the old project is in tradition Thumb syntax; for example, "CODE16" directive is used with ARM assembler, then

```
ADD R0, R1   ; This 16-bit Thumb instruction updates flag
ADD R0, #0x1 ; This 16-bit Thumb instruction updates flag
```

However, if you used the same code in UAL syntax; that is "THUMB" directive is used with ARM assembler, then

```
ADD R0, R1   ; This 16-bit Thumb instruction does not
             ; update flag
ADD R0, #0x1 ; This will become a 32-bit Thumb instruction
             ; that does not update flag
```

To make sure that the code works correctly with different tools, you should always use the S suffix if the flags need to be updated for conditional operations such as conditional branches.

The compare (CMP) instruction subtracts two values and updates the flags (just like SUBS), but the result is not stored in any registers. CMP can have the following formats:

```
CMP R0, R1    ; Calculate R0 - R1 and update flag
CMP R0, #0x12 ; Calculate R0 - 0x12 and update flag
```

A similar instruction is the CMN (compare negative). It compares one value to the negative (two's complement) of a second value; the flags are updated, but the result is not stored in any registers:

```
CMN R0, R1    ; Calculate R0 - (-R1) and update flag
CMN R0, #0x12 ; Calculate R0 - (-0x12) and update flag
```

The TST (test) instruction is more like the AND instruction. It ANDs two values and updates the flags. However, the result is not stored in any register. Similarly to CMP, it has two input formats:

```
TST R0, R1   ; Calculate R0 AND R1 and update flag
TST R0, #0x12 ; Calculate R0 AND 0x12 and update flag
```

4.3.6 **Assembler Language: Combined Compare and Conditional Branch**

With ARM architecture v7-M, two new instructions are provided on the Cortex-M3 to supply a simple compare with zero and conditional branch operations. These are CBZ (compare and branch if zero) and CBNZ (compare and branch if nonzero).

The compare and branch instructions only support forward branches. For example,

```
i = 5;
while (i != 0 ){
func1(); ; call a function
i--;
}
```

This can be compiled into the following:

```
        MOV R0, #5      ; Set loop counter
loop1 CBZ R0,loop1exit ; if loop counter = 0 then exit the loop
        BL   func1      ; call a function
        SUB R0, #1      ; loop counter decrement
        B    loop1      ; next loop
loop1exit
```

The usage of CBNZ is similar to CBZ, apart from the fact that the branch is taken if the Z flag is not set (result is not zero). For example,

```
status = strchr(email_address, '@');
if (status == 0){//status is 0 if @ is not in email_address
      show_error_message();
      exit(1);
      }
```

This can be compiled into the following:

```
      ...
      BL    strchr
      CBNZ R0, email_looks_okay ; Branch if result is not zero
      BL    show_error_message
      BL    exit
email_looks_okay
      ...
```

The APSR value is not affected by the CBZ and CBNZ instructions.

Assembler Language: Conditional Execution Using IT Instructions

The IT (IF-THEN) block is very useful for handling small conditional code. It avoids branch penalties because there is no change to program flow. It can provide a maximum of four conditionally executed instructions.

In IT instruction blocks, the first line must be the IT instruction, detailing the choice of execution, followed by the condition it checks. The first statement after the IT command must be

TRUE-THEN-EXECUTE, which is always written as *ITxyz*, where *T* means THEN and *E* means ELSE. The second through fourth statements can be either THEN (true) or ELSE (false):

```
IT<x><y><z> <cond>                      ; IT instruction (<x>, <y>,
                                        ; <z> can be T or E)
instr1<cond> <operands>                 ; 1st instruction (<cond>
                                        ; must be same as IT)
instr2<cond or not cond> <operands> ; 2nd instruction (can be
                                        ; <cond> or <!cond>
instr3<cond or not cond> <operands> ; 3rd instruction (can be
                                        ; <cond> or <!cond>
instr4<cond or not cond> <operands> ; 4th instruction (can be
                                        ; <cond> or <!cond>
```

If a statement is to be executed when *<cond>* is false, the suffix for the instruction must be the opposite of the condition. For example, the opposite of EQ is NE, the opposite of GT is LE, and so on. The following code shows an example of a simple conditional execution:

```
if (R1<R2) then
    R2=R2-R1
    R2=R2/2
else
    R1=R1-R2
    R1=R1/2
```

In assembly,

```
        CMP     R1, R2 ; If R1 < R2 (less then)
        ITTEE   LT     ; then execute instruction 1 and 2
                       ; (indicated by T)
                       ; else execute instruction 3 and 4
                       ; (indicated by E)
        SUBLT.W R2,R1  ; 1st instruction
        LSRLT.W R2,#1  ; 2nd instruction
        SUBGE.W R1,R2  ; 3rd instruction (notice the GE is
                       ; opposite of LT)
        LSRGE.W R1,#1  ; 4th instruction
```

You can have fewer than four conditionally executed instructions. The minimum is 1. You need to make sure the number of *T* and *E* occurrences in the IT instruction matches the number of conditionally executed instructions after the IT.

If an exception occurs during the IT instruction block, the execution status of the block will be stored in the stacked PSR (in the IT/Interrupt-Continuable Instruction [ICI] bit field). So, when the exception handler completes and the IT block resumes, the rest of the instructions in the block can continue the execution correctly. In the case of using multicycle instructions (for example, multiple load and store) inside an IT block, if an exception takes place during the execution, the whole instruction is abandoned and restarted after the interrupt process is completed.

4.3.7 **Assembler Language: Instruction Barrier and Memory Barrier Instructions**

The Cortex-M3 supports a number of barrier instructions. These instructions are needed as memory systems get more and more complex. In some cases, if memory barrier instructions are not used, race conditions could occur.

For example, if the memory map can be switched by a hardware register, after writing to the memory switching register you should use the DSB instruction. Otherwise, if the write to the memory switching register is buffered and takes a few cycles to complete, and the next instruction accesses the switched memory region immediately, the access could be using the old memory map. In some cases, this might result in an invalid access if the memory switching and memory access happen at the same time. Using DSB in this case will make sure that the write to the memory map switching register is completed before a new instruction is executed.

The following are the three barrier instructions in the Cortex-M3:

- DMB
- DSB
- ISB

These instructions are described in Table 4.27.

The memory barrier instructions can be accessed in C using Cortex Microcontroller Software Interface Standard (CMSIS) compliant device driver library as follows:

```
void __DMB(void); // Data Memory Barrier
void __DSB(void); // Data Synchronization Barrier
void __ISB(void); // Instruction Synchronization Barrier
```

The DSB and ISB instructions can be important for self-modifying code. For example, if a program changes its own program code, the next executed instruction should be based on the updated program. However, since the processor is pipelined, the modified instruction location might have already been fetched. Using DSB and then ISB can ensure that the modified program code is fetched again.

Architecturally, the ISB instruction should be used after updating the value of the CONTROL register. In the Cortex-M3 processor, this is not strictly required. But if you want to make sure your application is portable, you should ensure an ISB instruction is used after updating to CONTROL register.

DMB is very useful for multi-processor systems. For example, tasks running on separate processors might use shared memory to communicate with each other. In these environments, the order of memory accesses to the shared memory can be very important. DMB instructions can be inserted between accesses to the shared memory to ensure that the memory access sequence is exactly the same as expected.

Table 4.27 Barrier Instructions

Instruction	Description
DMB	Data memory barrier; ensures that all memory accesses are completed before new memory access is committed
DSB	Data synchronization barrier; ensures that all memory accesses are completed before next instruction is executed
ISB	Instruction synchronization barrier; flushes the pipeline and ensures that all previous instructions are completed before executing new instructions

More details about memory barriers can be found in the *ARM v7-M Architecture Application Level Reference Manual* [Ref. 2].

4.3.8 Assembly Language: Saturation Operations

The Cortex-M3 supports two instructions that provide signed and unsigned saturation operations: SSAT and USAT (for signed data type and unsigned data type, respectively). Saturation is commonly used in signal processing—for example, in signal amplification. When an input signal is amplified, there is a chance that the output will be larger than the allowed output range. If the value is adjusted simply by removing the unused MSB, an overflowed result will cause the signal waveform to be completely deformed (see Figure 4.3).

The saturation operation does not prevent the distortion of the signal, but at least the amount of distortion is greatly reduced in the signal waveform.

The instruction syntax of the SSAT and USAT instructions is outlined here and in Table 4.28.

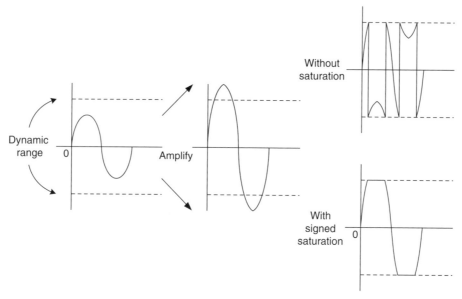

FIGURE 4.3

Signed Saturation Operation.

Table 4.28 Saturation Instructions

Instruction	Description
SSAT.W <Rd>, #<immed>, <Rn>, {,<shift>}	Saturation for signed value
USAT.W <Rd>, #<immed>, <Rn>, {,<shift>}	Saturation for a signed value into an unsigned value

Rn: Input value
Shift: Shift operation for input value before saturation; optional, can be #LSL N or #ASR N
Immed: Bit position where the saturation is carried out
Rd: Destination register

Besides the destination register, the Q-bit in the APSR can also be affected by the result. The Q flag is set if saturation takes place in the operation, and it can be cleared by writing to the APSR (see Table 4.29). For example, if a 32-bit signed value is to be saturated into a 16-bit signed value, the following instruction can be used:

```
SSAT.W R1, #16, R0
```

Similarly, if a 32-bit unsigned value is to saturate into a 16-bit unsigned value, the following instruction can be used:

```
USAT.W R1, #16, R0
```

This will provide a saturation feature that has the properties shown in Figure 4.4.

For the preceding 16-bit saturation example instruction, the output values shown in Table 4.30 can be observed.

Saturation instructions can also be used for data type conversions. For example, they can be used to convert a 32-bit integer value to 16-bit integer value. However, C compilers might not be able to directly use these instructions, so intrinsic function or assembler functions (or embedded/inline assembler code) for the data conversion could be required.

Table 4.29 Examples of Signed Saturation Results

Input (R0)	Output (R1)	Q Bit
0x00020000	0x00007FFF	Set
0x00008000	0x00007FFF	Set
0x00007FFF	0x00007FFF	Unchanged
0x00000000	0x00000000	Unchanged
0xFFFF8000	0xFFFF8000	Unchanged
0xFFFF7FFF	0xFFFF8000	Set
0xFFFE0000	0xFFFF8000	Set

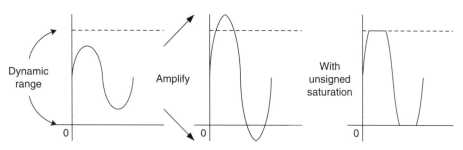

FIGURE 4.4

Unsigned Saturation Operation.

Table 4.30 Examples of Unsigned Saturation Results

Input (R0)	Output (R1)	Q Bit
0x00020000	0x0000FFFF	Set
0x00008000	0x00008000	Unchanged
0x00007FFF	0x00007FFF	Unchanged
0x00000000	0x00000000	Unchanged
0xFFFF8000	0x00000000	Set
0xFFFF8001	0x00000000	Set
0xFFFFFFFF	0x00000000	Set

4.4 SEVERAL USEFUL INSTRUCTIONS IN THE CORTEX-M3

Several useful Thumb-2 instructions from the architecture v7 and v6 are introduced here.

4.4.1 MSR and MRS

These two instructions provide access to the special registers in the Cortex-M3. Here is the syntax of these instructions:

```
MRS <Rn>, <SReg> ; Move from Special Register
MSR <SReg>, <Rn> ; Write to Special Register
```

where *<SReg>* could be one of the options shown in Table 4.31.

For example, the following code can be used to set up the process stack pointer:

```
LDR R0,=0x20008000 ; new value for Process Stack Pointer (PSP)
MSR PSP, R0
```

Unless accessing the APSR, the MRS and MSR instructions can be used in privileged mode only. Otherwise the operation will be ignored, and the returned read data (if MRS is used) will be zero.

After updating the value of the CONTROL register using MSR instruction, it is recommended to add an ISB instruction to ensure that the effect of the update takes place immediately. On the Cortex-M3 processor this is not strictly required, but for software portability (if the software code is to be used on other ARM processor) this is needed.

4.4.2 More on the IF-THEN Instruction Block

The IF-THEN instruction was introduced briefly in an earlier section in this chapter "Conditional Execution Using IT instruction." In here, we will cover more details about this instruction.

The IF-THEN (IT) instructions allow up to four succeeding instructions (called an *IT block*) to be conditionally executed. They are in the following formats as shown in Table 4.32, where,

- *<x>* specifies the execution condition for the second instruction
- *<y>* specifies the execution condition for the third instruction
- *<z>* specifies the execution condition for the fourth instruction
- *<cond>* specifies the base condition of the instruction block; the first instruction following IT executes if *<cond>* is true

Table 4.31 Special Register Names for MRS and MSR Instructions

Symbol	Description
IPSR	Interrupt status register
EPSR	Execution status register (read as zero)
APSR	Flags from previous operation
IEPSR	A composite of IPSR and EPSR
IAPSR	A composite of IPSR and APSR
EAPSR	A composite of EPSR and APSR
PSR	A composite of APSR, EPSR, and IPSR
MSP	Main stack pointer
PSP	Process stack pointer
PRIMASK	Normal exception mask register
BASEPRI	Normal exception priority mask register
BASEPRI_MAX	Same as normal exception priority mask register, with conditional write (new priority level must be higher than the old level)
FAULTMASK	Fault exception mask register (also disables normal interrupts)
CONTROL	Control register

Table 4.32 Various Length of IT Instruction Block

	IT Block (each of \<x>, \<y> and \<z> can either be T [true] or E [else])	Examples
Only one conditional instruction	IT \<cond> instr1\<cond>	IT EQ ADDEQ R0, R0, R1

Two conditional instructions	IT\<x> \<cond>
	instr1\<cond>
	instr2\<cond or ~(cond)>
		ITE GE
		ADDGE R0, R0, R1
		ADDLT R0, R0, R3

Three conditional instructions	IT\<x>\<y> \<cond>
	instr1\<cond>
	instr2\<cond or ~(cond)>
	instr3\<cond or ~(cond)>
		ITET GT
		ADDGT R0, R0, R1
		ADDLE R0, R0, R3
		ADDGT R2, R4, #1

Four conditional instructions	IT\<x>\<y>\<z> \<cond>
	instr1\<cond>
	instr2\<cond or ~(cond)>
	instr3\<cond or ~(cond)>
	instr4\<cond or ~(cond)>
		ITETT NE
		ADDNE R0, R0, R1
		ADDEQ R0, R0, R3
		ADDNE R2, R4, #1
		MOVNE R5, R3

The \<cond> part uses the same condition symbols as conditional branch. If "AL" is used as \<cond>, then you cannot use "E" in the condition control as it implies the instruction should never get executed.

Each of \<x>, \<y>, and \<z> can be either T (THEN) or E (ELSE), which refers to the base condition \<cond>, whereas \<cond> uses traditional syntax such as EQ, NE, GT, or the like.

Here is an example of IT use:

```
if (R0 equal R1) then {
  R3 = R4 + R5
  R3 = R3/2
  } else {
  R3 = R6 + R7
  R3 = R3/2
  }
```

This can be written as follows:

```
CMP R0, R1        ; Compare R0 and R1
ITTEE EQ          ; If R0 equal R1, Then-Then-Else-Else
ADDEQ R3, R4, R5  ; Add if equal
ASREQ R3, R3, #1  ; Arithmetic shift right if equal
ADDNE R3, R6, R7  ; Add if not equal
ASRNE R3, R3, #1  ; Arithmetic shift right if not equal
```

Aside from using the IT instruction directly, the IT instruction also helps porting of assembly application codes from ARM7TDMI to Cortex-M3. When ARM assembler (including KEIL RealView Microcontroller Development Kit, which is covered in Chapter 20) is used, and if a conditional execution instruction is used in assembly code without IT instruction, the assembler can insert the required IT instruction automatically. An example is shown in Table 4.33. This feature allows existing assembly code to be reused on Cortex-M3 without modifications.

Table 4.33 Automatic Insertion of IT Instruction in ARM Assembler

Original Assembly Code	Disassembled Assembly Code from Generated Object File

`CMP    R1, #2`
`ADDEQ  R0, R1, #1`
`...`
	`CMP    R1, #2`
	`IT     EQ`
	`ADDEQ  R0, R1, #1`

Note that 16-bit data processing instructions does not update APSR if they are used inside an IT instruction block. If you add the *S* suffix in the conditional executed instruction, the 32-bit version of the instruction would be used by the assembler.

4.4.3 SDIV and UDIV

The syntax for signed and unsigned divide instructions is as follows:

```
SDIV.W <Rd>, <Rn>, <Rm>
UDIV.W <Rd>, <Rn>, <Rm>
```

The result is Rd = Rn/Rm. For example,

```
LDR    R0,=300 ; Decimal 300
MOV    R1,#5
UDIV.W R2, R0, R1
```

This will give you an R2 result of 60 (0x3C).

You can set up the DIVBYZERO bit in the NVIC Configuration Control Register so that when a divide by zero occurs, a fault exception (usage fault) takes place. Otherwise, *<Rd>* will become 0 if a divide by zero takes place.

4.4.4 REV, REVH, and REVSH

REV reverses the byte order in a data word, and REVH reverses the byte order inside a half word. For example, if R0 is 0x12345678, in executing the following:

```
REV  R1, R0
REVH R2, R0
```

R1 will become 0x78563412, and R2 will be 0x34127856. REV and REVH are particularly useful for converting data between big endian and little endian.

REVSH is similar to REVH except that it only processes the lower half word, and then it sign extends the result. For example, if R0 is 0x33448899, running:

```
REVSH R1, R0
```

R1 will become 0xFFFF9988.

4.4.5 Reverse Bit

The RBIT instruction reverses the bit order in a data word. The syntax is as follows:

```
RBIT.W <Rd>, <Rn>
```

This instruction is very useful for processing serial bit streams in data communications. For example, if R0 is 0xB4E10C23 (binary value 1011_0100_1110_0001_0000_1100_0010_0011), executing:

```
RBIT.W R0, R1
```

R0 will become 0xC430872D (binary value 1100_0100_0011_0000_1000_0111_0010_1101).

4.4.6 SXTB, SXTH, UXTB, and UXTH

The four instructions SXTB, SXTH, UXTB, and UXTH are used to extend a byte or half word data into a word. The syntax of the instructions is as follows:

```
SXTB <Rd>, <Rn>
SXTH <Rd>, <Rn>
UXTB <Rd>, <Rn>
UXTH <Rd>, <Rn>
```

For SXTB/SXTH, the data are sign extended using bit[7]/bit[15] of Rn. With UXTB and UXTH, the value is zero extended to 32-bit.

For example, if R0 is 0x55AA8765:

```
SXTB R1, R0 ; R1 = 0x00000065
SXTH R1, R0 ; R1 = 0xFFFF8765
UXTB R1, R0 ; R1 = 0x00000065
UXTH R1, R0 ; R1 = 0x00008765
```

4.4.7 Bit Field Clear and Bit Field Insert

Bit Field Clear (BFC) clears 1–31 adjacent bits in any position of a register. The syntax of the instruction is as follows:

```
BFC.W <Rd>, <#lsb>, <#width>
```

For example,

```
LDR   R0,=0x1234FFFF
BFC.W R0, #4, #8
```

This will give R0 = 0x1234F00F.

Bit Field Insert (BFI) copies 1–31 bits (#width) from one register to any location (#lsb) in another register. The syntax is as follows:

```
BFI.W <Rd>, <Rn>, <#lsb>, <#width>
```

For example,

```
LDR   R0,=0x12345678
LDR   R1,=0x3355AACC
BFI.W R1, R0, #8, #16 ; Insert R0[15:0] to R1[23:8]
```

This will give R1 = 0x335678CC.

4.4.8 UBFX and SBFX

UBFX and SBFX are the unsigned and signed bit field extract instructions. The syntax of the instructions is as follows:

```
UBFX.W <Rd>, <Rn>, <#lsb>, <#width>
SBFX.W <Rd>, <Rn>, <#lsb>, <#width>
```

UBFX extracts a bit field from a register starting from any location (specified by #lsb) with any width (specified by #width), zero extends it, and puts it in the destination register. For example,

```
LDR    R0,=0x5678ABCD
UBFX.W R1, R0, #4, #8
```

This will give R1 = 0x000000BC.

Similarly, SBFX extracts a bit field, but its sign extends it before putting it in a destination register. For example,

```
LDR    R0,=0x5678ABCD
SBFX.W R1, R0, #4, #8
```

This will give R1 = 0xFFFFFFBC.

4.4.9 LDRD and STRD

The two instructions LDRD and STRD transfer two words of data from or into two registers. The syntax of the instructions is as follows:

```
LDRD.W <Rxf>, <Rxf2>, [Rn, #+/-offset]{!} ; Pre-indexed
```

```
LDRD.W <Rxf>, <Rxf2>, [Rn], #+/-offset     ; Post-indexed
STRD.W <Rxf>, <Rxf2>, [Rn, #+/-offset]{!} ; Pre-indexed
STRD.W <Rxf>, <Rxf2>, [Rn], #+/-offset     ; Post-indexed
```

where *<Rxf>* is the first destination/source register and *<Rxf2>* is the second destination/source register. Avoid using same register for *<Rn>* and *<Rxf>* when using LDRD because of an erratum in Cortex-M3 revision 0 to 2.

For example, the following code reads a 64-bit value located in memory address 0x1000 into R0 and R1:

```
LDR     R2,=0x1000
LDRD.W R0, R1, [R2] ; This will gives R0 = memory[0x1000],
                    ; R1 = memory[0x1004]
```

Similarly, we can use STRD to store a 64-bit value in memory. In the following example, preindexed addressing mode is used:

```
LDR     R2,=0x1000       ; Base address
STRD.W R0, R1, [R2, #0x20] ; This will gives memory[0x1020] = R0,
                          ; memory[0x1024] = R1
```

4.4.10 Table Branch Byte and Table Branch Halfword

Table Branch Byte (TBB) and Table Branch Halfword (TBH) are for implementing branch tables. The TBB instruction uses a branch table of byte size offset, and TBH uses a branch table of half word offset. Since the bit 0 of a program counter is always zero, the value in the branch table is multiplied by two before it's added to PC. Furthermore, because the PC value is the current instruction address plus four, the branch range for TBB is $(2 \times 255) + 4 = 514$, and the branch range for TBH is $(2 \times 65535) + 4 = 131074$. Both TBB and TBH support forward branch only.

TBB has this general syntax:

```
TBB.W [Rn, Rm]
```

where *Rn* is the base memory offset and *Rm* is the branch table index. The branch table item for TBB is located at Rn + Rm. Assuming we used PC for Rn, we can see the operation as shown in Figure 4.5.

For TBH instruction, the process is similar except the memory location of the branch table item is located at Rn + 2 x Rm and the maximum branch offset is higher. Again, we assume that Rn is set to PC, as shown in Figure 4.6.

If *Rn* in the table branch instruction is set to R15, the value used for *Rn* will be PC + 4 because of the pipeline in the processor. These two instructions are more likely to be used by a C compiler to generate code for switch (case) statements. Because the values in the branch table are relative to the current program counter, it is not easy to code the branch table content manually in assembler as the address offset value might not be able to be determined during assembly/compile stage, especially if the branch target is in a separate program code file. The coding syntax for calculating TBB/TBH branch table content could be dependent on the development tool. In ARM assembler (*armasm*), the TBB branch table can be created in the following way:

```
TBB.W [pc, r0] ; when executing this instruction, PC equal
               ; branchtable
```

```
branchtable
    DCB ((dest0 - branchtable)/2) ; Note that DCB is used because
                                  ; the value is 8-bit
    DCB ((dest1 - branchtable)/2)
    DCB ((dest2 - branchtable)/2)
    DCB ((dest3 - branchtable)/2)
dest0
    ... ; Execute if r0 = 0
dest1
    ... ; Execute if r0 = 1
dest2
    ... ; Execute if r0 = 2
dest3
    ... ; Execute if r0 = 3
```

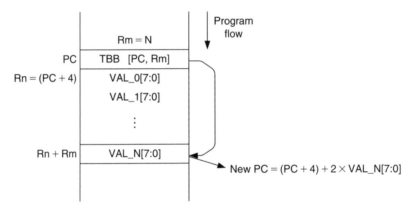

FIGURE 4.5

TBB Operation.

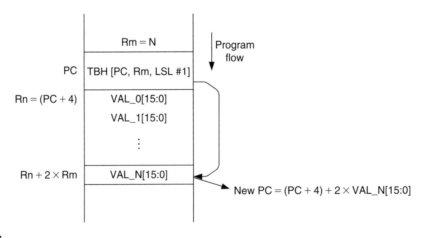

FIGURE 4.6

TBH Operation.

When the TBB instruction is executed, the current PC value is at the address labeled as *branchtable* (because of the pipeline in the processor). Similarly, for TBH instructions, it can be used as follows:

```
        TBH.W [pc, r0, LSL #1]
branchtable
        DCI ((dest0 - branchtable)/2) ; Note that DCI is used because
                                      ; the value is 16-bit
        DCI ((dest1 - branchtable)/2)
        DCI ((dest2 - branchtable)/2)
        DCI ((dest3 - branchtable)/2)
dest0
        ... ; Execute if r0 = 0
dest1
        ... ; Execute if r0 = 1
dest2
        ... ; Execute if r0 = 2
dest3
        ... ; Execute if r0 = 3
```

Memory Systems

IN THIS CHAPTER

Memory System Features Overview 79
Memory Maps 79
Memory Access Attributes 82
Default Memory Access Permissions 83
Bit-Band Operations 84
Unaligned Transfers 92
Exclusive Accesses 93
Endian Mode 95

5.1 MEMORY SYSTEM FEATURES OVERVIEW

The Cortex™-M3 processor has different memory architecture from that of traditional ARM processors. First, it has a predefined memory map that specifies which bus interface is to be used when a memory location is accessed. This feature also allows the processor design to optimize the access behavior when different devices are accessed.

Another feature of the memory system in the Cortex-M3 is the bit-band support. This provides atomic operations to bit data in memory or peripherals. The bit-band operations are supported only in special memory regions. This topic is covered in more detail later in this chapter.

The Cortex-M3 memory system also supports unaligned transfers and exclusive accesses. These features are part of the v7-M architecture. Finally, the Cortex-M3 supports both little endian and big endian memory configuration.

5.2 MEMORY MAPS

The Cortex-M3 processor has a fixed memory map (see Figure 5.1). This makes it easier to port software from one Cortex-M3 product to another. For example, components described in previous sections, such as Nested Vectored Interrupt Controller (NVIC) and Memory Protection Unit (MPU), have the

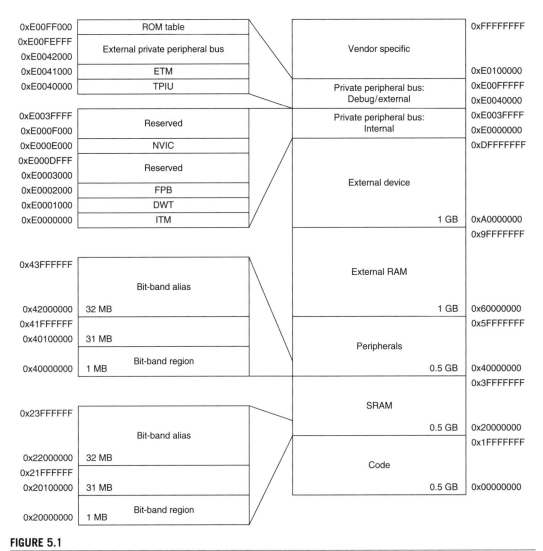

FIGURE 5.1

Cortex-M3 Predefined Memory Map.

same memory locations in all Cortex-M3 products. Nevertheless, the memory map definition allows great flexibility so that manufacturers can differentiate their Cortex-M3-based product from others.

Some of the memory locations are allocated for private peripherals such as debugging components. They are located in the private peripheral memory region. These debugging components include the following:

- Fetch Patch and Breakpoint Unit (FPB)
- Data Watchpoint and Trace Unit (DWT)

- Instrumentation Trace Macrocell (ITM)
- Embedded Trace Macrocell (ETM)
- Trace Port Interface Unit (TPIU)
- ROM table

The details of these components are discussed in later chapters on debugging features.

The Cortex-M3 processor has a total of 4 GB of address space. Program code can be located in the code region, the Static Random Access Memory (SRAM) region, or the external RAM region. However, it is best to put the program code in the code region because with this arrangement, the instruction fetches and data accesses are carried out simultaneously on two separate bus interfaces.

The SRAM memory range is for connecting internal SRAM. Access to this region is carried out via the system interface bus. In this region, a 32-MB range is defined as a bit-band alias. Within the 32-bit-band alias memory range, each word address represents a single bit in the 1-MB bit-band region. A data write access to this bit-band alias memory range will be converted to an atomic READ-MODIFY-WRITE operation to the bit-band region so as to allow a program to set or clear individual data bits in the memory. The bit-band operation applies only to data accesses not instruction fetches. By putting Boolean information (single bits) in the bit-band region, we can pack multiple Boolean data in a single word while still allowing them to be accessible individually via bit-band alias, thus saving memory space without the need for handling READ-MODIFY-WRITE in software. More details on bit-band alias can be found later in this chapter.

Another 0.5-GB block of address range is allocated to on-chip peripherals. Similar to the SRAM region, this region supports bit-band alias and is accessed via the system bus interface. However, instruction execution in this region is not allowed. The bit-band support in the peripheral region makes it easy to access or change control and status bits of peripherals, making it easier to program peripheral control.

Two slots of 1-GB memory space are allocated for external RAM and external devices. The difference between the two is that program execution in the external device region is not allowed, and there are some differences with the caching behaviors.

The last 0.5-GB memory is for the system-level components, internal peripheral buses, external peripheral bus, and vendor-specific system peripherals. There are two segments of the private peripheral bus (PPB):

- Advanced High-Performance Bus (AHB) PPB, for Cortex-M3 internal AHB peripherals only; this includes NVIC, FPB, DWT, and ITM
- Advance Peripheral Bus (APB) PPB, for Cortex-M3 internal APB devices as well as external peripherals (external to the Cortex-M3 processor); the Cortex-M3 allows chip vendors to add additional on-chip APB peripherals on this private peripheral bus via an APB interface

The NVIC is located in a memory region called the system control space (SCS) (see Figure 5.2). Besides providing interrupt control features, this region also provides the control registers for SYSTICK, MPU, and code debugging control.

The remaining unused vendor-specific memory range can be accessed via the system bus interface. However, instruction execution in this region is not allowed.

The Cortex-M3 processor also comes with an optional MPU. Chip manufacturers can decide whether to include the MPU in their products.

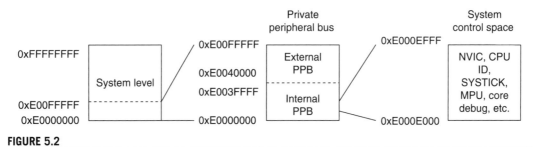

FIGURE 5.2

The System Control Space.

What we have shown in the memory map is merely a template; individual semiconductor vendors provide detailed memory maps including the actual location and size of ROM, RAM, and peripheral memory locations.

5.3 MEMORY ACCESS ATTRIBUTES

The memory map shows what is included in each memory region. Aside from decoding which memory block or device is accessed, the memory map also defines the memory attributes of the access. The memory attributes you can find in the Cortex-M3 processor include the following:

- *Bufferable*: Write to memory can be carried out by a write buffer while the processor continues on next instruction execution.
- *Cacheable*: Data obtained from memory read can be copied to a memory cache so that next time it is accessed the value can be obtained from the cache to speed up the program execution.
- *Executable*: The processor can fetch and execute program code from this memory region.
- *Sharable*: Data in this memory region could be shared by multiple bus masters. Memory system needs to ensure coherency of data between different bus masters in shareable memory region.

The Cortex-M3 bus interfaces output the memory access attributes information to the memory system for each instruction and data transfer. The default memory attribute settings can be overridden if MPU is present and the MPU region configurations are programmed differently from the default. Though the Cortex-M3 processor does not have a cache memory or cache controller, a cache unit can be added on the microcontroller which can use the memory attribute information to define the memory access behaviors. In addition, the cache attributes might also affect the operation of memory controllers for on-chip memory and off-chip memory, depending on the memory controllers used by the chip manufacturers.

The memory access attributes for each memory region are as follows:

- *Code memory region* (0x00000000–0x1FFFFFFF): This region is executable, and the cache attribute is write through (WT). You can put data memory in this region as well. If data operations are carried out for this region, they will take place via the data bus interface. Write transfers to this region are bufferable.

- *SRAM memory region* (0x20000000–0x3FFFFFFF): This region is intended for on-chip RAM. Write transfers to this region are bufferable, and the cache attribute is write back, write allocated (WB-WA). This region is executable, so you can copy program code here and execute it.

- *Peripheral region* (0x40000000–0x5FFFFFFF): This region is intended for peripherals. The accesses are noncacheable. You cannot execute instruction code in this region (Execute Never, or XN in ARM documentation, such as the Cortex-M3 TRM).

- *External RAM region* (0x60000000–0x7FFFFFFF): This region is intended for either on-chip or off-chip memory. The accesses are cacheable (WB-WA), and you can execute code in this region.

- *External RAM region* (0x80000000–0x9FFFFFFF): This region is intended for either on-chip or off-chip memory. The accesses are cacheable (WT), and you can execute code in this region.

- *External devices* (0xA0000000–0xBFFFFFFF): This region is intended for external devices and/or shared memory that needs ordering/nonbuffered accesses. It is also a nonexecutable region.

- *External devices* (0xC0000000–0xDFFFFFFF): This region is intended for external devices and/or shared memory that needs ordering/nonbuffered accesses. It is also a nonexecutable region.

- *System region* (0xE0000000–0xFFFFFFFF): This region is for private peripherals and vendor-specific devices. It is nonexecutable. For the PPB memory range, the accesses are strongly ordered (noncacheable, nonbufferable). For the vendor-specific memory region, the accesses are bufferable and noncacheable.

Note that from Revision 1 of the Cortex-M3, the code region memory attribute export to external memory system is hardwired to cacheable and nonbufferable. This cannot be overridden by MPU configuration. This update only affects the memory system outside the processor (e.g., level 2 cache and certain types of memory controllers with cache features). Within the processor, the internal write buffer can still be used for write transfers accessing the code region.

5.4 DEFAULT MEMORY ACCESS PERMISSIONS

The Cortex-M3 memory map has a default configuration for memory access permissions. This prevents user programs (non-privileged) from accessing system control memory spaces such as the NVIC. The default memory access permission is used when either no MPU is present or MPU is present but disabled.

If MPU is present and enabled, the access permission in the MPU setup will determine whether user accesses are allowed.

The default memory access permissions are shown in Table 5.1.

Table 5.1 Default Memory Access Permissions

Memory Region	Address	Access in User Program
Vendor specific	0xE0100000–0xFFFFFFFF	Full access
ROM table	0xE00FF000–0xE00FFFFF	Blocked; user access results in bus fault
External PPB	0xE0042000–0xE00FEFFF	Blocked; user access results in bus fault
ETM	0xE0041000–0xE0041FFF	Blocked; user access results in bus fault

Continued

Table 5.1 Default Memory Access Permissions *Continued*

Memory Region	Address	Access in User Program
TPIU	0xE0040000–0xE0040FFF	Blocked; user access results in bus fault
Internal PPB	0xE000F000–0xE003FFFF	Blocked; user access results in bus fault
NVIC	0xE000E000–0xE000EFFF	Blocked; user access results in bus fault, except Software Trigger Interrupt Register that can be programmed to allow user accesses
FPB	0xE0002000–0xE0003FFF	Blocked; user access results in bus fault
DWT	0xE0001000–0xE0001FFF	Blocked; user access results in bus fault
ITM	0xE0000000–0xE0000FFF	Read allowed; write ignored except for stimulus ports with user access enabled
External device	0xA0000000–0xDFFFFFFF	Full access
External RAM	0x60000000–0x9FFFFFFF	Full access
Peripheral	0x40000000–0x5FFFFFFF	Full access
SRAM	0x20000000–0x3FFFFFFF	Full access
Code	0x00000000–0x1FFFFFFF	Full access

When a user access is blocked, the fault exception takes place immediately.

5.5 BIT-BAND OPERATIONS

Bit-band operation support allows a single load/store operation to access (read/write) to a single data bit. In the Cortex-M3, this is supported in two predefined memory regions called bit-band regions. One of them is located in the first 1 MB of the SRAM region, and the other is located in the first 1 MB of the peripheral region. These two memory regions can be accessed like normal memory, but they can also be accessed via a separate memory region called the bit-band alias (see Figure 5.3). When the bit-band alias address is used, each individual bit can be accessed separately in the least significant bit (LSB) of each word-aligned address.

For example, to set bit 2 in word data in address 0x20000000, instead of using three instructions to read the data, set the bit, and then write back the result, this task can be carried out by a single instruction (see Figure 5.4). The assembler sequence for these two cases could be like the one shown in Figure 5.5.

Similarly, bit-band support can simplify application code if we need to read a bit in a memory location. For example, if we need to determine bit 2 of address 0x20000000, we use the steps outlined in Figure 5.6. The assembler sequence for these two cases could be like the one shown in Figure 5.7.

Bit-band operation is not a new idea; in fact, a similar feature has existed for more than 30 years on 8-bit microcontrollers such as the 8051. Although the Cortex-M3 does not have special instructions for bit operation, special memory regions are defined so that data accesses to these regions are automatically converted into bit-band operations.

Note that the Cortex-M3 uses the following terms for the bit-band memory addresses:

- *Bit-band region*: This is a memory address region that supports bit-band operation.
- *Bit-band alias*: Access to the bit-band alias will cause an access (a bit-band operation) to the bit-band region. (*Note:* A memory remapping is performed.)

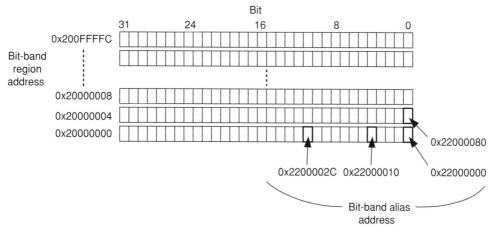

FIGURE 5.3

Bit Accesses to Bit-Band Region via the Bit-Band Alias.

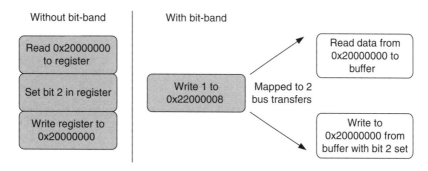

FIGURE 5.4

Write to Bit-Band Alias.

```
           Without bit-band                          With bit-band
LDR    R0,=0x20000000 ; Setup address     LDR    R0,=0x22000008 ; Setup add
LDR    R1, [R0]       ; Read              MOV    R1, #1         ; Setup dat
ORR.W  R1, #0x4       ; Modify bit        STR    R1, [R0]       ; Write
STR    R1, [R0]       ; Write back result
```

FIGURE 5.5

Example Assembler Sequence to Write a Bit with and without Bit-Band.

Within the bit-band region, each word is represented by an LSB of 32 words in the bit-band alias address range. What actually happens is that when the bit-band alias address is accessed, the address is remapped into a bit-band address. For read operations, the word is read and the chosen bit location is shifted to the LSB of the read return data. For write operations, the written bit data are shifted to the required bit position, and a READ-MODIFY-WRITE is performed.

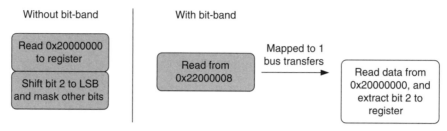

FIGURE 5.6

Read from the Bit-Band Alias.

```
               Without bit-band                    |                With bit-band
   LDR    R0,=0x20000000 ; Setup address          |    LDR    R0,=0x22000008 ; Setup address
   LDR    R1, [R0]       ; Read                    |    LDR    R1, [R0]       ; Read
   UBFX.W R1,  R1, #2, #1 ; Extract bit[2]         |
```

FIGURE 5.7

Read from the Bit-Band Alias.

Table 5.2 Remapping of Bit-Band Addresses in SRAM Region

Bit-Band Region	Aliased Equivalent
0x20000000 bit[0]	0x22000000 bit[0]
0x20000000 bit[1]	0x22000004 bit[0]
0x20000000 bit[2]	0x22000008 bit[0]
...	...
0x20000000 bit[31]	0x2200007C bit[0]
0x20000004 bit[0]	0x22000080 bit[0]
...	...
0x20000004 bit[31]	0x220000FC bit[0]
...	...
0x200FFFFC bit[31]	0x23FFFFFC bit[0]

There are two regions of memory for bit-band operations:

- 0x20000000–0x200FFFFF (SRAM, 1 MB)
- 0x40000000–0x400FFFFF (peripherals, 1 MB)

For the SRAM memory region, the remapping of the bit-band alias is shown in Table 5.2.

Similarly, the bit-band region of the peripheral memory region can be accessed via bit-band aliased addresses, as shown in Table 5.3.

Table 5.3 Remapping of Bit-Band Addresses in Peripheral Memory Region

Bit-Band Region	Aliased Equivalent
0x40000000 bit[0]	0x42000000 bit[0]
0x40000000 bit[1]	0x42000004 bit[0]
0x40000000 bit[2]	0x42000008 bit[0]
…	…
0x40000000 bit[31]	0x4200007C bit[0]
0x40000004 bit[0]	0x42000080 bit[0]
…	…
0x40000004 bit[31]	0x420000FC bit[0]
…	…
0x400FFFFC bit[31]	0x43FFFFFC bit[0]

Here's a simple example:

1. Set address 0x20000000 to a value of 0x3355AACC.
2. Read address 0x22000008. This read access is remapped into read access to 0x20000000. The return value is 1 (bit[2] of 0x3355AACC).
3. Write 0x0 to 0x22000008. This write access is remapped into a READ-MODIFY-WRITE to 0x20000000. The value 0x3355AACC is read from memory, bit 2 is cleared, and a result of 0x3355AAC8 is written back to address 0x20000000.
4. Now, read 0x20000000. That gives you a return value of 0x3355AAC8 (bit[2] cleared).

When you access bit-band alias addresses, only the LSB (bit[0]) in the data is used. In addition, accesses to the bit-band alias region should not be unaligned. If an unaligned access is carried out to bit-band alias address range, the result is unpredictable.

5.5.1 Advantages of Bit-Band Operations

So, what are the uses of bit-band operations? We can use them to, for example, implement serial data transfers in general-purpose input/output (GPIO) ports to serial devices. The application code can be implemented easily because access to serial data and clock signals can be separated.

BIT-BAND VERSUS BIT-BANG

In the Cortex-M3, we use the term *bit-band* to indicate that the feature is a special memory band (region) that provides bit accesses. *Bit-bang* commonly refers to driving I/O pins under software control to provide serial communication functions. The bit-band feature in the Cortex-M3 can be used for bit-banging implementations, but the definitions of these two terms are different.

Bit-band operation can also be used to simplify branch decisions. For example, if a branch should be carried out based on 1 single bit in a status register in a peripheral, instead of

- Reading the whole register
- Masking the unwanted bits
- Comparing and branching

you can simplify the operations to

- Reading the status bit via the bit-band alias (get 0 or 1)
- Comparing and branching

Besides providing faster bit operations with fewer instructions, the bit-band feature in the Cortex-M3 is also essential for situations in which resources are being shared by more than one process. One of the most important advantages or properties of a bit-band operation is that it is atomic. In other words, the READ-MODIFY-WRITE sequence cannot be interrupted by other bus activities. Without this behavior in, for example, using a software READ-MODIFY-WRITE sequence, the following problem can occur: consider a simple output port with bit 0 used by a main program and bit 1 used by an interrupt handler. A software-based READ-MODIFY-WRITE operation can cause data conflicts, as shown in Figure 5.8.

With the Cortex-M3 bit-band feature, this kind of race condition can be avoided because the READ-MODIFY-WRITE is carried out at the hardware level and is atomic (the two transfers cannot be pulled apart) and interrupts cannot take place between them (see Figure 5.9).

Similar issues can be found in multitasking systems. For example, if bit 0 of the output port is used by Process A and bit 1 is used by Process B, a data conflict can occur in software-based READ-MODIFY-WRITE (see Figure 5.10).

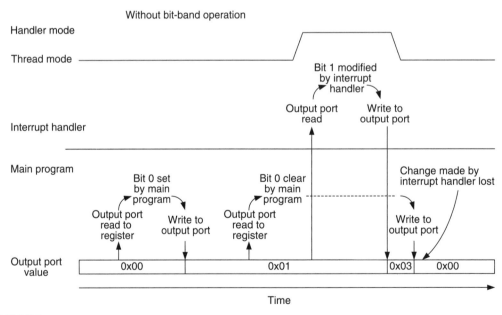

FIGURE 5.8

Data Are Lost When an Exception Handler Modifies a Shared Memory Location.

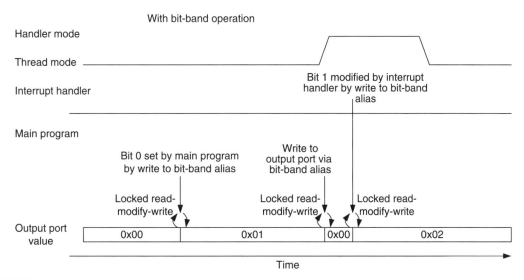

FIGURE 5.9

Data Loss Prevention with Locked Transfers Using the Bit-Band Feature.

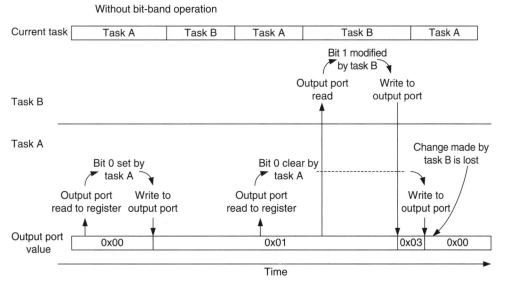

FIGURE 5.10

Data Are Lost When a Different Task Modifies a Shared Memory Location.

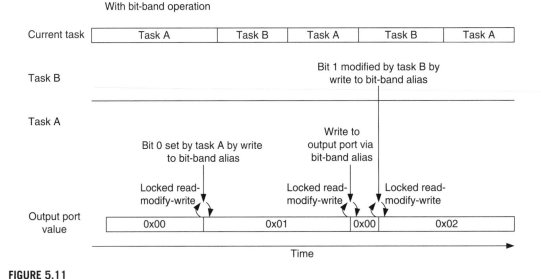

FIGURE 5.11

Data Loss Prevention with Locked Transfers Using the Bit-Band Feature.

Again, the bit-band feature can ensure that bit accesses from each task are separated so that no data conflicts occur (see Figure 5.11).

Besides I/O functions, the bit-band feature can be used for storing and handling Boolean data in the SRAM region. For example, multiple Boolean variables can be packed into one single memory location to save memory space, whereas the access to each bit is still completely separated when the access is carried out via the bit-band alias address range.

For system-on-chip (SoC) designers designing a bit-band-capable device, the device's memory address should be located within the bit-band memory, and the lock (HMASTLOCK) signal from the AHB interface must be checked to make sure that writable register contents will not be changed except by the bus when a locked transfer is carried out.

5.5.2 Bit-Band Operation of Different Data Sizes

Bit-band operation is not limited to word transfers. It can be carried out as byte transfers or half word transfers as well. For example, when a byte access instruction (LDRB/STRB) is used to access a bit-band alias address range, the accesses generated to the bit-band region will be in byte size. The same applies to half word transfers (LDRH/STRH). When you use nonword transfers to bit-band alias addresses, the address value should still be word aligned.

5.5.3 Bit-Band Operations in C Programs

There is no native support of bit-band operation in most C compilers. For example, C compilers do not understand that the same memory can be accessed using two different addresses, and they do not know that accesses to the bit-band alias will only access the LSB of the memory location. To use the bit-band feature in C, the simplest solution is to separately declare the address and the bit-band alias of a memory location. For example:

```
#define  DEVICE_REG0       *((volatile unsigned long *)  (0x40000000))
#define  DEVICE_REG0_BIT0  *((volatile unsigned long *)  (0x42000000))
#define  DEVICE_REG0_BIT1  *((volatile unsigned long *)  (0x42000004))
...
DEVICE_REG0 = 0xAB; // Accessing the hardware register by normal
                    // address
...
DEVICE_REG0 = DEVICE_REG0 | 0x2; // Setting bit 1 without using
                                 // bitband feature
...
DEVICE_REG0_BIT1 = 0x1; // Setting bit 1 using bitband feature
                        // via the bit band alias address
```

It is also possible to develop C macros to make accessing the bit-band alias easier. For example, we could set up one macro to convert the bit-band address and the bit number into the bit-band alias address and set up another macro to access the memory location by taking the address value as a pointer:

```
// Convert bit band address and bit number into
// bit band alias address
#define BITBAND(addr,bitnum) ((addr & 0xF0000000)+0x2000000+((addr &
    0xFFFFF)<<5)+(bitnum <<2))
// Convert the address as a pointer
#define MEM_ADDR(addr) *((volatile unsigned long *) (addr))
```

Based on the previous example, we rewrite the code as follows:

```
#define DEVICE_REG0 0x40000000
#define BITBAND(addr,bitnum) ((addr & 0xF0000000)+0x02000000+((addr &
    0xFFFFF)<<5)+(bitnum<<2))
#define MEM_ADDR(addr) *((volatile unsigned long *) (addr))
...
MEM_ADDR(DEVICE_REG0) = 0xAB; // Accessing the hardware
                              // register by normal address
...
// Setting bit 1 without using bitband feature
MEM_ADDR(DEVICE_REG0) = MEM_ADDR(DEVICE_REG0) | 0x2;
...
// Setting bit 1 with using bitband feature
MEM_ADDR(BITBAND(DEVICE_REG0,1)) = 0x1;
```

Note that when the bit-band feature is used, the variables being accessed might need to be declared as volatile. The C compilers do not know that the same data could be accessed in two different addresses, so the volatile property is used to ensure that each time a variable is accessed, the memory location is accessed instead of a local copy of the data inside the processor.

Starting from ARM RealView Development Suite version 4.0 and Keil MDK-ARM 3.80, bit band support is provided by __attribute__((bitband)) language extension and __bitband command line option (see reference 6). You can find further examples of bit-band accesses with C macros using ARM RealView Compiler Tools in the *ARM Application Note 179* [Ref. 7].

5.6 UNALIGNED TRANSFERS

The Cortex-M3 supports unaligned transfers on single accesses. Data memory accesses can be defined as aligned or unaligned. Traditionally, ARM processors (such as the ARM7/ARM9/ARM10) allow only aligned transfers. That means in accessing memory, a word transfer must have address bit[1] and bit[0] equal to 0, and a half word transfer must have address bit[0] equal to 0. For example, word data can be located at 0x1000 or 0x1004, but it cannot be located in 0x1001, 0x1002, or 0x1003. For half word data, the address can be 0x1000 or 0x1002, but it cannot be 0x1001.

So, what does an unaligned transfer look like? Figures 5.12 through 5.16 show some examples. Assuming that the memory infrastructure is 32-bit (4 bytes) wide, an unaligned transfer can be any word size read/write such that the address is not a multiple of 4, as shown in Figures 5.12–5.14, or when the transfer is in half word size, and the address is not a multiple of 2, as shown in Figures 5.15 and 5.16.

All the byte-size transfers are aligned on the Cortex-M3 because the minimum address step is 1 byte.

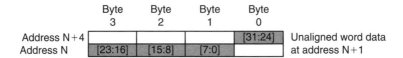

FIGURE 5.12

Unaligned Transfer Example 1.

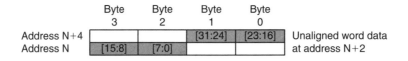

FIGURE 5.13

Unaligned Transfer Example 2.

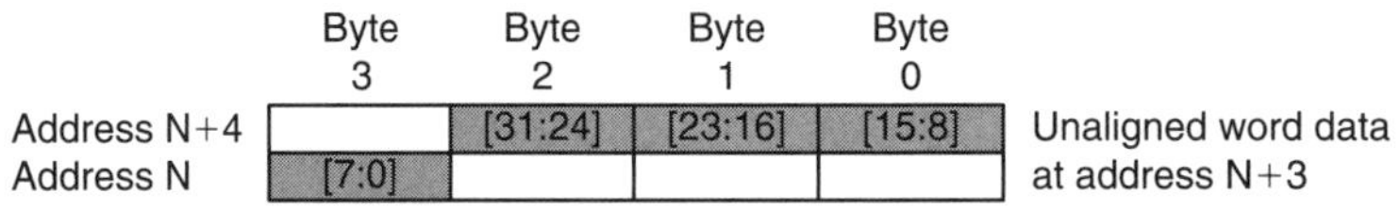

FIGURE 5.14

Unaligned Transfer Example 3.

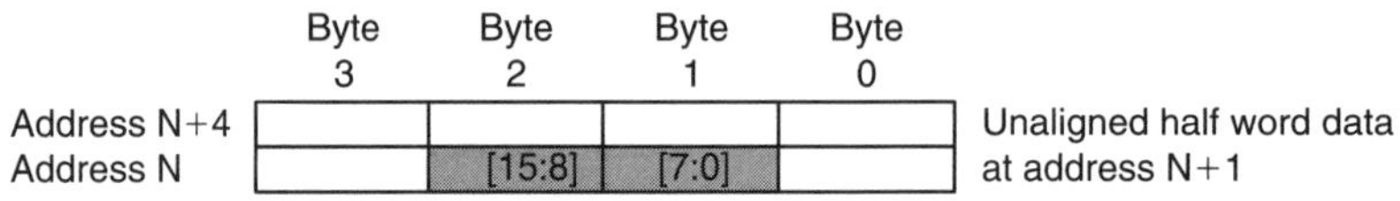

FIGURE 5.15

Unaligned Transfer Example 4.

FIGURE 5.16

Unaligned Transfer Example 5.

In the Cortex-M3, unaligned transfers are supported in normal memory accesses (such as LDR, LDRH, STR, and STRH instructions). There are a number of limitations:

- Unaligned transfers are not supported in Load/Store multiple instructions.
- Stack operations (PUSH/POP) must be aligned.
- Exclusive accesses (such as LDREX or STREX) must be aligned; otherwise, a fault exception (usage fault) will be triggered.
- Unaligned transfers are not supported in bit-band operations. Results will be unpredictable if you attempt to do so.

When unaligned transfers are used, they are actually converted into multiple aligned transfers by the processor's bus interface unit. This conversion is transparent, so application programmers do not have to worry about it. However, when an unaligned transfer takes place, it is broken into separate transfers, and as a result, it takes more clock cycles for a single data access and might not be good for situations in which high performance is required. To get the best performance, it's worth making sure that data are aligned properly.

It is also possible to set up the NVIC so that an exception is triggered when an unaligned transfer takes place. This is done by setting the UNALIGN_TRP (unaligned trap) bit in the configuration control register in the NVIC (0xE000ED14). In this way, the Cortex-M3 generates usage fault exceptions when unaligned transfers take place. This is useful during software development to test whether an application produces unaligned transfers.

5.7 EXCLUSIVE ACCESSES

You might have noticed that the Cortex-M3 has no SWP instruction (swap), which was used for semaphore operations in traditional ARM processors like ARM7TDMI. This is now being replaced by exclusive access operations. Exclusive accesses were first supported in architecture v6 (for example, in the ARM1136).

Semaphores are commonly used for allocating shared resources to applications. When a shared resource can only service one client or application processor, we also call it Mutual Exclusion (MUTEX). In such cases, when a resource is being used by one process, it is locked to that process and cannot serve another process until the lock is released. To set up a MUTEX semaphore, a memory location is defined as the lock flag to indicate whether a shared resource is locked by a process. When a process or application wants to use the resource, it needs to check whether the resource has been locked first. If it is not being used, it can set the lock flag to indicate that the resource is now locked. In traditional ARM processors, the access to the lock flag is carried out by the SWP instruction. It allows

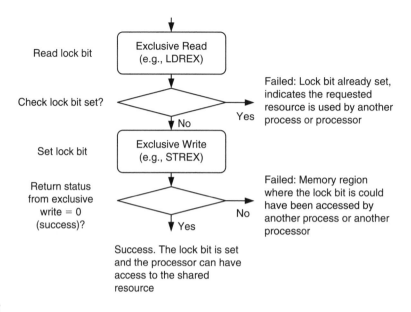

Read lock bit — Exclusive Read (e.g., LDREX)

Check lock bit set? — Yes → Failed: Lock bit already set, indicates the requested resource is used by another process or processor

No

Set lock bit — Exclusive Write (e.g., STREX)

Return status from exclusive write = 0 (success)? — No → Failed: Memory region where the lock bit is could have been accessed by another process or another processor

Yes

Success. The lock bit is set and the processor can have access to the shared resource

FIGURE 5.17

Using Exclusive Access in MUTEX Semaphores.

the lock flag read and write to be atomic, preventing the resource from being locked by two processes at the same time.

In newer ARM processors, the read/write access can be carried out on separated buses. In such situations, the SWP instructions can no longer be used to make the memory access atomic because the read and write in a locked transfer sequence must be on the same bus. Therefore, the locked transfers are replaced by exclusive accesses. The concept of exclusive access operation is quite simple but different from SWP; it allows the possibility that the memory location for a semaphore could be accessed by another bus master or another process running on the same processor (see Figure 5.17).

To allow exclusive access to work properly in a multiple processor environment, an additional hardware called "exclusive access monitor" is required. This monitor checks the transfers toward shared address locations and replies to the processor if an exclusive access is success. The processor bus interface also provides additional control signals[1] to this monitor to indicate if the transfer is an exclusive access.

If the memory device has been accessed by another bus master between the exclusive read and the exclusive write, the exclusive access monitor will flag an exclusive failed through the bus system when the processor attempts the exclusive write. This will cause the return status of the exclusive write to be 1. In the case of failed exclusive write, the exclusive access monitor also blocks the write transfer from getting to the exclusive access address.

[1]Exclusive access signals are available on the system bus and the D-Code bus of the Cortex-M3 processor. They are EXREQD and EXRESPD for the D-Code bus and EXREQS and EXRESPS for the system bus. The I-Code bus that is used for instruction fetch cannot generate exclusive accesses.

Exclusive access instructions in the Cortex-M3 include LDREX (word), LDREXB (byte), LDREXH (half word), STREX (word), STREXB (byte), and STREXH (half word). A simple example of the syntax is as follows:

```
LDREX  <Rxf>, [Rn, #offset]
STREX  <Rd>, <Rxf>,[Rn, #offset]
```

Where *Rd* is the return status of the exclusive write (0 = success and 1 = failure).

Example code for exclusive accesses can be found in Chapter 10. You can also access exclusive access instructions in C using intrinsic functions provided in Cortex Microcontroller Software Interface Standard (CMSIS) compliant device driver libraries from microcontroller vendors: __LDREX, __LEDEXH, __LDREXB, __STREX, __STREXH, __STREXB. More details of these functions are covered in Appendix G.

When exclusive accesses are used, the internal write buffers in the Cortex-M3 bus interface will be bypassed, even when the MPU defines the region as bufferable. This ensures that semaphore information on the physical memory is always up to date and coherent between bus masters. SoC designers using Cortex-M3 on multiprocessor systems should ensure that the memory system enforces data coherency when exclusive transfers occur.

5.8 ENDIAN MODE

The Cortex-M3 supports both little endian and big endian modes. However, the supported memory type also depends on the design of the rest of the microcontroller (bus connections, memory controllers, peripherals, and so on). Make sure that you check your microcontroller datasheets in detail before developing your software. In most cases, Cortex-M3-based microcontrollers will be little endian. With little endian mode, the first byte of a word size data is stored in the least significant byte of the 32-bit memory location (see Table 5.4).

There are some microcontrollers that use big endian mode. In such a case, the first byte of a word size data is stored in the most significant byte of the 32-bit address memory location (see Table 5.5).

The definition of big endian in the Cortex-M3 is different from the ARM7. In the ARM7TDMI, the big endian scheme is called *word-invariant big endian*, also referred as BE-32 in ARM documentation, whereas in the Cortex-M3, the big endian scheme is called *byte-invariant big endian*, also referred as BE-8 (byte-invariant big endian is supported on ARM architecture v6 and v7). The memory view of both schemes is the same, but the byte lane usage on the bus interface during data transfers is different (see Tables 5.6 and 5.7).

Note that the data transfer on the AHB bus in BE-8 mode uses the same data byte lanes as in little endian. However, the data byte inside the half word or word data is reversely ordered compared to little endian (see Table 5.8).

Table 5.4 The Cortex-M3 Little Endian Memory View Example

Address	Bits 31 – 24	Bits 23 – 16	Bits 15 – 8	Bits 7 – 0
0x1003 – 0x1000	Byte – 0x1003	Byte – 0x1002	Byte – 0x1001	Byte – 0x1000
0x1007 – 0x1004	Byte – 0x1007	Byte – 0x1006	Byte – 0x1005	Byte – 0x1004
…	Byte – 4xN+3	Byte – 4xN+2	Byte – 4xN+1	Byte – 4xN

Table 5.5 The Cortex-M3 Big Endian Memory View Example

Address	Bits 31 – 24	Bits 23 – 16	Bits 15 – 8	Bits 7 – 0
0x1003 – 0x1000	Byte – 0x1000	Byte – 0x1001	Byte – 0x1002	Byte – 0x1003
0x1007 – 0x1004	Byte – 0x1004	Byte – 0x1005	Byte – 0x1006	Byte – 0x1007
...	Byte – 4xN	Byte – 4xN+1	Byte – 4xN+2	Byte – 4xN+3

Table 5.6 The Cortex-M3 (Byte-Invariant Big Endian, BE-8)—Data on the AHB Bus

Address, Size	Bits 31 – 24	Bits 23 – 16	Bits 15 – 8	Bits 7 – 0
0x1000, word	Data bit [7:0]	Data bit [15:8]	Data bit [23:16]	Data bit [31:24]
0x1000, half word	—	—	Data bit [7:0]	Data bit [15:8]
0x1002, half word	Data bit [7:0]	Data bit [15:8]	—	—
0x1000, byte	—	—	—	Data bit [7:0]
0x1001, byte	—	—	Data bit [7:0]	—
0x1002, byte	—	Data bit [7:0]	—	—
0x1003, byte	Data bit [7:0]	—	—	—

Table 5.7 ARM7TDMI (Word-Invariant Big Endian, BE-32)—Data on the AHB Bus

Address, Size	Bits 31 – 24	Bits 23 – 16	Bits 15 – 8	Bits 7 – 0
0x1000, word	Data bit [7:0]	Data bit [15:8]	Data bit [23:16]	Data bit [31:24]
0x1000, half word	Data bit [7:0]	Data bit [15:8]	—	—
0x1002, half word	—	—	Data bit [7:0]	Data bit [15:8]
0x1000, byte	Data bit [7:0]	—	—	—
0x1001, byte	—	Data bit [7:0]	—	—
0x1002, byte	—	—	Data bit [7:0]	—
0x1003, byte	—	—	—	Data bit [7:0]

Table 5.8 The Cortex-M3 Little Endian—Data on the AHB Bus

Address, Size	Bits 31 – 24	Bits 23 – 16	Bits 15 – 8	Bits 7 – 0
0x1000, word	Data bit [31:24]	Data bit [23:16]	Data bit [15:8]	Data bit [7:0]
0x1000, half word	—	—	Data bit [15:8]	Data bit [7:0]
0x1002, half word	Data bit [15:8]	Data bit [7:0]	—	—
0x1000, byte	—	—	—	Data bit [7:0]
0x1001, byte	—	—	Data bit [7:0]	—
0x1002, byte	—	Data bit [7:0]	—	—
0x1003, byte	Data bit [7:0]	—	—	—

In the Cortex-M3 processor, the endian mode is set when the processor exits reset. The endian mode cannot be changed afterward. (There is no dynamic endian switching, and the SETEND instruction is not supported.) Instruction fetches are always in little endian as are data accesses in the system control memory space (such as NVIC and FPB) and the external PPB memory range (memory range from 0xE0000000 to 0xE00FFFFF is always little endian).

In case your SoC does not support big endian but one or some of the peripherals you are using contain big endian data, you can easily convert the data between little endian and big endian using some of the data type conversion instructions in the Cortex-M3. For example, REV and REV16 are very useful for this kind of conversion.

Cortex-M3 Implementation Overview

6

IN THIS CHAPTER

The Pipeline 99
A Detailed Block Diagram 101
Bus Interfaces on the Cortex-M3 104
Other Interfaces on the Cortex-M3 105
The External PPB 105
Typical Connections 106
Reset Types and Reset Signals 107

This chapter is mainly written for system-on-chip (SoC) designers who are interested in using the Cortex™-M3 processor in their project. Normal microcontroller users do not need to learn these details. However, for those who are interested in understanding the internal operations of the Cortex-M3 processor, this chapter provides a good overview of the design.

6.1 THE PIPELINE

The Cortex-M3 processor has a three-stage pipeline. The pipeline stages are instruction fetch, instruction decode, and instruction execution (see Figure 6.1).

Some people might argue that there are four stages because of the pipeline behavior in the bus interface when it accesses memory, but this stage is outside the processor, so the processor itself still has only three stages.

When running programs with mostly 16-bit instructions, you will find that the processor might not fetch instructions in every cycle. This is because the processor fetches up to two instructions (32-bit) in one go, so after one instruction is fetched, the next one is already inside the processor. In this case, the processor bus interface may try to fetch the instruction after the next or, if the buffer is full, the bus interface could be idle. Some of the instructions take multiple cycles to execute; in this case, the pipeline will be stalled.

In executing a branch instruction, the pipeline will be flushed. The processor will have to fetch instructions from the branch destination to fill up the pipeline again. However, the Cortex-M3 processor

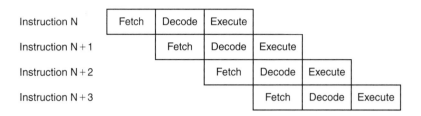

FIGURE 6.1

The Three-Stage Pipeline in the Cortex-M3.

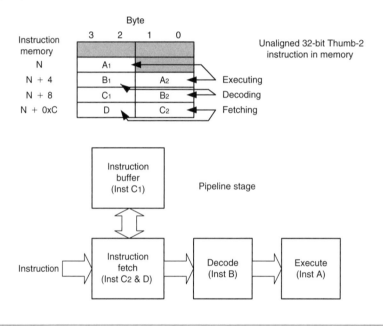

FIGURE 6.2

Use of a Buffer in the Instruction Fetch Unit to Improve 32-Bit Instruction Handling.

supports a number of instructions in v7-M architecture, so some of the short-distance branches can be avoided by replacing them with conditional execution codes.[1]

Because of the pipeline nature of the processor and to ensure that the program is compatible with Thumb® codes, the read value will be the address of the instruction plus 4, when the program counter is read during instruction execution. If the program counter is used for address generation for memory accesses, the word aligned value of the instruction address plus 4 would be used. This offset is constant, independent of the combination of 16-bit Thumb instructions and 32-bit Thumb-2 instructions. This ensures consistency between Thumb and Thumb-2.

Inside the instruction prefetch unit of the processor core, there is also an instruction buffer (see Figure 6.2). This buffer allows additional instructions to be queued before they are needed. This buffer

---

[1]For more information, refer to the "IF-THEN Instructions" section of Chapter 4.

prevents the pipeline being stalled when the instruction sequence contains 32-bit Thumb-2 instructions that are not word aligned. However, this buffer does not add an extra stage to the pipeline, so it does not increase the branch penalty.

6.2 A DETAILED BLOCK DIAGRAM

The Cortex-M3 processor contains not only the processor core but also a number of components for system management, as well as debugging support components (see Figure 6.3). These components are linked together using an Advanced High-Performance Bus (AHB), and an Advanced Peripheral Bus (APB). The AHB and APB are part of the Advanced Microcontroller Bus Architecture (AMBA) standards [Ref. 4].

Note that the MPU, WIC, and ETM blocks are optional blocks that can be included in the microcontroller system at the time of implementation. A number of new components are shown in Table 6.1.

The Cortex-M3 processor is released as a processor subsystem (see Figure 6.3). The CPU core itself is closely coupled to the interrupt controller (NVIC) and various debug logic blocks:

- *CM3Core*: The Cortex-M3 core contains the registers, ALU, data path, and bus interface.
- *NVIC*: The NVIC is a built-in interrupt controller. The number of interrupts is customized by chip manufacturers. The NVIC is closely coupled to the CPU core and contains a number of system

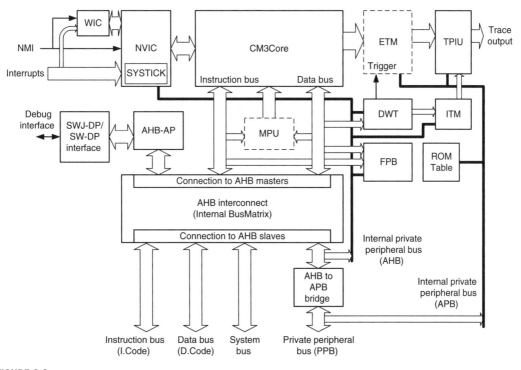

FIGURE 6.3

The Cortex-M3 Processor System Block Diagram.

Name	Description
	Table 6.1 Block Diagram Acronyms and Definitions
CM3Core	Central processing core of the Cortex-M3 processor
NVIC	Nested Vectored Interrupt Controller
SYSTICK timer	A simple timer that can be used by the operating system
WIC	Wakeup Interrupt Controller (optional)
MPU	Memory Protection Unit (optional)
BusMatrix	Internal AHB interconnection
AHB to APB	Bus bridge to convert AHB to APB
SW-DP/SWJ-DP interface	Serial Wire/Serial Wire Joint Test Action Group (JTAG) debug port (DP) interface; debug interface connection implemented using either Serial Wire Protocol or traditional JTAG Protocol (for SWJ-DP)
AHB-AP	AHB Access Port; converts commands from SW/SWJ interface into AHB transfers
ETM	Embedded Trace Macrocell; a module to handle instruction trace for debug (optional)
DWT	Data Watchpoint and Trace unit; a module to handle the data watchpoint function for debug
ITM	Instrumentation Trace Macrocell
TPIU	Trace Port Interface Unit; an interface block to send debug data to external trace capture hardware
FPB	Flash Patch and Breakpoint unit
ROM table	A small lookup table that stores configuration information

control registers. It supports the nested interrupt handling, which means that with the Cortex-M3, nested interrupt handling is very simple. It also comes with a vectored interrupt feature so that when an interrupt occurs, it can enter the corresponding interrupt handler routine directly, without using a shared handler to determine which interrupt has occurred.

- *SYSTICK Timer*: The System Tick (SYSTICK) Timer is a basic countdown timer that can be used to generate interrupts at regular time intervals, even when the system is in sleep mode. It makes OS porting between Cortex-M3 devices much easier because there is no need to change the OS's system timer code. The SYSTICK Timer is implemented as part of the NVIC.

- *WIC*: A module interface with NVIC but separated from the main processor design to allow the system to wake up from interrupt events while the processor (including the NVIC) is completely stopped or powered down. This module is new from the Cortex-M3 revision 2 and is optional.

- *MPU*: The MPU block is optional. This means that some versions of the Cortex-M3 might have the MPU and some might not. If it is included, the MPU can be used to protect memory contents by, for example, making memory regions read-only or preventing user applications from accessing privileged applications data.

- *BusMatrix*: A BusMatrix is used as the heart of the Cortex-M3 internal bus system. It is an AHB interconnection network, allowing transfer to take place on different buses simultaneously unless both bus masters are trying to access the same memory region. The BusMatrix also provides additional data transfer management, including a write buffer as well as bit-oriented operations (bit-band).

- *AHB to APB*: An AHB-to-APB bus bridge is used to connect a number of APB devices such as debugging components to the private peripheral bus in the Cortex-M3 processor. In addition, the Cortex-M3 allows chip manufacturers to attach additional APB devices to the external private peripheral bus (PPB) using this APB bus.

The rest of the components in the block diagram are for debugging support and normally should not be used by application code.

- *SW-DP/SWJ-DP*: The Serial Wire Debug Port (SW-DP)/Serial Wire JTAG Debug Port (SWJ-DP) work together with the AHB Access Port (AHB-AP) so that external debuggers can generate AHB transfers to control debug activities. There is no JTAG scan chain inside the processor core of the Cortex-M3; most debugging functions are controlled by the NVIC registers through AHB accesses. SWJ-DP supports both the Serial Wire Protocol and the JTAG Protocol, whereas SW-DP can support only the Serial Wire Protocol.

- *AHB-AP*: The AHB-AP provides access to the whole Cortex-M3 memory through a few registers. This block is controlled by the SW-DP/SWJ-DP through a generic debug interface called the Debug Access Port (DAP). To carry out debugging functions, the external debugging hardware needs to access the AHB-AP through the SW-DP/SWJ-DP to generate the required AHB transfers.

- *ETM*: The ETM is an optional component for instruction trace, so some Cortex-M3 products might not have real-time instruction trace capability. Trace information is output to the trace port through TPIU. The ETM control registers are memory mapped, which can be controlled by the debugger through the DAP.

- *DWT*: The DWT allows data watchpoints to be set up. When a data address or data value match is found, the match hit event can be used to generate watchpoint events to activate the debugger, generate data trace information, or activate the ETM.

- *ITM*: The ITM can be used in several ways. Software can write to this module directly to output information to TPIU, or the DWT matching events can be used to generate data trace packets through ITM for output into a trace data stream.

- *TPIU*: The TPIU is used to interface with external trace hardware such as trace port analyzers. Internal to the Cortex-M3, trace information is formatted as Advanced Trace Bus (ATB) packets, and the TPIU reformats the data to allow data to be captured by external devices.

- *FPB*: The FPB is used to provide Flash Patch and Breakpoint functionalities. Flash Patch means that if an instruction access by the CPU matches a certain address, the address can be remapped to a different location so that a different value is fetched. Alternatively, the matched address can be used to trigger a breakpoint event. The Flash Patch feature is very useful for testing, such as adding diagnosis program code to a device that cannot be used in normal situations unless the FPB is used to change the program control.

- *ROM table*: A small ROM table is provided. This is simply a small lookup table to provide memory map information for various system devices and debugging components. Debugging systems use this table to locate the memory addresses of debugging components. In most cases, the memory map should be fixed to the standard memory location, as documented in the *Cortex-M3 Technical Reference Manual* (TRM) [Ref. 1], but because some of the debugging components are optional and additional components can be added, individual chip manufacturers might want to customize their chip's debugging features. In this case, the ROM table must be customized and used for debugging software to determine the correct memory map and hence detect the type of debugging components available.

6.3 BUS INTERFACES ON THE CORTEX-M3

Unless you are designing an SoC product using the Cortex-M3 processor, it is unlikely that you can directly access the bus interface signals described here. Normally, the chip manufacturer will hook up all the bus signals to memory blocks and peripherals, and in a few cases, you might find that the chip manufacturer connected the bus to a bus bridge and allows external bus systems to be connected off-chip. The bus interfaces on the Cortex-M3 processor are based on AHB-Lite and APB protocols, which are documented in the AMBA Specification [Ref. 4].

6.3.1 The I-Code Bus

The I-Code bus is a 32-bit bus based on the AHB-Lite bus protocol for instruction fetches in memory regions from 0x00000000 to 0x1FFFFFFF. Instruction fetches are performed in word size, even for 16-bit Thumb instructions. Therefore, during execution, the CPU core could fetch up to two Thumb instructions at a time.

6.3.2 The D-Code Bus

The D-Code bus is a 32-bit bus based on the AHB-Lite bus protocol; it is used for data access in memory regions from 0x00000000 to 0x1FFFFFFF. Although the Cortex-M3 processor supports unaligned transfers, you won't get any unaligned transfer on this bus, because the bus interface on the processor core converts the unaligned transfers into aligned transfers for you. Therefore, devices (such as memory) that attach to this bus need only support AHB-Lite (AMBA 2.0) aligned transfers.

6.3.3 The System Bus

The system bus is a 32-bit bus based on the AHB-Lite bus protocol; it is used for instruction fetch and data access in memory regions from 0x20000000 to 0xDFFFFFFF and 0xE0100000 to 0xFFFFFFFF. Similar to the D-Code bus, all the transfers on the system bus are aligned.

6.3.4 The External PPB

The External PPB is a 32-bit bus based on the APB bus protocol. This is intended for private peripheral accesses in memory regions 0xE0040000 to 0xE00FFFFF. However, since some part of this APB memory is already used for TPIU, ETM, and the ROM table, the memory region that can be used for attaching extra peripherals on this bus is only 0xE0042000 to 0xE00FF000. Transfers on this bus are word aligned.

6.3.5 **The DAP Bus**

The DAP bus interface is a 32-bit bus based on an enhanced version of the APB specification. This is for attaching debug interface blocks such as SWJ-DP or SW-DP. Do not use this bus for other purposes. More information on this interface can be found in Chapter 15, or in the ARM document *CoreSight Technology System Design Guide* [Ref. 3].

6.4 **OTHER INTERFACES ON THE CORTEX-M3**

Apart from bus interfaces, the Cortex-M3 processor has a number of other interfaces for various purposes. These signals are unlikely to appear on the pins of the silicon chip, because they are mostly for connecting to various parts of the SoC or are unused. The details of the signals are contained in the *Cortex-M3 Technical Reference Manual* [Ref. 1]. Table 6.2 contains a short summary of some of them.

6.5 **THE EXTERNAL PPB**

The Cortex-M3 processor has an External PPB interface. The External PPB interface is based on the APB protocol in AMBA specification 2.0 (for Cortex-M3 revision 0 and revision 1) or 3.0 (for Cortex-M3 revision 2). It is intended for system devices that should not be shared, such as debugging components.

Table 6.2 Miscellaneous Interface Signals

Signal Group	Function
Multiprocessor communication (TXEV, RXEV)	Simple task synchronization signals between multiple processors
Sleep signals (SLEEPING, SLEEPDEEP)	Sleep status for power management
Interrupt status signals (ETMINTNUM, ETMINTSTATE, CURRPRI)	Status of interrupt operation, for ETM operation and debug usage
Reset request (SYSRESETREQ)	Resets request output from NVIC
Lockup[2] and Halted status (LOCKUP, HALTED)	Indicate that the processor core has entered a lockup state (caused by error conditions within hard fault handler or Nonmaskable Interrupt handler) or a halted state (for debug operations)
Endian input (ENDIAN)	Sets the endian of the Cortex-M3 when the core is reset
ETM interface	Connects to ETM for instruction trace
ITM's ATB interface	ATB is a bus protocol in ARM's CoreSight debug architecture for trace data transfer; here this interface provides trace data output from Cortex-M3's ITM, which is connected to the TPIU

[2]More information on lockup is included in Chapter 12.

This bus interface supports the use of CoreSight compliant debug components. To achieve this, this interface is slightly different from normal APB—it contains an extra signal called PADDR31 that indicates the source of a transfer. If this signal is 0, it means that the transfer is generated from software running on the Cortex-M3. If this signal is 1, it means that the transfer is generated by debugging hardware. Based on this signal, a peripheral can be designed so that only a debugger can use it, or when being used by software, only some of the features are allowed.

This bus is not intended for general use, as in peripherals. Although there is nothing to stop chip designers from designing and attaching general peripherals on this bus, users might find it a problem for programming later, because of privileged access-level management—for example, to program the device in the user state or to separate the devices from other memory regions when the MPU is used.

The External PPB does not support unaligned accesses. Because the data width of the bus is 32-bit and APB based, when you're designing peripherals for this memory region, it is necessary to make sure that all register addresses in the peripheral are word aligned. In addition, when writing software accessing devices in this region, it is recommended that you make sure that all the accesses are in word size. The PPB accesses are always in little endian.

6.6 TYPICAL CONNECTIONS

Because there are a number of bus interfaces on the Cortex-M3 processor, you might find it confusing to see how it will connect with other devices such as memory or peripherals. Figure 6.4 shows a simplified example.

Since the Code memory region can be accessed by the instruction bus (if it is an instruction fetch) and from the data bus (if it is a data access), an AHB bus switch called the *BusMatrix*[3] or an AHB bus multiplexer is needed. With the BusMatrix, the Flash memory and the additional Static Random Access Memory (SRAM) (if implemented) can be accessed by either bus interface. The BusMatrix is available from ARM in the AMBA Design Kit[4] (ADK). When both data bus and instruction bus are trying to access the same memory device at the same time, the data bus access could be given higher priority for best performance.

Using the AHB BusMatrix, if the instruction bus and the data bus are accessing different memory devices at the same time (for example, an instruction fetch from fetch and a data bus reading data from the additional SRAM), the transfers can be carried out simultaneously. If a bus multiplexer is used, however, the transfers cannot take place at the same time, but the circuit size would be smaller. Common Cortex-M3 microcontroller designs use system bus for SRAM connection.

The main SRAM block should be connected through the system bus interface, using the SRAM memory address region. This allows data access to be carried out at the same time as instruction access. It also allows setting up of Boolean data types by using the bit-band feature.

Some microcontrollers might have an external memory interface. That requires an external memory controller because you cannot connect off chip memory devices directly to AHB. The external memory

---

[3]The BusMatrix required here is different from the internal BusMatrix inside the Cortex-M3 shown in Figure 6.4. The Cortex-M3 internal BusMatrix is specially designed and is different from standard AMBA Design Kit (ADK) version.

[4]ADK is a collection of AMBA components and example systems in VHDL/Verilog.

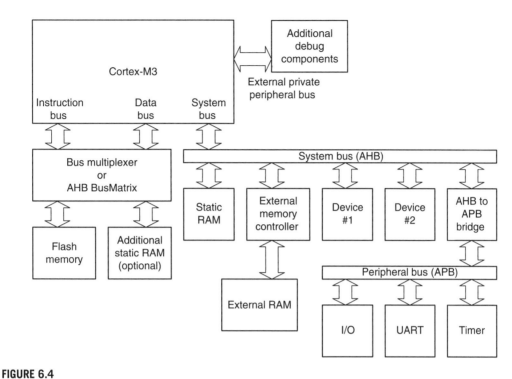

FIGURE 6.4

The Cortex-M3 Processor System Block Diagram.

controller can be connected to the system bus of the Cortex-M3. Additional AHB devices can also be easily connected to the system bus without the need for a BusMatrix.

Simple peripherals can be connected to the Cortex-M3 through an AHB-to-APB bridge. This allows the use of the simpler bus protocol APB for peripherals.

The diagram shown in Figure 6.4 is just a very simple example; chip designers might choose different bus connection designs. For software/firmware development, you will only need to know the memory map.

Design blocks shown in the diagram, such as the BusMatrix, AHB-to-APB bus bridge, memory controller, I/O interface, timer, and universal asynchronous receiver/transmitter (UART), are all available from ARM and a number of Internet Protocol providers. Because microcontrollers can have different providers for the peripherals, you need to access your microcontroller's datasheet for the correct programmer model when you're developing software for Cortex-M3 systems.

6.7 **RESET TYPES AND RESET SIGNALS**

There are a number of different reset types on a Cortex-M3 system. Some Cortex-M3 product might have more reset types depending on the design of reset circuitry on the Cortex-M3 microcontroller or SoC (see Figure 6.5). In general, there are at least three types of reset as shown in Table 6.3.

Table 6.3 Common Reset Types on Cortex-M3 Microcontrollers

Reset Type	Reset Signal on the Cortex-M3 Processor	Description
Power on reset	PORESETn	Reset that should be asserted when the device is powered up; resets processor core, peripherals, and debugging system Activate by power up sequence of the device
System reset	SYSRESETn	System reset; affects the whole system including processor core, NVIC (except debug control registers), MPU, peripherals but not the debugging system; activate by power up sequence of the device, reset request from debugger through NVIC register "AIRCR"
Processor reset	VECTRESET bit in the NVIC AIRCR register	Reset processor core only; affect the processor system including processor core, NVIC (except debug control registers), MPU, but not the debugging system; activate reset request from debugger through NVIC register "AIRCR"—intended to be used by debugger
JTAG reset	nTRST	Reset for JTAG tap controller (only if JTAG interface is available)

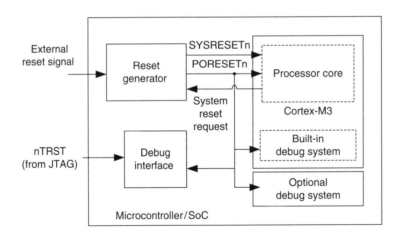

FIGURE 6.5

Generation of Internal Reset Signals in a Typical Cortex-M3 Microcontroller.

The details of the reset signals on the processor can be found in the *Cortex-M3 Technical Reference Manual* [Ref. 1]. The reset signals on the processors are connected to the reset generator inside the microcontroller or SoC. Externally you may find only one or two reset signals.

Exceptions

IN THIS CHAPTER

Exception Types 109
Definitions of Priority 111
Vector Tables 117
Interrupt Inputs and Pending Behavior 118
Fault Exceptions 120
Supervisor Call and Pendable Service Call 126

7.1 EXCEPTION TYPES

The Cortex™-M3 provides a feature-packed exception architecture that supports a number of system exceptions and external interrupts. Exceptions are numbered 1–15 for system exceptions and 16 and above for external interrupt inputs. Most of the exceptions have programmable priority, and a few have fixed priority.

Cortex-M3 chips can have different numbers of external interrupt inputs (from 1 to 240) and different numbers of priority levels. This is because chip designers can configure the Cortex-M3 design source code for different needs.

Exception types 1–15 are system exceptions (there is no exception type 0), as outlined in Table 7.1. Exceptions of type 16 or above are external interrupt inputs (see Table 7.2).

The value of the current running exception is indicated by the special register Interrupt Program Status register (IPSR), or from the Nested Vectored Interrupt Controllers (NVICs) Interrupt Control State register (the VECTACTIVE field).

Note that here the interrupt number (e.g., Interrupt #0) refers to the interrupt inputs to the Cortex-M3 NVIC. In actual microcontroller products or system-on-chips (SoCs), the external interrupt input pin number might not match the interrupt input number on the NVIC. For example, some of the first few interrupt inputs might be assigned to internal peripherals, and external interrupt pins could be assigned to the next couple of interrupt inputs. Therefore, you need to check the chip manufacturer's datasheets to determine the numbering of the interrupts.

Table 7.1 List of System Exceptions

Exception Number	Exception Type	Priority	Description
1	Reset	−3 (Highest)	Reset
2	NMI	−2	Nonmaskable interrupt (external NMI input)
3	Hard fault	−1	All fault conditions if the corresponding fault handler is not enabled
4	MemManage fault	Programmable	Memory management fault; Memory Protection Unit (MPU) violation or access to illegal locations
5	Bus fault	Programmable	Bus error; occurs when Advanced High-Performance Bus (AHB) interface receives an error response from a bus slave (also called *prefetch abort* if it is an instruction fetch or *data abort* if it is a data access)
6	Usage fault	Programmable	Exceptions resulting from program error or trying to access coprocessor (the Cortex-M3 does not support a coprocessor)
7–10	Reserved	NA	—
11	SVC	Programmable	Supervisor Call
12	Debug monitor	Programmable	Debug monitor (breakpoints, watchpoints, or external debug requests)
13	Reserved	NA	—
14	PendSV	Programmable	Pendable Service Call
15	SYSTICK	Programmable	System Tick Timer

Table 7.2 List of External Interrupts

Exception Number	Exception Type	Priority
16	External Interrupt #0	Programmable
17	External Interrupt #1	Programmable
...	...	...
255	External Interrupt #239	Programmable

When an enabled exception occurs but cannot be carried out immediately (for instance, if a higher-priority interrupt service routine is running or if the interrupt mask register is set), it will be pended (except for some fault exceptions[1]). This means that a register (pending status) will hold the exception request until the exception can be carried out. This is different from traditional ARM processors. Previously, the

---

[1]There are a few exceptions for the exception-pending behavior. If a fault takes place and the corresponding fault handler cannot be executed immediately because a higher-priority handler is running, the hard fault handler (highest priority fault handler) might be executed instead. More details on this topic are covered later in this chapter, where we look at fault exceptions; full details can be found in the *ARM v7-M Architecture Application Level Reference Manual*.

devices that generate interrupts, such as interrupt request (IRQ)/fast interrupt request (FIQ), must hold the request until they are served. Now, with the pending registers in the NVIC, an occurred interrupt will be handled even if the source requesting the interrupt deasserts its request signal.

7.2 DEFINITIONS OF PRIORITY

In the Cortex-M3, whether and when an exception can be carried out can be affected by the priority of the exception. A higher-priority (smaller number in priority level) exception can preempt a lower-priority (larger number in priority level) exception; this is the nested exception/interrupt scenario. Some of the exceptions (reset, NMI, and hard fault) have fixed priority levels. They are negative numbers to indicate that they are of higher priority than other exceptions. Other exceptions have programmable priority levels.

The Cortex-M3 supports three fixed highest-priority levels and up to 256 levels of programmable priority (a maximum of 128 levels of preemption). However, most Cortex-M3 chips have fewer supported levels—for example, 8, 16, 32, and so on. When a Cortex-M3 chip or SoC is being designed, designers can customize it to obtain the number of levels required. This reduction of levels is implemented by cutting out the Least Significant Bit (LSB) part of the priority configuration registers.

For example, if only 3 bits of priority level are implemented in the design, a priority-level configuration register will look like Figure 7.1.

Because bit 4 to bit 0 are not implemented, they are always read as zero, and writes to these bits will be ignored. With this setup, we have possible priority levels of 0x00 (high priority), 0x20, 0x40, 0x60, 0x80, 0xA0, 0xC0, and 0xE0 (the lowest).

Similarly, if 4 bits of priority level are implemented in the design, a priority-level configuration register will look like Figure 7.2.

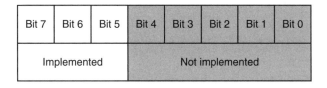

FIGURE 7.1

A Priority Level Register with 3 Bits Implemented.

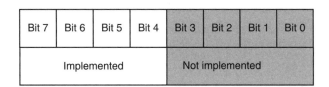

FIGURE   7.2

A Priority Level Register with 4 Bits Implemented.

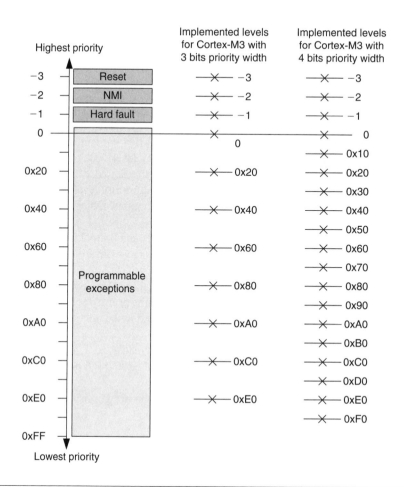

FIGURE 7.3

Available Priority Levels with 3-Bit or 4-Bit Priority Width.

If more bits are implemented, more priority levels will be available (see Figure 7.3). However, more priority bits can also increase gate counts and hence the power consumption. For the Cortex-M3, the minimum number of implemented priority register widths is 3 bits (eight levels).

The reason for removing the LSB of the register instead of the Most Significant Bit (MSB) is to make it easier to port software from one Cortex-M3 device to another. In this way, a program written for devices with 4-bit priority configuration registers is likely to be able to run on devices with 3-bit priority configuration registers. If the MSB is removed instead of the LSB, you might get an inversion of priority arrangement when porting an application from one Cortex-M3 chip to another. For example, if an application uses priority level 0x05 for IRQ #0 and level 0x03 for IRQ #1, IRQ #1 should have higher priority. But when MSB bit 2 is removed, IRQ #0 will become level 0x01 and have a higher priority than IRQ #1.

Examples of available exception priority levels for devices with 3-bit, 5-bit, and 8-bit priority registers are shown in Table 7.3.

Some readers might wonder whether, if the priority level configuration registers are 8 bits wide, why there are only 128 preemption levels? This is because the 8-bit register is further divided into two parts: *preempt priority* and *subpriority*.

Using a configuration register in the NVIC called *Priority Group* (a part of the Application Interrupt and Reset Control register in the NVIC, see Table 7.4), the priority-level configuration registers for each exception with programmable priority levels is divided into two halves. The upper half (left bits) is the preempt priority, and the lower half (right bits) is the subpriority (see Table 7.5).

Table 7.3 Available Priority Levels for Devices with 3-Bit, 5-Bit, and 8-Bit Priority Level Registers

Priority Level	Exception Type	Devices with 3-Bit Priority Configuration Registers	Devices with 5-Bit Priority Configuration Registers	Devices with 8-Bit Priority Configuration Registers
−3 (Highest)	Reset	−3	−3	−3
−2	NMI	−2	−2	−2
−1	Hard fault	−1	−1	−1
0, 1, … 0xFF	Exceptions with programmable priority level	0x00, 0x20, … 0xE0	0x00, 0x08, … 0xF8	0x00, 0x01, 0x02, 0x03, … 0xFE, 0xFE

Table 7.4 Application Interrupt and Reset Control Register (Address 0xE000ED0C)

Bits	Name	Type	Reset Value	Description
31:16	VECTKEY	R/W	—	Access key; 0x05FA must be written to this field to write to this register, otherwise the write will be ignored; the read-back value of the upper half word is 0xFA05
15	ENDIANNESS	R	—	Indicates endianness for data: 1 for big endian (BE8) and 0 for little endian; this can only change after a reset
10:8	PRIGROUP	R/W	0	Priority group
2	SYSRESETREQ	W	—	Requests chip control logic to generate a reset
1	VECTCLRACTIVE	W	—	Clears all active state information for exceptions; typically used in debug or OS to allow system to recover from system error (Reset is safer)
0	VECTRESET	W	—	Resets the Cortex-M3 processor (except debug logic), but this will not reset circuits outside the processor

Table 7.5 Definition of Preempt Priority Field and Subpriority Field in a Priority Level Register in Different Priority Group Settings

Priority Group	Preempt Priority Field	Subpriority Field
0	Bit [7:1]	Bit [0]
1	Bit [7:2]	Bit [1:0]
2	Bit [7:3]	Bit [2:0]
3	Bit [7:4]	Bit [3:0]
4	Bit [7:5]	Bit [4:0]
5	Bit [7:6]	Bit [5:0]
6	Bit [7]	Bit [6:0]
7	None	Bit [7:0]

The *preempt priority level* defines whether an interrupt can take place when the processor is already running another interrupt handler. The *subpriority level* value is used only when two exceptions with the same preempt priority level occurred at the same time. In this case, the exception with higher subpriority (lower value) will be handled first.

As a result of the priority grouping, the maximum width of preempt priority is 7, so there can be 128 levels. When the priority group is set to 7, all exceptions with a programmable priority level will be in the same level, and no preemption between these exceptions will take place, except that hard fault, NMI, and reset, which have priority of −1, −2, and −3, respectively, can preempt these exceptions.

When deciding the effective preempt priority level and subpriority level, you must take the following factors into account:

- Implemented priority-level configuration registers
- Priority group setting

For example, if the width of the configuration registers is 3 (bit 7 to bit 5 are available) and priority group is set to 5, you can have four levels of preempt priority levels (bit 7 to bit 6), and inside each preempt level there are two levels of subpriority (bit 5).

With the setting as shown in Figure 7.4, the available priority levels are illustrated in Figure 7.5. For the same design, if the priority group is set to 0x1, there can be only eight preempt priority levels and no further subpriority levels inside each preempt level. (Bit [1:0] of preempt priority is always 0.) The definition of the priority level configuration registers is shown in Figure 7.6, and the available priority levels are illustrated in Figure 7.7.

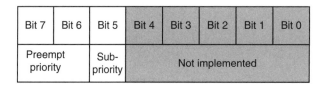

FIGURE 7.4

Definition of Priority Fields in a 3-Bit Priority Level Register with Priority Group Set to 5.

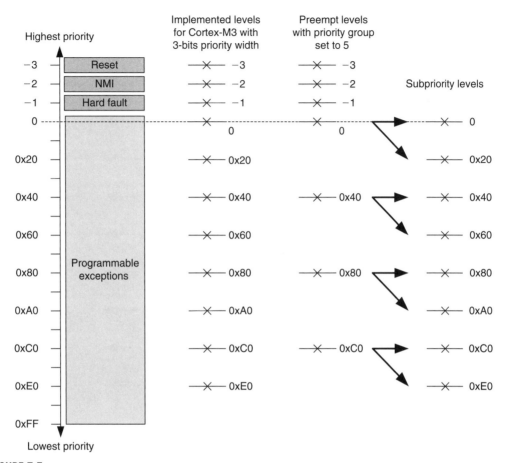

FIGURE 7.5

Available Priority Levels with 3-Bit Priority Width and Priority Group Set to 5.

Bit 7	Bit 6	Bit 5	Bit 4	Bit 3	Bit 2	Bit 1	Bit 0
Preempt priority [5:3]			Preempt priority [2:0] (always 0)			Sub- priority [1:0] (always 0)	

FIGURE 7.6

Definition of Priority Fields in an 8-Bit Priority Level Register with Priority Group Set to 1.

If a Cortex-M3 device has implemented all 8 bits in the priority-level configuration registers, the maximum number of preemption levels it can have is only 128, using a priority group setting of 0. The priority fields definition is shown in Figure 7.8.

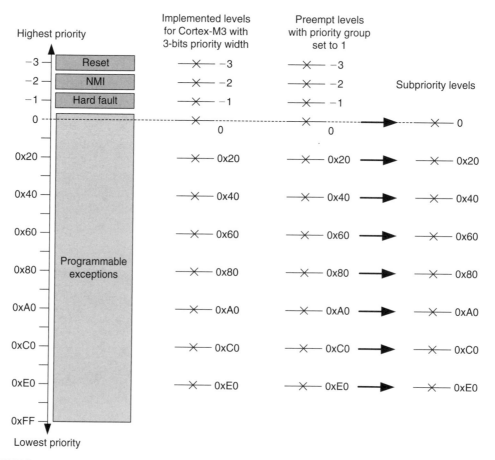

FIGURE 7.7

Available Priority Levels with 3-Bit Priority Width and Priority Group Set to 1.

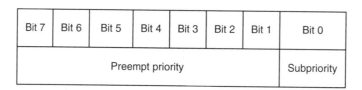

FIGURE 7.8

Definition of Priority Fields in an 8-Bit Priority Level Register with Priority Group Set to 0.

When two interrupts are asserted at the same time with exactly the same preempt priority level as well as subpriority level, the interrupt with the smaller exception number has higher priority. (IRQ #0 has higher priority than IRQ #1.)

To avoid unexpected changes of priority levels for interrupts, be careful when writing to the Application Interrupt and Reset Control register (address 0xE000ED0C). In most cases, after the priority group is configured, there is no need to use this register except to generate a reset (see Table 7.4).

7.3 VECTOR TABLES

When an exception takes place and is being handled by the Cortex-M3, the processor will need to locate the starting address of the exception handler. This information is stored in the vector table in the memory. By default, the vector table starts at memory address 0, and the vector address is arranged according to the exception number times four (see Table 7.6).

Since the address 0x0 should be boot code, usually it will be either Flash memory or ROM devices, and the value cannot be changed at run time. However, the vector table can be relocated to other memory locations in the code or Random Access Memory (RAM) region where the RAM is so that we can change the handlers during run time. This is done by setting a register in the NVIC called the *vector table offset register* (address 0xE000ED08). The address offset should be aligned to the vector table size, extended to the next larger power of 2. For example, if there are 32 IRQ inputs, the total number of exceptions will be 32 + 16 (system exceptions) = 48. Extending it to the power of 2 makes it 64. Multiplying it by 4 (4 bytes per vector) makes it 256 bytes (0x100). Therefore, the vector table offset can be programmed as 0x0, 0x100, 0x200, and so on. The vector table offset register contains the items shown in Table 7.7.

In applications where you want to allow dynamic changing of exception handlers, in the beginning of the boot image, you need to have the following (at a minimum):

- Initial main stack pointer value
- Reset vector
- NMI vector
- Hard fault vector

Table 7.6 Exception Vector Table After Power Up

Address	Exception Number	Value (Word Size)
0x00000000	—	MSP initial value
0x00000004	1	Reset vector (program counter initial value)
0x00000008	2	NMI handler starting address
0x0000000C	3	Hard fault handler starting address
…	…	Other handler starting address

Table 7.7 Vector Table Offset Register (Address 0xE000ED08)

Bits	Name	Type	Reset Value	Description
29	TBLBASE	R/W	0	Table base in code (0) or RAM (1)
28:7	TBLOFF	R/W	0	Table offset value from code region or RAM region

These are required because the NMI and hard fault can potentially occur during your boot process. Other exceptions cannot take place until they are enabled.

When the booting process is done, you can define a part of your Static Random Access Memory as the new vector table and relocate the vector table to the new one, which is writable.

7.4 INTERRUPT INPUTS AND PENDING BEHAVIOR

This section describes the behavior of IRQ inputs and pending behavior. It also applies to NMI input, except that an NMI will be executed immediately in most cases, unless the core is already executing an NMI handler, halted by a debugger, or locked up because of some serious system error.

When an interrupt input is asserted, it will be pended, which means it is put into a state of waiting for the processor to process the request. Even if the interrupt source deasserts the interrupt, the pended interrupt status will still cause the interrupt handler to be executed when the priority is allowed. Once the interrupt handler is started, the pending status is cleared automatically. This is shown in Figure 7.9.

However, if the pending status is cleared before the processor starts responding to the pended interrupt (for example, the interrupt was not taken immediately because PRIMASK/FAULTMASK is set to 1, and the pending status was cleared by software writing to NVIC interrupt control registers), the interrupt can be cancelled (Figure 7.10). The pending status of the interrupt can be accessed in the NVIC and is writable, so you can clear a pending interrupt or use software to pend a new interrupt by setting the pending register.

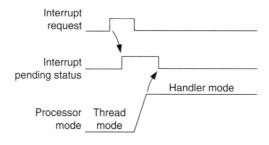

FIGURE 7.9

Interrupt Pending.

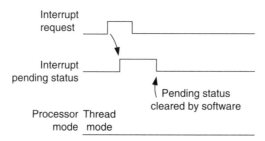

FIGURE 7.10

Interrupt Pending Cleared Before Processor Takes Action.

When the processor starts to execute an interrupt, the interrupt becomes active and the pending bit will be cleared automatically (Figure 7.11). When an interrupt is active, you cannot start processing the same interrupt again, until the interrupt service routine is terminated with an interrupt return (also called an *exception exit*, as discussed in Chapter 9). Then the active status is cleared, and the interrupt can be processed again if the pending status is 1. It is possible to repend an interrupt before the end of the interrupt service routine.

If an interrupt source continues to hold the interrupt request signal active, the interrupt will be pended again at the end of the interrupt service routine as shown in Figure 7.12. This is just like the traditional ARM7TDMI.

If an interrupt is pulsed several times before the processor starts processing it, it will be treated as one single interrupt request as illustrated in Figure 7.13. If an interrupt is deasserted and then pulsed again during the interrupt service routine, it will be pended again as shown in Figure 7.14.

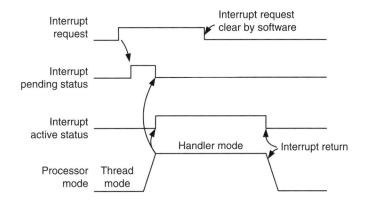

FIGURE 7.11

Interrupt Active Status Set as Processor Enters Handler.

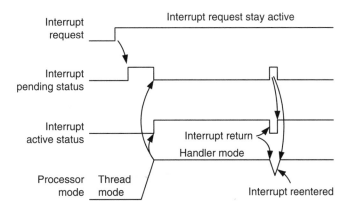

FIGURE 7.12

Continuous Interrupt Request Pends Again After Interrupt Exit.

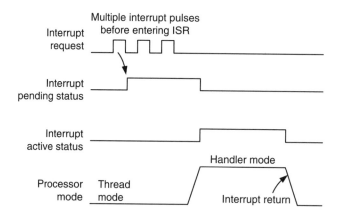

FIGURE 7.13

Interrupt Pending Only Once, Even with Multiple Pulses Before the Handler.

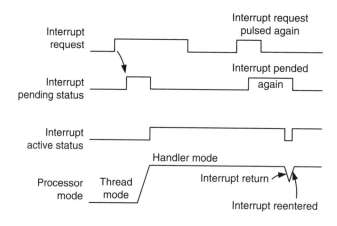

FIGURE 7.14

Interrupt Pending Occurs Again during the Handler.

Pending of an interrupt can happen even if the interrupt is disabled; the pended interrupt can then trigger the interrupt sequence when the enable is set later. As a result, before enabling an interrupt, it could be useful to check whether the pending register has been set. The interrupt source might have been activated previously and have set the pending status. If necessary, you can clear the pending status before you enable an interrupt.

7.5 FAULT EXCEPTIONS

A number of system exceptions are useful for fault handling. There are several categories of faults:

- Bus faults
- Memory management faults
- Usage faults
- Hard faults

7.5.1 **Bus Faults**

Bus faults are produced when an error response is received during a transfer on the AHB interfaces. It can happen at these stages:

* Instruction fetch, commonly called *prefetch abort*
* Data read/write, commonly called *data abort*

In the Cortex-M3, bus faults can also occur during the following:

* Stack PUSH in the beginning of interrupt processing, called a *stacking error*
* Stack POP at the end of interrupt processing, called an *unstacking error*
* Reading of an interrupt vector address (vector fetch) when the processor starts the interrupt-handling sequence (a special case classified as a hard fault)

When these types of bus faults (except vector fetches) take place and if the bus fault handler is enabled and no other exceptions with the same or higher priority are running, the bus fault handler will be executed. If the bus fault handler is enabled but at the same time the core receives another exception handler with higher priority, the bus fault exception will be pending. Finally, if the bus fault handler is not enabled or when the bus fault happens in an exception handler that has the same or higher priority than the bus fault handler, the hard fault handler will be executed instead. If another bus fault takes place when running the hard fault handler, the core will enter a lockup state.[2]

WHAT CAN CAUSE AHB ERROR RESPONSES?

Bus faults occur when an error response is received on the AHB bus. The common causes are as follows:

* Attempts to access an invalid memory region (for example, a memory location with no memory attached)
* The device is not ready to accept a transfer (for example, trying to access SDRAM without initializing the SDRAM controller)
* Attempts to carry out a transfer with a transfer size not supported by the target device (for example, doing a byte access to a peripheral register that must be accessed as a word)
* The device does not accept the transfer for various reasons (for example, a peripheral that can only be programmed at the privileged access level)

To enable the bus fault handler, you need to set the BUSFAULTENA bit in the System Handler Control and State register in the NVIC. Before doing that, make sure that the bus fault handler starting address is set up in the vector table if the vector table has been relocated to RAM.

Hence, how do you find out what went wrong when the processor entered the bus fault handler? The NVIC has a number of Fault Status registers (FSRs). One of them is the Bus Fault Status register (BFSR). From this register, the bus fault handler can find out if the fault was caused by data/instruction access or an interrupt stacking or unstacking operation.

For precise bus faults, the offending instruction can be located by the stacked program counter, and if the BFARVALID bit in BFSR is set, it is also possible to determine the memory location that caused the bus fault. This is done by reading another NVIC register called the *Bus Fault Address*

---

[2]More information on the lockup state is covered in Chapter 12.

register (BFAR). However, the same information is not available for imprecise bus faults because by the time the processor receives the error, the processor could have already executed a number of other instructions.

PRECISE AND IMPRECISE BUS FAULTS

Bus faults caused by data accesses can be further classified as precise or imprecise. In imprecise bus faults, the fault is caused by an already completed operation (such as a buffered write) that might have occurred a number of clock cycles ago. Precise bus faults are caused by the last completed operation—for example, a memory read is precise on the Cortex-M3 because the instruction cannot be completed until it receives the data.

The programmer's model for BFSR is as follows: It is 8 bits wide and can be accessed through byte transfer to address 0xE000ED29 or with a word transfer to address 0xE000ED28 with BFSR in the second byte (see Table 7.8). The error indication bit is cleared when a 1 is written to it.

Table 7.8 Bus Fault Status Register (0xE000ED29)

Bits	Name	Type	Reset Value	Description
7	BFARVALID	—	0	Indicates BFAR is valid
6:5	—	—	—	—
4	STKERR	R/Wc	0	Stacking error
3	UNSTKERR	R/Wc	0	Unstacking error
2	IMPRECISERR	R/Wc	0	Imprecise data access violation
1	PRECISERR	R/Wc	0	Precise data access violation
0	IBUSERR	R/Wc	0	Instruction access violation

7.5.2 Memory Management Faults

Memory management faults can be caused by memory accesses that violate the setup in the MPU or by certain illegal accesses (for example, trying to execute code from nonexecutable memory regions), which can trigger the fault, even if no MPU is presented.

Some of the common MPU faults include the following:

- Access to memory regions not defined in MPU setup
- Writing to read-only regions
- An access in the user state to a region defined as privileged access only

When a memory management fault occurs and if the memory management handler is enabled, the memory management fault handler will be executed. If the fault occurs at the same time a higher-priority exception takes place, the other exceptions will be handled first and the memory management fault will be pended. If the processor is already running an exception handler with the same or higher priority or if the memory management fault handler is not enabled, the hard fault handler will be executed instead. If a memory management fault takes place inside the hard fault handler or the NMI handler, the processor will enter the lockup state.

Like the bus fault handler, the memory management fault handler needs to be enabled. This is done by the MEMFAULTENA bit in the System Handler Control and State register in the NVIC. If the vector table has been relocated to RAM, the memory management fault handler starting address should be set up in the vector table first.

The NVIC contains a Memory Management Fault Status register (MFSR) to indicate the cause of the memory management fault. If the status register indicates that the fault is a data access violation (DACCVIOL bit) or an instruction access violation (IACCVIOL bit), the offending code can be located by the stacked program counter. If the MMARVALID bit in the MFSR is set, it is also possible to determine the memory address location that caused the fault from the Memory Management Address register (MMAR) in the NVIC.

The programmer's model for the MFSR is shown in Table 7.9. It is 8 bits wide and can be accessed through byte transfer or with a word transfer to address 0xE000ED28, with the MFSR in the lowest byte. As with other FSRs, the fault status bit can be cleared by writing 1 to the bit.

Table 7.9 Memory Management Fault Status Register (0xE000ED28)

Bits	Name	Type	Reset Value	Description
7	MMARVALID	—	0	Indicates the MMAR is valid
6:5	—	—	—	—
4	MSTKERR	R/Wc	0	Stacking error
3	MUNSTKERR	R/Wc	0	Unstacking error
2	—	—	—	—
1	DACCVIOL	R/Wc	0	Data access violation
0	IACCVIOL	R/Wc	0	Instruction access violation

7.5.3 Usage Faults

Usage faults can be caused by a number of things:

- Undefined instructions
- Coprocessor instructions (the Cortex-M3 processor does not support a coprocessor, but it is possible to use the fault exception mechanism to run software compiled for other Cortex processors through coprocessor emulation)
- Trying to switch to the ARM state (software can use this faulting mechanism to test whether the processor it is running on supports the ARM code; because the Cortex-M3 does not support the ARM state, a usage fault takes place if there's an attempt to switch)
- Invalid interrupt return (link register contains invalid/incorrect values)
- Unaligned memory accesses using multiple load or store instructions

It is also possible, by setting up certain control bits in the NVIC, to generate usage faults for the following:

- Divide by zero
- Any unaligned memory accesses

When a usage fault occurs and if the usage fault handler is enabled, normally the usage fault handler will be executed. However, if at the same time a higher-priority exception takes place, the usage fault will be pended. If the processor is already running an exception handler with the same or higher priority or if the usage fault handler is not enabled, the hard fault handler will be executed instead. If a usage fault takes place inside the hard fault handler or the NMI handler, the processor will enter the lockup state.

The usage fault handler is enabled by setting the USGFAULTENA bit in the System Handler Control and State register in the NVIC. If the vector table has been relocated to RAM, the usage fault handler starting address should be set up in the vector table first.

The NVIC provides a Usage Fault Status register (UFSR) for the usage fault handler to determine the cause of the fault. Inside the handler, the program code that causes the error can also be located using the stacked program counter value.

ACCIDENTALLY SWITCHING TO THE ARM STATE

One of the most common causes of usage faults is accidentally trying to switch the processor to ARM mode. This can happen if you load a new value to PC with the LSB equal to 0—for example, if you try to branch to an address in a register using the BX or BLX instruction without setting the LSB of the target address, have zero in the LSB of a vector in the exception vector table, or the stacked PC value to be read by POP {PC} is modified manually, leaving the LSB cleared. When these situations happen, the usage fault exception will take place with the INVSTATE bit in the UFSR set.

The UFSR is shown in Table 7.10. It occupies 2 bytes and can be accessed by half word transfer to address 0xE000ED2A, or as a word transfer to address 0xE000ED28 with the UFSR in the upper half word. As with other FSRs, the fault status bit can be cleared by writing 1 to the bit.

Table 7.10 Usage Fault Status Register (0xE000ED2A)

Bits	Name	Type	Reset Value	Description
9	DIVBYZERO	R/Wc	0	Indicates a divide by zero has taken place (can be set only if DIV_0_TRP is set)
8	UNALIGNED	R/Wc	0	Indicates that an unaligned access fault has taken place
7:4	—	—	—	—
3	NOCP	R/Wc	0	Attempts to execute a coprocessor instruction
2	INVPC	R/Wc	0	Attempts to do an exception with a bad value in the EXC_RETURN number
1	INVSTATE	R/Wc	0	Attempts to switch to an invalid state (e.g., ARM)
0	UNDEFINSTR	R/Wc	0	Attempts to execute an undefined instruction

7.5.4 **Hard Faults**

The hard fault handler can be caused by usage faults, bus faults, and memory management faults if their handler cannot be executed. In addition, it can also be caused by a bus fault during vector fetch (reading of a vector table during exception handling). In the NVIC, there is a hard fault status register that can be used to determine whether the fault was caused by a vector fetch. If not, the hard fault handler will need to check the other FSRs to determine the cause of the hard fault.

Details of the Hard Fault Status register (HFSR) are shown in Table 7.11. As with other FSRs, the fault status bit can be cleared by writing 1 to the bit.

Table 7.11 Hard Fault Status Register (0xE000ED2C)				
Bits	**Name**	**Type**	**Reset Value**	**Description**
31	DEBUGEVT	R/Wc	0	Indicates hard fault is triggered by debug event
30	FORCED	R/Wc	0	Indicates hard fault is taken because of bus fault, memory management fault, or usage fault
29:2	—	—	—	—
1	VECTBL	R/Wc	0	Indicates hard fault is caused by failed vector fetch
0	—	—	—	—

7.5.5 **Dealing with Faults**

During software development, we can use the FSRs to determine the causes of errors in the program and correct them. A troubleshooting guide is included in Appendix E of this book for common causes of various faults. In a real running system, the situation is different. After the cause of a fault is determined, the software will have to decide what to do next. In systems that run an OS, the offending tasks or applications could be terminated. In some other cases, the system might need a reset. The requirements of fault recovery depend on the target application. Doing it properly could make the product more robust, but it is best to prevent the faults from happening in the first place. The following are some fault-handling methods:

- *Reset*: This can be carried out using the SYSRESETREQ control bit in the Application Interrupt and Reset Control register in the NVIC. This will reset most parts of the system apart from the debug logic. Depending on the application, if you do not want to reset the whole system, you could reset just the processor using the VECTRESET bit.
- *Recovery*: In some cases, it might be possible to resolve the problem that caused the fault exception. For example, in the case of coprocessor instructions, the problem can be resolved using coprocessor emulation software.
- *Task termination*: For systems running an OS, it is likely that the task that caused the fault will be terminated and restarted if needed.

The FSRs retain their status until they are cleared manually. Fault handlers should clear the fault status bit they have dealt with. Otherwise, the next time another fault takes place, the fault handler will be invoked again and could mistake that the first fault still exists and so will try to deal with it again. The FSRs use a write-to-clear mechanism (clear by writing 1 to the bits that need to be cleared).

Chip manufacturers can also include an auxiliary FSR in the chip to indicate other fault situations. The implementation of an AFSR depends on individual chip design requirements.

7.6 SUPERVISOR CALL AND PENDABLE SERVICE CALL

Supervisor Call (SVC) and Pendable Service Call (PendSV) are two exceptions targeted at software and operating systems. SVC is for generating system function calls. For example, instead of allowing user programs to directly access hardware, an operating system may provide access to hardware through an SVC. So when a user program wants to use certain hardware, it generates the SVC exception using SVC instructions, and then the software exception handler in the operating system is executed and provides the service the user application requested. In this way, access to hardware is under the control of the OS, which can provide a more robust system by preventing the user applications from directly accessing the hardware.

SVC can also make software more portable because the user application does not need to know the programming details of the hardware. The user program will only need to know the application programming interface (API) function ID and parameters; the actual hardware-level programming is handled by device drivers (see Figure 7.15).

SVC exception is generated using the SVC instruction. An immediate value is required for this instruction, which works as a parameter-passing method. The SVC exception handler can then extract the parameter and determine what action it needs to perform. For example,

```
SVC #0x3 ; Call SVC function 3
```

The traditional syntax for SVC is also acceptable (without the "#"):

```
SVC 0x3 ; Call SVC function 3
```

For C language development, the SVC instruction can be generated using __svc function (for ARM RealView C Compiler or KEIL Microcontroller Development Kit for ARM), or using inline assembly in other C compilers.

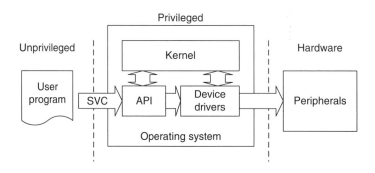

FIGURE 7.15

SVC as a Gateway for OS Functions.

When the SVC handler is executed, you can determine the immediate data value in the SVC instruction by reading the stacked program counter value, then reading the instruction from that address and masking out the unneeded bits. If the system uses a Process Stack Pointer for user applications, you might need to determine which stack was used first. This can be determined from the link register value when the handler is entered. (This topic is covered in more depth in Chapter 8).

SVC AND SOFTWARE INTERRUPT INSTRUCTION (ARM7)

If you have used traditional ARM processors (such as the ARM7), you might know that they have a software interrupt instruction (SWI). The SVC has a similar function, and in fact the binary encoding of SVC instructions is the same as SWI in ARM7. However, since the exception model has changed, this instruction is renamed to make sure that programmers will properly port software code from ARM7 to the Cortex-M3.

Because of the interrupt priority model in the Cortex-M3, you cannot use SVC inside an SVC handler (because the priority is the same as the current priority). Doing so will result in a usage fault. For the same reason, you cannot use SVC in an NMI handler or a hard fault handler.

PendSV (Pendable Service Call) works with SVC in the OS. Although SVC (by SVC instruction) cannot be pended (an application calling SVC will expect the required task to be done immediately), PendSV can be pended and is useful for an OS to pend an exception so that an action can be performed after other important tasks are completed. PendSV is generated by writing 1 to the PENDSVSET bit in the NVIC Interrupt Control State register.

A typical use of PendSV is context switching (switching between tasks). For example, a system might have two active tasks, and context switching can be triggered by the following:

- Calling an SVC function
- The system timer (SYSTICK)

Let's look at a simple example of having only two tasks in a system, and a context switch is triggered by SYSTICK exceptions (see Figure 7.16).

If an interrupt request takes place before the SYSTICK exception, the SYSTICK exception will preempt the IRQ handler. In this case, the OS should not carry out the context switching. Otherwise the

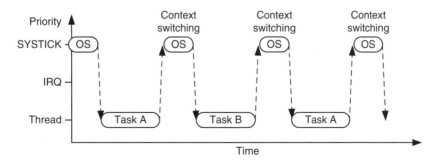

FIGURE 7.16

A Simple Scenario Using SYSTICK to Switch between Two Tasks.

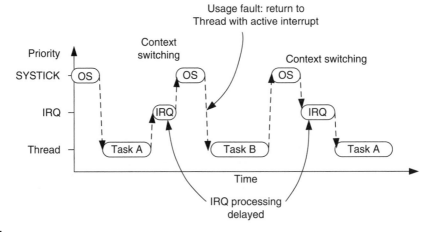

FIGURE 7.17

Problem with Context Switching at the IRQ.

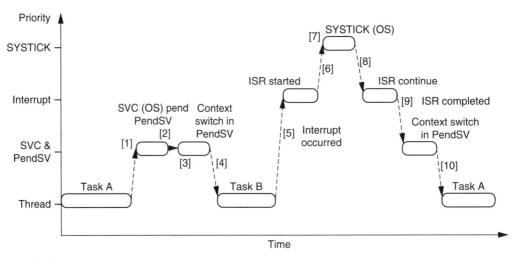

FIGURE 7.18

Example Context Switching with PendSV.

IRQ handler process will be delayed, and for the Cortex-M3, a usage fault could be generated if the OS tries to switch to thread mode when an interrupt is active (see Figure 7.17).

To avoid the problem of delaying the IRQ processing, some OS implementations carry out only context switching if they detect that none of the IRQ handlers are being executed. However, this can result in a very long delay for task switching, especially if the frequency of an interrupt source is close to that of the SYSTICK exception.

The PendSV exception solves the problem by delaying the context-switching request until all other IRQ handlers have completed their processing. To do this, the PendSV is programmed as the lowest priority exception. If the OS detects that an IRQ is currently active (IRQ handler running and preempted by SYSTICK), it defers the context switching by pending the PendSV exception. Figure 7.18 illustrates a context switching example with the following event sequence:

1. Task A calls SVC for task switching (for example, waiting for some work to complete).
2. The OS receives the request, prepares for context switching, and pends the PendSV exception.
3. When the CPU exits SVC, it enters PendSV immediately and does the context switch.
4. When PendSV finishes and returns to the thread level, it executes Task B.
5. An interrupt occurs and the interrupt handler is entered.
6. While running the interrupt handler routine, a SYSTICK exception (for OS tick) takes place.
7. The OS carries out the essential operation, then pends the PendSV exception and gets ready for the context switch.
8. When the SYSTICK exception exits, it returns to the interrupt service routine.
9. When the interrupt service routine completes, the PendSV starts and does the actual context switch operations.
10. When PendSV is complete, the program returns to the thread level; this time it returns to Task A and continues the processing.

The Nested Vectored Interrupt Controller and Interrupt Control

IN THIS CHAPTER

Nested Vectored Interrupt Controller Overview 131
The Basic Interrupt Configuration 132
Example Procedures in Setting Up an Interrupt 138
Software Interrupts 140
The SYSTICK Timer 141

8.1 NESTED VECTORED INTERRUPT CONTROLLER OVERVIEW

As we've seen, the Nested Vectored Interrupt Controller (NVIC) is an integrated part of the Cortex™-M3 processor. It is closely linked to the Cortex-M3 CPU core logic. Its control registers are accessible as memory-mapped devices. Besides control registers and control logic for interrupt processing, the NVIC unit also contains control registers for the SYSTICK Timer, and debugging controls. In this chapter, we'll examine the control logic for interrupt processing. Memory Protection Unit and debugging control logic are discussed in later chapters.

The NVIC supports 1–240 external interrupt inputs (commonly known as interrupt request [IRQs]). The exact number of supported interrupts is determined by the chip manufacturers when they develop their Cortex-M3 chips. In addition, the NVIC also has a Nonmaskable Interrupt (NMI) input. The actual function of the NMI is also decided by the chip manufacturer. In some cases, this NMI cannot be controlled from an external source.

The NVIC can be accessed in the System Control Space (SCS) address range, which is memory location 0xE000E000. Most of the interrupt control/status registers are accessible only in privileged mode, except the Software Trigger Interrupt register (STIR), which can be set up to be accessible in user mode. The interrupt control/status register can be accessed in word, half word, or byte transfers.

In addition, a few other interrupt-masking registers are also involved in the interrupts. They are the "special registers" covered in Chapter 3 and are accessed through special registers access instructions: move special register to general-purpose register (MRS) and move to special register from general-purpose register (MSR) instructions.

8.2 THE BASIC INTERRUPT CONFIGURATION

Each external interrupt has several registers associated with it.

- Enable and Clear Enable registers
- Set-Pending and Clear-Pending registers
- Priority level
- Active status

In addition, a number of other registers can also affect the interrupt processing:

- Exception-masking registers (PRIMASK, FAULTMASK, and BASEPRI)
- Vector Table Offset register
- STIR
- Priority group

8.2.1 Interrupt Enable and Clear Enable

The Interrupt Enable register is programmed through two addresses. To set the enable bit, you need to write to the SETENA register address; to clear the enable bit, you need to write to the CLRENA register address. In this way, enabling or disabling an interrupt will not affect other interrupt enable states. The SETENA/CLRENA registers are 32 bits wide; each bit represents one interrupt input.

As there could be more than 32 external interrupts in the Cortex-M3 processor, you might find more than one SETENA and CLRENA register—for example, SETENA0, SETENA1, and so on (see Table 8.1). Only the enable bits for interrupts that exist are implemented. So, if you have only 32 interrupt inputs, you will only have SETENA0 and CLRENA0. The SETENA and CLRENA registers can be accessed as word, half word, or byte. As the first 16 exception types are system exceptions, external Interrupt #0 has a start exception number of 16 (see Table 7.2).

8.2.2 Interrupt Set Pending and Clear Pending

If an interrupt takes place but cannot be executed immediately (for instance, if another higher-priority interrupt handler is running), it will be pended. The interrupt-pending status can be accessed through the Interrupt Set Pending (SETPEND) and Interrupt Clear Pending (CLRPEND) registers. Similarly to the enable registers, the pending status controls might contain more than one register if there are more than 32 external interrupt inputs.

The values of pending status registers can be changed by software, so you can cancel a current pended exception through the CLRPEND register, or generate software interrupts through the SETPEND register (see Table 8.2).

8.2.3 Priority Levels

Each external interrupt has an associated priority-level register, which has a maximum width of 8 bits and a minimum width of 3 bits. As described in the previous chapter, each register can be further divided into preempt priority level and subpriority level based on priority group settings. The priority-level registers can be accessed as byte, half word, or word. The number of priority-level registers depends on how many external interrupts the chip contains (see Table 8.3). The priority level configuration registers details can be found in Appendix D, Table D.19.

Table 8.1 Interrupt Set Enable Registers and Interrupt Clear Enable Registers
(0xE000E100–0xE000E11C, 0xE000E180–0xE000E19C)

Address	Name	Type	Reset Value	Description
0xE000E100	SETENA0	R/W	0	Enable for external Interrupt #0–31 bit[0] for Interrupt #0 (exception #16) bit[1] for Interrupt #1 (exception #17) … bit[31] for Interrupt #31 (exception #47) Write 1 to set bit to 1; write 0 has no effect Read value indicates the current status
0xE000E104	SETENA1	R/W	0	Enable for external Interrupt #32–63 Write 1 to set bit to 1; write 0 has no effect Read value indicates the current status
0xE000E108	SETENA2	R/W	0	Enable for external Interrupt #64–95 Write 1 to set bit to 1; write 0 has no effect Read value indicates the current status
…	—	—	—	—
0xE000E180	CLRENA0	R/W	0	Clear enable for external Interrupt #0–31 bit[0] for Interrupt #0 bit[1] for Interrupt #1 … bit[31] for Interrupt #31 Write 1 to clear bit to 0; write 0 has no effect Read value indicates the current enable status
0xE000E184	CLRENA1	R/W	0	Clear enable for external Interrupt #32–63 Write 1 to clear bit to 0; write 0 has no effect Read value indicates the current enable status
0xE000E188	CLRENA2	R/W	0	Clear enable for external Interrupt #64–95 Write 1 to clear bit to 0; write 0 has no effect Read value indicates the current enable status
…	—	—	—	—

Table 8.2 Interrupt Set-Pending Registers and Interrupt Clear-Pending Registers
(0xE000E200–0xE000E21C, 0xE000E280–0xE000E29C)

Address	Name	Type	Reset Value	Description
0xE000E200	SETPEND0	R/W	0	Pending for external Interrupt #0–31 bit[0] for Interrupt #0 (exception #16) bit[1] for Interrupt #1 (exception #17) … bit[31] for Interrupt #31 (exception #47) Write 1 to set bit to 1; write 0 has no effect Read value indicates the current status
0xE000E204	SETPEND1	R/W	0	Pending for external Interrupt #32–63 Write 1 to set bit to 1; write 0 has no effect Read value indicates the current status

Continued

Table 8.2 Interrupt Set-Pending Registers and Interrupt Clear-Pending Registers (0xE000E200-0xE000E21C, 0xE000E280-0xE000E29C) *Continued*

Address	Name	Type	Reset Value	Description
0xE000E208	SETPEND2	R/W	0	Pending for external Interrupt #64–95 Write 1 to set bit to 1; write 0 has no effect Read value indicates the current status
...	—	—	—	—
0xE000E280	CLRPEND0	R/W	0	Clear pending for external Interrupt #0–31 bit[0] for Interrupt #0 (exception #16) bit[1] for Interrupt #1 (exception #17) ... bit[31] for Interrupt #31 (exception #47) Write 1 to clear bit to 0; write 0 has no effect Read value indicates the current pending status
0xE000E284	CLRPEND1	R/W	0	Clear pending for external Interrupt #32–63 Write 1 to clear bit to 0; write 0 has no effect Read value indicates the current pending status
0xE000E288	CLRPEND2	R/W	0	Clear pending for external Interrupt #64–95 Write 1 to clear bit to 0; write 0 has no effect Read value indicates the current pending status
...	—	—	—	—

Table 8.3 Interrupt Priority-Level Registers (0xE000E400-0xE000E4EF)

Address	Name	Type	Reset Value	Description
0xE000E400	PRI_0	R/W	0 (8 bit)	Priority-level external Interrupt #0
0xE000E401	PRI_1	R/W	0 (8 bit)	Priority-level external Interrupt #1
...	—	—	—	—
0xE000E41F	PRI_31	R/W	0 (8 bit)	Priority-level external Interrupt #31
...	—	—	—	—

8.2.4 Active Status

Each external interrupt has an active status bit. When the processor starts the interrupt handler, the bit is set to 1 and cleared when the interrupt return is executed. However, during an Interrupt Service Routine (ISR) execution, a higher-priority interrupt might occur and cause preemption. During this period, although the processor is executing another interrupt handler, the previous interrupt is still defined as active. The active registers are 32 bit but can also be accessed using half word or byte-size transfers. If there are more than 32 external interrupts, there will be more than one active register. The active status registers for external interrupts are read-only (see Table 8.4).

Table 8.4 Interrupt Active Status Registers (0xE000E300-0xE000E31C)

Address	Name	Type	Reset Value	Description
0xE000E300	ACTIVE0	R	0	Active status for external Interrupt #0–31 bit[0] for Interrupt #0 bit[1] for Interrupt #1 … bit[31] for Interrupt #31
0xE000E304	ACTIVE1	R	0	Active status for external Interrupt #32–63
…	—	—	—	—

8.2.5 **PRIMASK and FAULTMASK Special Registers**

The PRIMASK register is used to disable all exceptions except NMI and hard fault. It effectively changes the current priority level to 0 (highest programmable level). In C programming, you can use the intrinsic functions provided in Cortex Microcontroller Software Interface Standard (CMSIS) compliant device driver libraries or provided in the compiler to set and clear PRIMASK:

```
void __enable_irq(); // Clear PRIMASK
void __disable_irq(); // Set PRIMASK
void __set_PRIMASK(uint32_t priMask); // Set PRIMASK to value
uint32_t __get_PRIMASK(void); // Read the PRIMASK value
```

For assembly language users, you can change the current status of PRIMASK using Change Process State (CPS) instructions:

```
CPSIE I ; Clear PRIMASK (Enable interrupts)
CPSID I ; Set PRIMASK (Disable interrupts)
```

This register is also programmable using MRS and MSR instructions. For example,

```
MOV   R0, #1
MSR   PRIMASK, R0 ; Write 1 to PRIMASK to disable all
                  ; interrupts
```

and

```
MOV   R0, #0
MSR   PRIMASK, R0 ; Write 0 to PRIMASK to allow interrupts
```

PRIMASK is useful for temporarily disabling all interrupts for critical tasks. When PRIMASK is set, if a fault takes place, the hard fault handler will be executed.

FAULTMASK is just like PRIMASK except that it changes the effective current priority level to -1, so that even the hard fault handler is blocked. Only the NMI can be executed when FAULT-MASK is set. It can be used by fault handlers to raise its priority to -1, so that they can have access to some features for hard fault exception (more information on this is provided in Chapter 12). In C programming with CMSIS compliant driver libraries, you can use the intrinsic functions provided in device driver libraries to set and clear FAULTMASK as follows:

```
void __set_FAULTMASK(uint32_t faultMask);
uint32_t __get_FAULTMASK(void);
```

For assembly language users, you can change the current status of FAULTMASK using CPS instructions as follows:

```
CPSIE F ; Clear FAULTMASK
CPSID F ; Set FAULTMASK
```

You can also access the FAULTMASK register using MRS and MSR instructions.

FAULTMASK is cleared automatically upon exiting the exception handler except return from NMI handler. Both FAULTMASK and PRIMASK registers cannot be set in the user state.

8.2.6 The BASEPRI Special Register

In some cases, you might want to disable interrupts only with priority lower than a certain level. In this case, you could use the BASEPRI register. To do this, simply write the required masking priority level to the BASEPRI register. For example, if you want to block all exceptions with priority level equal to or lower than 0x60, you can write the value to BASEPRI:

```
__set_BASEPRI(0x60); // Disable interrupts with priority
                     // 0x60-0xFF using CMSIS
```

Or in assembly language:

```
MOV   R0, #0x60
MSR   BASEPRI, R0 ; Disable interrupts with priority
                  ; 0x60-0xFF
```

You can also read back the value of BASEPRI:

```
x = __get_BASEPRI(void); // Read value of BASEPRI
```

Or in assembly language:

```
MRS   R0, BASEPRI
```

To cancel the masking, just write 0 to the BASEPRI register:

```
__set_BASEPRI(0x0); // Turn off BASEPRI masking
```

Or in assembly language:

```
MOV   R0, #0x0
MSR   BASEPRI, R0 ; Turn off BASEPRI masking
```

The BASEPRI register can also be accessed using the BASEPRI_MAX register name. It is actually the same register, but when you use it with this name, it will give you a conditional write operation. (As far as hardware is concerned, BASEPRI and BASEPRI_MAX are the same register, but in the assembler code they use different register name coding.) When you use BASEPRI_MAX as a register, the processor hardware automatically compares the current value and the new value and only allows the update if it is to be changed to a higher priority level; it cannot be changed to lower priority levels. For example, consider the following instruction sequence:

```
MOV   R0, #0x60
MSR   BASEPRI_MAX, R0 ; Disable interrupts with priority
                      ; 0x60, 0x61,...., etc
```

```
MOV   R0, #0xF0
MSR   BASEPRI_MAX, R0 ; This write will be ignored because
                      ; it is lower
                      ; level than 0x60
MOV   R0, #0x40
MSR   BASEPRI_MAX, R0 ; This write is allowed and change the
                      ; masking level to 0x40
```

To change to a lower masking level or disable the masking, the BASEPRI register name should be used. The BASEPRI/ BASEPRI_MAX register cannot be set in the user state.

As with other priority-level registers, the formatting of the BASEPRI register is affected by the number of implemented priority register widths. For example, if only 3 bits are implemented for priority-level registers, BASEPRI can be programmed as 0x00, 0x20, 0x40 ... 0xC0, and 0xE0.

8.2.7 Configuration Registers for Other Exceptions

Usage faults, memory management faults, and bus fault exceptions are enabled by the System Handler Control and State register (0xE000ED24). The pending status of faults and active status of most system exceptions are also available from this register (see Table 8.5).

Table 8.5 The System Handler Control and State Register (0xE000ED24)				
Bits	**Name**	**Type**	**Reset Value**	**Description**
18	USGFAULTENA	R/W	0	Usage fault handler enable
17	BUSFAULTENA	R/W	0	Bus fault handler enable
16	MEMFAULTENA	R/W	0	Memory management fault handler enable
15	SVCALLPENDED	R/W	0	SVC pended; SVC was started but was replaced by a higher-priority exception
14	BUSFAULTPENDED	R/W	0	Bus fault pended; bus fault handler was started but was replaced by a higher-priority exception
13	MEMFAULTPENDED	R/W	0	Memory management fault pended; memory management fault started but was replaced by a higher-priority exception
12	USGFAULTPENDED	R/W	0	Usage fault pended; usage fault started but was replaced by a higher-priority exception
11	SYSTICKACT	R/W	0	Read as 1 if SYSTICK exception is active
10	PENDSVACT	R/W	0	Read as 1 if PendSV exception is active
8	MONITORACT	R/W	0	Read as 1 if debug monitor exception is active
7	SVCALLACT	R/W	0	Read as 1 if SVC exception is active
3	USGFAULTACT	R/W	0	Read as 1 if usage fault exception is active
1	BUSFAULTACT	R/W	0	Read as 1 if bus fault exception is active
0	MEMFAULTACT	R/W	0	Read as 1 if memory management fault is active
Note: Bit 12 (USGFAULTPENDED) is not available on revision 0 of Cortex-M3.				

Table 8.6 Interrupt Control and State Register (0xE000ED04)

Bits	Name	Type	Reset Value	Description
31	NMIPENDSET	R/W	0	NMI pended
28	PENDSVSET	R/W	0	Write 1 to pend system call Read value indicates pending status
27	PENDSVCLR	W	0	Write 1 to clear PendSV pending status
26	PENDSTSET	R/W	0	Write 1 to pend SYSTICK exception Read value indicates pending status
25	PENDSTCLR	W	0	Write 1 to clear SYSTICK pending status
23	ISRPREEMPT	R	0	Indicates that a pending interrupt is going to be active in the next step (for debug)
22	ISRPENDING	R	0	External interrupt pending (excluding system exceptions such as NMI for fault)
21:12	VECTPENDING	R	0	Pending ISR number
11	RETTOBASE	R	0	Set to 1 when the processor is running an exception handler; will return to thread level if interrupt return and no other exceptions pending
9:0	VECTACTIVE	R	0	Current running ISR

Be cautious when writing to this register; make sure that the active status bits of system exceptions are not changed accidentally. Otherwise, if an activated system exception has its active state cleared by accident, a fault exception will be generated when the system exception handler generates an exception exit.

Pending for NMI, the SYSTICK Timer, and PendSV is programmable through the Interrupt Control and State register. In this register, quite a number of the bit fields are for debugging purposes. In most cases, only the pending bits would be useful for application development (see Table 8.6).

8.3 EXAMPLE PROCEDURES IN SETTING UP AN INTERRUPT

For most simple applications, the application is stored in ROM and there is no need to change the exception handlers, we can have the whole vector table coded in the beginning of ROM in the Code region (0x00000000). This way, the vector table offset will always be 0 and the interrupt vector is already in ROM. The only steps required to set up an interrupt will be as follows:

1. Set up the priority group setting. This step is optional. By default priority group setting is zero—only bit 0 of the priority level register is used for subpriority.
2. Set up the priority level of the interrupt. This step is optional. By default, all interrupts are at priority level 0 (highest).
3. Enable the interrupt.

Here is a simple example procedure for setting up an interrupt:

```
NVIC_SetPriorityGrouping(5);
NVIC_SetPriority(7, 0xC0); // Set IRQ#7 priority level to 0xC0
NVIC_EnableIRQ(7);
```

In addition, make sure that you have enough stack memory if you allow a large number of nested interrupt levels. Because exception handlers always use the Main Stack Pointer, the main stack memory should contain enough space for the largest number of nesting interrupts.

If the interrupt handlers need to be changed at different stage of the application, we might need to relocate the vector table to Static Random Access Memory (SRAM), so that we can modify the exception vectors. In this case, the following extra steps would be required:

1. When the system boots up, the priority group register might need to be set up. By default, the priority group 0 is used (bit[7:1] of priority level is the preemption level and bit[0] is the subpriority level).
2. Copy the hard fault, NMI handlers and other required vector to a new vector table location in SRAM.
3. Set up the Vector Table Offset register (Table 7.7) to point to the new vector table.
4. Set up the interrupt vector for the interrupt in the new vector table.
5. Set up the priority level for the interrupt.
6. Enable the interrupt.

For example, this can be done in C programming with a CMSIS compliant device driver library, assume the starting address of the new vector table is defined as "NEW_VECT_TABLE":

```
// HW_REG is a macro to convert address value to pointer
#define HW_REG(addr) (*((volatile unsigned long *)(addr)))
#define NEW_VECT_TABLE 0x20008000 // An SRAM region for vector table
    NVIC_SetPriorityGrouping(5);
    ...
    HW_REG((NEW_VECT_TABLE +0x8)) = HW_REG(0x8); // Copy NMI vector
    HW_REG((NEW_VECT_TABLE +0xC)) = HW_REG(0xC); // Copy HardFault
    ...
    SCB->VTOR = NEW_VECT_TABLE; // Relocate vector table to SRAM
    ...
    HW_REG(4*(7+16)) = (unsigned) IRQ7_Handler; // Setup vector
    ...
    NVIC_SetPriority(7, 0xC0); // Set IRQ#7 priority level to 0xC0
    ...
    NVIC_EnableIRQ(7);
```

The program in assembly might be something like this:

```
    LDR R0, =0xE000ED0C      ; Application Interrupt and Reset
                             ; Control Register
    LDR R1, =0x05FA0500      ; Priority Group 5 (2/6)
    STR R1, [R0]             ; Set Priority Group
    ...
    MOV R4,#8                ; Vector Table in ROM
    LDR R5,=(NEW_VECT_TABLE+8)
    LDMIA R4!,{R0-R1}        ; Read vectors address for NMI and
                             ; Hard Fault
    STMIA R5!,{R0-R1}        ; Copy vectors to new vector table
    ...
    LDR R0,=0xE000ED08       ; Vector Table Offset Register
    LDR R1,=NEW_VECT_TABLE
```

```
STR R1,[R0]                 ; Set vector table to new location
...
LDR R0,=IRQ7_Handler        ; Get starting address of IRQ#7 handler
LDR R1,=0xE000ED08          ; Vector Table Offset Register
LDR R1,[R1]
ADD R1, R1, #(4*(7+16))     ; Calculate IRQ#7 handler vector
                            ; address
STR R0,[R1]                 ; Setup vector for IRQ#7
...
LDR R0,=0xE000E400          ; External IRQ priority base
MOV R1, #0x0
STRB R1,[R0,#7]             ; Set IRQ#7 priority to 0x0
...
LDR R0,=0xE000E100          ; SETEN register
MOV R1,#(1<<7)              ; IRQ#7 enable bit (value 0x1 shifted
                            ; by 7 bits)
STR R1,[R0]                 ; Enable the interrupt
```

In cases where the software needs to be able to run on a number of hardware devices, it might be necessary to determine the following:

- The number of interrupts supported in the design
- The number of bits in priority-level registers

The Cortex-M3 has an Interrupt Controller Type register that gives the number of interrupt inputs supported, in granularities of 32 (see Table 8.7). Alternatively, you can detect the exact number of external interrupts by performing a read/write test to interrupt configuration registers such as SETEN or priority registers.

To determine the number of bits implemented for interrupt priority-level registers, you can write 0xFF to one of the priority-level registers, then read it back and see how many bits are set. The minimum number is three. In that case you should get a read-back value of 0xE0.

8.4 SOFTWARE INTERRUPTS

Software interrupts can be generated in more than one way. The first one is to use the SETPEND register; the second solution is to use the STIR, outlined in Table 8.8.

For example, you can generate Interrupt #3 by writing the following code in C:

```
NVIC->STIR = 3;     /* NVIC->STIR is defined in CMSIS compliant device driver
                       library */
```

This is functionally equivalent to using SETPEND register using CMSIS function:

```
NVIC_SetPendingIRQ(3);
```

System exceptions (NMI, faults, PendSV, and so on) cannot be pended using this register. By default, a user program cannot write to the NVIC; however, if it is necessary for a user program to write to this register, the bit 1 (USERSETMPEND) of the NVIC Configuration Control register (0xE000ED14) can be set to allow user access to the NVIC's STIR.

Table 8.7 Interrupt Controller Type Register (0xE000E004)

Bits	Name	Type	Reset Value	Description
4:0	INTLINESNUM	R	—	Number of interrupt inputs in step of 32 0 = 1 to 32 1 = 33 to 64 …

Table 8.8 Software Trigger Interrupt Register (0xE000EF00)

Bits	Name	Type	Reset Value	Description
8:0	INTID	W	—	Writing the interrupt number sets the pending bit of the interrupt; for example, write 0 to pend external Interrupt #0

8.5 THE SYSTICK TIMER

The SYSTICK Timer is integrated with the NVIC and can be used to generate a SYSTICK exception (exception type #15). In many operating systems, a hardware timer is used to generate interrupts so that the OS can carry out task management—for example, to allow multiple tasks to run at different time slots and to make sure that no single task can lock up the whole system. To do that, the timer needs to be able to generate interrupts, and if possible, it should be protected from user tasks so that user applications cannot change the timer behavior.

The Cortex-M3 processor includes a simple timer. Because all Cortex-M3 chips have the same timer, porting software between different Cortex-M3 products is simplified. The timer is a 24-bit down counter. It can use the internal free running processor clock signal on the Cortex-M3 processor or an external reference clock (documented as the STCLK signal on the Cortex-M3 TRM). However, the source of the STCLK will be decided by chip designers, so the clock frequency might vary between products. You should check the chip's datasheet carefully when selecting a clock source.

The SYSTICK Timer can be used to generate interrupts. It has a dedicated exception type and exception vector. It makes porting operating systems and software easier because the process will be the same across different Cortex-M3 products. The SYSTICK Timer is controlled by four registers, shown in Tables 8.9–8.12.

The Calibration Value register provides a solution for applications to generate the same SYS-TICK interrupt interval when running on various Cortex-M3 products. To use it, just write the value in TENMS to the reload value register. This will give an interrupt interval of about 10 ms. For other interrupt timing intervals, the software code will need to calculate a new suitable value from the calibration value. However, the TENMS field might not be available in all Cortex-M3 products (the calibration input signals to the Cortex-M3 might have been tied low), so check with your manufacturer's datasheets before using this feature.

Aside from being a system tick timer for operating systems, the SYSTICK Timer can be used in a number of ways: as an alarm timer, for timing measurement, and more. Note that the SYSTICK

Table 8.9 SYSTICK Control and Status Register (0xE000E010)

Bits	Name	Type	Reset Value	Description
16	COUNTFLAG	R	0	Read as 1 if counter reaches 0 since last time this register is read; clear to 0 automatically when read or when current counter value is cleared
2	CLKSOURCE	R/W	0	0 = External reference clock (STCLK) 1 = Use processor free running clock
1	TICKINT	R/W	0	1 = Enable SYSTICK interrupt generation when SYSTICK Timer reaches 0 0 = Do not generate interrupt
0	ENABLE	R/W	0	SYSTICK Timer enable

Table 8.10 SYSTICK Reload Value Register (0xE000E014)

Bits	Name	Type	Reset Value	Description
23:0	RELOAD	R/W	0	Reload value when timer reaches 0

Table 8.11 SYSTICK Current Value Register (0xE000E018)

Bits	Name	Type	Reset Value	Description
23:0	CURRENT	R/Wc	0	Read to return current value of the timer. Write to clear counter to 0. Clearing of current value also clears COUNTFLAG in SYSTICK Control and Status register

Table 8.12 SYSTICK Calibration Value Register (0xE000E01C)

Bits	Name	Type	Reset Value	Description
31	NOREF	R	—	1 = No external reference clock (STCLK not available) 0 = External reference clock available
30	SKEW	R	—	1 = Calibration value is not exactly 10 ms 0 = Calibration value is accurate
23:0	TENMS	R/W	0	Calibration value for 10 ms; chip designer should provide this value through Cortex-M3 input signals. If this value is read as 0, calibration value is not available

Timer stops counting when the processor is halted during debugging. Depending on the design of the microcontroller, the SYSTICK Timer could also be stopped when the processor enters certain type of sleep modes.

To set up the SYSTICK Timer, the recommended programming sequence is as follows:

- Disable SYSTICK by writing 0 to the SYSTICK Control and Status register.
- Write new reload value to the SYSTICK Reload Value register.
- Write to the SYSTICK Current Value register to clear the current value to 0.
- Write to the SYSTICK Control and Status register to start the SYSTICK timer.

This programming sequence can be used on all Cortex-M3 processors. More details of the SYSTICK setup is covered in Chapter 14.

Interrupt Behavior

IN THIS CHAPTER

Interrupt/Exception Sequences 145
Exception Exits 147
Nested Interrupts 148
Tail-Chaining Interrupts.. 148
Late Arrivals... 149
More on the Exception Return Value 149
Interrupt Latency 152
Faults Related to Interrupts 152

9.1 INTERRUPT/EXCEPTION SEQUENCES

When an exception takes place, a number of things happen, such as

- Stacking (pushing eight registers' contents to stack)
- Vector fetch (reading the exception handler starting address from the vector table)
- Update of the stack pointer, link register (LR), and program counter (PC)

9.1.1 Stacking

When an exception takes place, the registers R0–R3, R12, LR, PC, and Program Status (PSR) are pushed to the stack. If the code that is running uses the Process Stack Pointer (PSP), the process stack will be used; if the code that is running uses the Main Stack Pointer (MSP), the main stack will be used. Afterward, the main stack will always be used during the handler, so all nested interrupts will use the main stack.

The block of eight words of data being pushed to the stack is commonly called a stack frame. Prior to Cortex™-M3 revision 2, the stack frame was started in any word address by default. In Cortex-M3 revision 2, the stack frame is aligned to double word address by default, although the alignment feature can be turned off by programming the STKALIGN bit in Nested Vectored Interrupt Controller (NVIC)

Configuration Control register to zero. The stack frame feature is also available in Cortex-M3 revision 1, but it needs to be enabled by writing 1 to the STKALIGN bit. More details on this register can be found in Chapter 12.

The data arrangement inside an exception stack frame is shown in Figure 9.1. The order of stacking is shown in Figure 9.2 (assuming that the stack pointer [SP] value is N after the exception). Due to the pipeline nature of the Advanced High-Performance Bus (AHB) interface, the address and data are offset by one pipeline state.

The values of PC and PSR are stacked first so that instruction fetch can be started early (which requires modification of PC) and the Interrupt Program Status register (IPSR) can be updated early. After stacking, SP will be updated, and the stacked data arrangement in the stack memory will look like Figure 9.1.

The reason the registers R0–R3, R12, LR, PC, and PSR are stacked is that these are caller-saved registers, according to C standards (C/C++ standard *Procedure Call Standard for the ARM Architecture*,

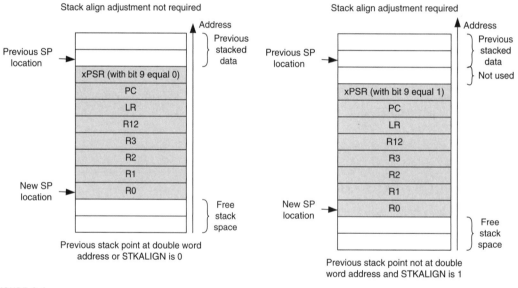

FIGURE 9.1

Exception Stack Frame.

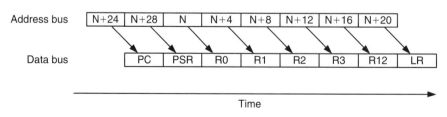

FIGURE 9.2

Stacking Sequence.

AAPCS [Ref. 5]). This arrangement allows the interrupt handler to be a normal C function because registers that could be changed by the exception handler are saved in the stack.

The general registers (R0–R3 and R12) are located at the end of the stack frame so that they can be easily accessed using SP-related addressing. As a result, it's easy to pass parameters to software interrupts using stacked registers.

9.1.2 Vector Fetches

Although the data bus is busy stacking the registers, the instruction bus carries out another important task of the interrupt sequence: It fetches the exception vector (the starting address of the exception handler) from the vector table. Since the stacking and vector fetch are performed on separate bus interfaces, they can be carried out at the same time.

9.1.3 Register Updates

After the stacking and vector fetch are completed, the exception vector will start to execute. On entry of the exception handler, a number of registers will be updated. They are as follows:

- *SP*: The SP (either the MSP or the PSP) will be updated to the new location during stacking. During execution of the interrupt service routine, the MSP will be used if the stack is accessed.
- *PSR*: The IPSR (the lowest part of the PSR) will be updated to the new exception number.
- *PC*: This will change to the vector handler as the vector fetch completes and starts fetching instructions from the exception vector.
- *LR*: The LR will be updated to a special value called EXC_RETURN.[1] This special value drives the interrupt return operation. The last 4 bits of the LR is used to provide exception return information. This is covered later in this chapter.

A number of other NVIC registers will also be updated. For example, the pending status of the exception will be cleared and the active bit of the exception will be set.

9.2 EXCEPTION EXITS

At the end of the exception handler, an exception exit (known as an *interrupt return* in some processors) is required to restore the system status so that the interrupted program can resume normal execution. There are three ways to trigger the interrupt return sequence; all of them use the special value stored in the LR in the beginning of the handler (see Table 9.1).

Some microprocessor architectures use special instructions for interrupt returns (for example, *reti* in 8051). In the Cortex-M3, a normal return instruction is used so that the whole interrupt handler can be implemented as a C subroutine.

When the interrupt return instruction is executed, the unstacking and the NVIC registers update processes that are listed in Table 9.1 are carried out.

---

[1]EXC_RETURN has bit 31 to 4 all set to one (i.e., 0xfffffffx). The last 4 bits define the return information. More information on the EXC_RETURN value is covered later in this chapter.

Table 9.1 Instructions That Can Be Used for Triggering Exception Return

Return Instruction	Description
BX *reg*	If the EXC_RETURN value is still in LR, we can use the *BX LR* instruction to perform the interrupt return.
POP {PC}, or POP {..., PC}	Very often the value of LR is pushed to the stack after entering the exception handler. We can use the POP instruction, either a single POP or multiple POPs, to put the EXC_RETURN value to the program counter. This will cause the processor to perform the interrupt return.
Load (LDR) or Load multiple (LDM)	It is possible to produce an interrupt return using the LDR or LDM instruction with PC as the destination register.

1. *Unstacking*: The registers pushed to the stack will be restored. The order of the POP will be the same as in stacking. The stack pointer will also be changed back.
2. *NVIC register update*: The active bit of the exception will be cleared. For external interrupts, if the interrupt input is still asserted, the pending bit will be set again, causing it to reenter the interrupt handler.

9.3 NESTED INTERRUPTS

Nested interrupt support is built into the Cortex-M3 processor core and the NVIC. There is no need to use assembler wrapper code to enable nested interrupts. In fact, you do not have to do anything apart from setting up the appropriate priority level for each interrupt source. First, the NVIC in the Cortex-M3 processor sorts out the priority decoding for you. So when the processor is handling an exception, all other exceptions with the same or lower priority will be blocked. Second, the automatic hardware stacking and unstacking allow the nested interrupt handler to execute without risk of losing data in registers.

However, one thing needs to be taken care of: Make sure that there is enough space in the main stack if several nested interrupts are allowed. Since each exception level will use eight words of stack space and the exception handler code might require extra stack space, it might end up using more stack memory than expected.

Reentrant exceptions are not allowed in the Cortex-M3. Since each exception has a priority level assigned and, during exception handling, exceptions with the same or lower priority will be blocked, the same exception cannot be carried out until the handler is ended. For this reason, Supervisor Call (SVC) instructions cannot be used inside an SVC handler, since doing so will cause a fault exception.

9.4 TAIL-CHAINING INTERRUPTS

The Cortex-M3 uses a number of methods to improve interrupt latency. The first one we'll look at is *tail chaining* (see Figure 9.3).

When an exception takes place but the processor is handling another exception of the same or higher priority, the exception will enter pending state. When the processor has finished executing the current

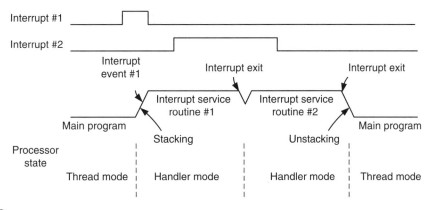

FIGURE 9.3

Tail Chaining of Exceptions.

exception handler, it can then process the pended interrupt. Instead of restoring the registers back from the stack (unstacking) and then pushing them onto the stack again (stacking), the processor skips the unstacking and stacking steps and enters the exception handler of the pended exception as soon as possible. In this way, the timing gap between the two exception handlers is considerably reduced.

9.5 LATE ARRIVALS

Another feature that improves interrupt performance is *late arrival* exception handling. When an exception takes place and the processor has started the stacking process, and if during this delay a new exception arrives with higher preemption priority, the late arrival exception will be processed first.

For example, if Exception #1 (lower priority) takes place a few cycles before Exception #2 (higher priority), the processor will behave as shown in Figure 9.4, such that Handler #2 is executed as soon as the stacking completes.

9.6 MORE ON THE EXCEPTION RETURN VALUE

When entering an exception handler, the LR is updated to a special value called EXC_RETURN, with the upper 28 bits all set to 1. This value, when loaded into the PC at the end of the exception handler execution, will cause the processor to perform an exception return sequence.

The instructions that can be used to generate exception returns are as follows:

- POP/LDM
- LDR with PC as a destination
- BX with any register

The EXC_RETURN value has bit (31:4) all set to 1, and bit (3:0) provides information required by the exception return operation (see Table 9.2). When the exception handler is entered, the LR value is updated automatically, so there is no need to generate these values manually.

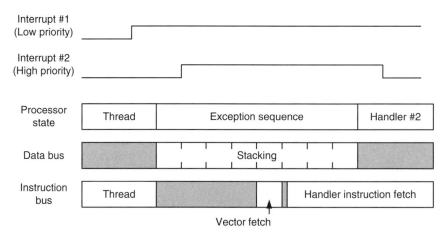

FIGURE 9.4

Late Arrival Exception Behavior.

Table 9.2 Description of Bit Fields in EXC_RETURN Value

Bits	31:4	3	2	1	0
Descriptions	0xFFFFFFF	Return mode (thread/handler)	Return stack	Reserved; must be 0	Process state (Thumb/ARM)

Table 9.3 Allowed EXC_RETURN Values on Cortex-M3

Value	Condition
0xFFFFFFF1	Return to handler mode
0xFFFFFFF9	Return to thread mode and on return use the main stack
0xFFFFFFFD	Return to thread mode and on return use the process stack

Bit 0 indicates the process state being used after the exception return. Since the Cortex-M3 supports only the Thumb® state, bit 0 must be 1. The valid values (for the Cortex-M3) are shown in Table 9.3.

If the thread is using the MSP (main stack), the value of LR will be set to 0xFFFFFFF9 when it enters an exception, and 0xFFFFFFF1 when a nested exception is entered, as shown in Figure 9.5. If the thread is using PSP (process stack), the value of LR would be 0xFFFFFFFD when entering the first exception and 0xFFFFFFF1 for entering a nested exception, as shown in Figure 9.6.

As a result of the EXC_RETURN number format, you cannot perform interrupt returns to an address in the 0xFFFFFFF0–0xFFFFFFFF memory range. However, since this address is in a nonexecutable region anyway, it is not a problem.

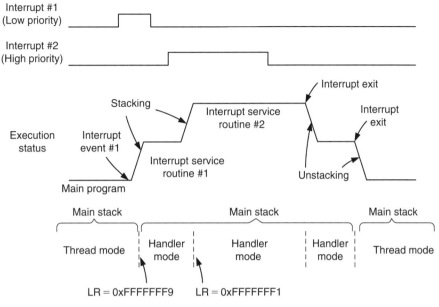

FIGURE 9.5

LR Set to EXC_RETURN at Exception (Main Stack Used in Thread Mode).

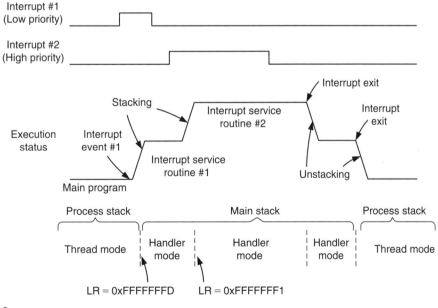

FIGURE 9.6

LR Set to EXC_RETURN at Exception (Process Stack Used in Thread Mode).

9.7 INTERRUPT LATENCY

The term *interrupt latency* refers to the delay from the start of the interrupt request to the start of interrupt handler execution. In the Cortex-M3 processor, if the memory system has zero latency, and provided that the bus system design allows vector fetch and stacking to happen at the same time, the interrupt latency can be as low as 12 cycles. This includes stacking the registers, vector fetch, and fetching instructions for the interrupt handler. However, this depends on memory access wait states and a few other factors.

For tail-chaining interrupts, since there is no need to carry out stacking operations, the latency of switching from one exception handler to another exception handler can be as low as six cycles.

When the processor is executing a multicycle instruction, such as divide, the instruction could be abandoned and restarted after the interrupt handler completes. This also applies to load double (LDRD) and store double (STRD) instructions.

To reduce exception latency, the Cortex-M3 processor allows exceptions in the middle of Multiple Load and Store instructions (LDM/STM). If the LDM/STM instruction is executing, the current memory accesses will be completed, and the next register number will be saved in the stacked xPSR (Interrupt-Continuable Instruction [ICI] bits). After the exception handler completes, the multiple load/store will resume from the point at which the transfer stopped. There is a corner case: If the multiple load/store instruction being interrupted is part of an IF-THEN (IT) instruction block, the load/store instruction will be cancelled and restarted when the interrupt is completed. This is because the ICI bits and IT execution status bits share the same space in the Execution Program Status Register (EPSR).

In addition, if there is an outstanding transfer on the bus interface, such as a buffered write, the processor will wait until the transfer is completed. This is necessary to ensure that a bus fault handler preempts the correct process.

Of course, the interrupt could be blocked if the processor is already executing another exception handler of the same or higher priority or if the Interrupt Mask register was masking the interrupt request. In these cases, the interrupt will be pended and will not be processed until the blocking is removed.

9.8 FAULTS RELATED TO INTERRUPTS

Various faults can be caused by exception handling. Let's take a look at these now.

9.8.1 Stacking

If a bus fault takes place during stacking, the stacking sequence will be terminated and the bus fault exception will be triggered or pended. If the bus fault is disabled, the hard fault handler will be executed. Otherwise, if the bus fault handler has higher priority than the original exception, the bus fault handler will be executed; if not, it will be pended until the original exception is completed. This scenario, called a *stacking error*, is indicated by the STKERR (bit 4) in the Bus Fault Status register (0xE000ED29).

If the stacking error is caused by a Memory Protection Unit (MPU) violation, the memory management fault handler will be executed and the MSTKERR (bit 4) in the Memory Management Fault Status register (0xE000ED28) will be set to indicate the problem. If the memory management fault is disabled, the hard fault handler will be executed.

9.8.2 **Unstacking**

If a bus fault takes place during unstacking (an interrupt return), the unstacking sequence will be terminated and the bus fault exception will be triggered or pended. If the bus fault is disabled, the hard fault handler will be executed. Otherwise, if the bus fault handler has higher priority than the current priority of the executing task (the core could already be executing another exception in a nested interrupt case), the bus fault handler will be executed. This scenario, called an *unstacking error*, is indicated by the UNSTKERR (bit 3) in the Bus Fault Status register (0xE000ED29).

Similarly, if the stacking error is caused by an MPU violation, the memory management fault handler will be executed and the MUNSTKERR (bit 3) in the Memory Management Fault Status register (0xE000ED28) will be set to indicate the problem. If the memory management fault is disabled, the hard fault handler will be executed.

9.8.3 **Vector Fetches**

If a bus fault or memory management fault takes place during a vector fetch, the hard fault handler will be executed. This is indicated by VECTTBL (bit 1) in the Hard Fault Status register (0xE000ED2C).

9.8.4 **Invalid Returns**

If the EXC_RETURN number is invalid or does not match the state of the processor (as in using 0xFFFFFFF1 to return to thread mode), it will trigger the usage fault. If the usage fault handler is not enabled, the hard fault handler will be executed instead. The INVPC bit (bit 2) or INVSTATE (bit 1) bit in the Usage Fault Status register (0xE000ED2A) will be set, depending on the actual cause of the fault.

Cortex-M3 Programming

10

IN THIS CHAPTER

Overview 155
A Typical Development Flow 155
Using C 156
CMSIS 164
Using Assembly 169
Using Exclusive Access for Semaphores 177
Using Bit Band for Semaphores 179
Working with Bit Field Extract and Table Branch 181

10.1 OVERVIEW

The Cortex™-M3 can be programmed using either assembly language, C language, or other high-level languages like National Instruments LabVIEW. For most embedded applications using the Cortex-M3 processor, the software can be written entirely in C language. There are of course some people who prefer to use assembly language or a combination of C and assembly language in their projects. The procedure of building and downloading the resultant image files to the target device is largely dependent on the tool chain used. Although this is not the main focus of this book, some simple examples showing how to use the Gnu's Not Unix (GNU) and Keil tool chains are provided in Chapters 19 and 20, and an introduction of using LabVIEW on Cortex-M3 is covered in Chapter 21.

10.2 A TYPICAL DEVELOPMENT FLOW

Various software programs are available for developing Cortex-M3 applications. The concepts of code generation flow in terms of these tools are similar. For the most basic uses, you will need assembler, a C compiler, a linker, and binary file generation utilities. For ARM solutions, the RealView Development Suite (RVDS) or RealView Compiler Tools (RVCT) provide a file generation flow, as shown in

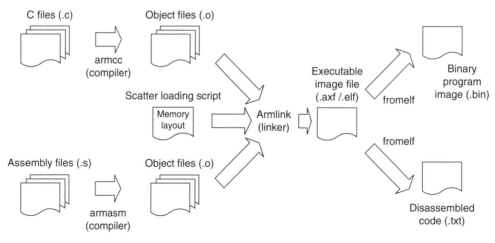

FIGURE 10.1

Example Flow Using ARM Development Tools.

Figure 10.1. The scatter-loading script is optional but often required when the memory map becomes more complex.

Besides these basic tools, RVDS also contains a large number of utilities, including an Integrated Development Environment (IDE) and debuggers. Please visit the ARM web site (*www.arm.com*) for details.

10.3 USING C

For beginners in embedded programming, using C language for software development on the Cortex-M3 processor is the best choice. Programming in C with the Cortex-M3 processor is made even easier as most microcontroller vendors provide device driver libraries written in C to control peripherals. These can then be included into your project. Since modern C compilers can generate very efficient code, it is better to program in C than spending a lot of time to try to develop complex routines in assembly language, which is error prone and less portable.

In this chapter, we will have a quick look at a simple example of using C language to create a simple program image. Then, we will have a look at some C language development areas including using device driver libraries and the Cortex Microcontroller Software Interface Standard (CMSIS).

C has the advantage of being portable and easier for implementing complex operations, compared with assembly language. Since it's a generic computer language, C does not specify how the processor is initialized. For these areas, tool chains can have different approaches. The best way to get started is to look at example codes. For users of ARM C compiler products, such as RVDS or Keil RealView Microcontroller Development Kit (MDK-ARM), a number of Cortex-M3 program examples are already included in the installation. For users of the GNU tool chain, Chapter 19 provides a simple C example based on the CodeSourcery GNU tool chain for ARM.

10.3.1 Example of a Simple C Program Using RealView Development Site

A normal program for the Cortex-M3 contains at least the "main" program and a vector table. Let's start with the most basic main program that toggles an Light Emitting Diode (LED):

```c
#define LED *((volatile unsigned int *)(0xDFFF000C))

int main (void)
{
int i;             /* loop counter for delay function */
volatile int j; /* dummy volatile variable to prevent
            C compiler from optimize the delay away */

  while (1) {
    LED = 0x00; /* toogle LED */
    for (i=0;i<10;i++) {j=0;} /* delay */
    LED = 0x01; /* toogle LED */
    for (i=0;i<10;i++) {j=0;} /* delay */
    }
  return 0;
}
```

This file is named "blinky.c." For the vector table, we create a separate C program called "vectors.c." The file "vectors.c" contains the vector table, as well as a number of dummy exception handlers (these can be customized for target application later on):

```c
typedef void(* const ExecFuncPtr)(void) __irq;
extern int __main(void);

/*
 * Dummy handlers Exception Handlers
 */
__irq void NMI_Handler(void)
{   while(1); }
__irq void HardFault_Handler(void)
{   while(1); }
__irq void SVC_Handler(void)
{   while(1); }
__irq void DebugMon_Handler(void)
{   while(1); }
__irq void PendSV_Handler(void)
{   while(1); }
__irq void SysTick_Handler(void)
{   while(1); }
__irq void ExtInt0_IRQHandler(void)
{   while(1); }
__irq void ExtInt1_IRQHandler(void)
{   while(1); }
__irq void ExtInt2_IRQHandler(void)
{   while(1); }
__irq void ExtInt3_IRQHandler(void)
{   while(1); }

#pragma arm section rodata="exceptions_area"
```

```
ExecFuncPtr exception_table[] = { /* vector table */
    (ExecFuncPtr)0x20002000,
    (ExecFuncPtr)__main,
    NMI_Handler, /* NMI */
    HardFault_Handler,
    0, /* MemManage_Handler in Cortex-M3 */
    0, /* BusFault_Handler in Cortex-M3 */
    0, /* UsageFault_Handler in Cortex-M3 */

    0, /* Reserved */
    0, /* Reserved */
    0, /* Reserved */
    0, /* Reserved */
    SVC_Handler,
    0, /* DebugMon_Handler in Cortex-M3 */
    0, /* Reserved */
    PendSV_Handler,
    SysTick_Handler,

    /* External Interrupts*/
    ExtInt0_IRQHandler,
    ExtInt1_IRQHandler,
    ExtInt2_IRQHandler,
    ExtInt3_IRQHandler
};
#pragma arm section
```

Assuming you are using RVDS, you can compile the program using the following command line:

```
$> armcc -c -g -W blinky.c -o blinky.o
$> armcc -c -g -W vectors.c -o vectors.o
```

Then the linker can be used to generate the program image. A scatter loading file "led.scat" is used to tell the linker the memory layout and to put the vector table in the starting of the program image. The "led.scat" is

```
#define HEAP_BASE 0x20001000
#define STACK_BASE 0x20002000
#define HEAP_SIZE ((STACK_BASE-HEAP_BASE)/2)
#define STACK_SIZE ((STACK_BASE-HEAP_BASE)/2)

LOAD_REGION 0x00000000 0x00200000
{
  VECTORS 0x0 0xC0
  {
    ; Provided by the user in vectors.c
    * (exceptions_area)
  }

  CODE 0xC0 FIXED
  {
    * (+RO)
  }

  DATA 0x20000000 0x00010000
  {
    * (+RW, +ZI)
  }
```

```
;; Heap starts at 4KB and grows upwards
ARM_LIB_HEAP HEAP_BASE EMPTY HEAP_SIZE
{
}

;; Stack starts at the end of the 8KB of RAM
;; And grows downwards for 2KB
ARM_LIB_STACK STACK_BASE EMPTY -STACK_SIZE
{
}
}
```

And the command line for the linker is

```
$> armlink -scatter led.scat "--keep=vectors.o(exceptions_area)"
  blinky.o vectors.o -o blinky.elf
```

The executable image `blinky.elf` is now generated. We can convert it to binary file and disassembly file using fromelf.

```
/* create binary file */
$> fromelf --bin blinky.elf -output blinky.bin
/* Create disassembly output */
$> fromelf -c blinky.elf > list.txt
```

Previously in ARM processors, because there is a Thumb® state and an ARM state, the code for different states has to be compiled differently. In the Cortex-M3, there is no such need because everything is in the Thumb state, and project file management is much simpler.

When you're developing applications in C, it is recommended that you use the double word stack alignment function (configured by the STKALIGN bit in the Nested Vectored Interrupt Controller [NVIC] Configuration Control register). For users of Cortex-M3 revision 2 or future products, the STKALIGN bit is set by default at reset so there is no need to set up this bit in the software. Users of Cortex-M3 revision 1 can enable this feature by setting this bit in the beginning of their applications, for example. The details of STKALIGN feature are covered in Chapter 9.

```
SCB->CCR = SCB->CCR | 0x200; /* Set STKALIGN */
/* SCB->CCR is defined in device driver library. */
```

If you are not using a CMSIS compliant device driver, you can use the following code instead.

```
#define NVIC_CCR *((volatile unsigned long *)(0xE000ED14))
NVIC_CCR = NVIC_CCR | 0x200; /* Set STKALIGN */
```

Using this feature ensures that the system conforms to Procedure Call Standards for the ARM Architecture (AAPCS). Additional information on this subject is covered in Chapter 12.

10.3.2 Compile the Same Example Using Keil MDK-ARM

For users of Keil MDK-ARM, it is possible to compile the same program as in RVDS. However, the command line options and a few symbols in the linker script (scatter loading file) have to be modified. Based on the example in Section 10.3.1, scatter loading file "led.scat" needed to be modified to

```
#define HEAP_BASE 0x20001000
#define STACK_BASE 0x20002000
#define HEAP_SIZE ((STACK_BASE-HEAP_BASE)/2)
#define STACK_SIZE ((STACK_BASE-HEAP_BASE)/2)

LOAD_REGION 0x00000000 0x00200000
{
  VECTORS 0x0 0xC0
  {
    ; Provided by the user in vectors.c
    * (exceptions_area)
  }

  CODE 0xC0 FIXED
  {
    * (+RO)
  }

  DATA 0x20000000 0x00010000
  {
    * (+RW, +ZI)
  }

  ;; Heap starts at 4KB and grows upwards
  Heap_Mem HEAP_BASE EMPTY HEAP_SIZE
  {
  }

  ;; Stack starts at the end of the 8KB of RAM
  ;; And grows downwards for 2KB
  Stack_Mem STACK_BASE EMPTY -STACK_SIZE
  {
  }
}
```

And the compile sequence can be created in a DOS batch file

```
SET PATH=C:\Keil\ARM\BIN40\;%PATH%
SET RVCT40INC=C:\Keil\ARM\RV31\INC
SET RVCT40LIB=C:\Keil\ARM\RV31\LIB
SET CPU_TYPE=Cortex-M3
SET CPU_VENDOR=ARM
SET UV2_TARGET=Target 1
SET CPU_CLOCK=0x00000000
C:\Keil\ARM\BIN40\armcc -c -O3 -W -g -Otime --device DLM vectors.c
C:\Keil\ARM\BIN40\armcc -c -O3 -W -g -Otime --device DLM blinky.c
C:\Keil\ARM\BIN40\armlink --device DLM "--keep=Startup.o(RESET)"
  "--first=Startup.o(RESET)" -scatter led.scat --map vectors.o
  blinky.o -o blinky.elf
C:\Keil\ARM\BIN40\fromelf --bin blinky.elf -o blinky.bin
```

In general, it is much easier to use the μVision IDE to create and compile projects rather than using command lines. Chapter 20 is ideal for beginners who want to start using the Cortex-M3 microcontrollers with the Keil Microcontroller Development Kit for ARM (MDK-ARM).

10.3.3 **Accessing Memory-Mapped Registers in C**

There are various ways to access memory-mapped peripheral registers in C language. For illustration, we will use the System Tick (SYSTICK) Timer in the Cortex-M3 as an example peripheral to demonstrate different access methods in C language. The SYSTICK is a 24-bit timer which contains only four registers. The functionality of the SYSTICK will be covered in Chapter 14. In the previous examples, we have already illustrated the easiest method—defining each register as a pointer. To apply the same solution to the SYSTICK, we can define each register separately. This is illustrated in Figure 10.2.

Based on the same method, we can define a macro to convert address values to C pointer. The C-code looks a bit different, but the generated code is the same as previous implementation. This is illustrated in Figure 10.3.

Method 2 is to define the registers as a data structure, and then define a pointer of the defined structure. This is the method used in CMSIS compliant device driver libraries. This is illustrated in Figure 10.4.

Method 3 also uses data structure, but the base address of the peripheral is defined using a scatter loading file (or linker script) during linking stage. This is illustrated in Figure 10.5.

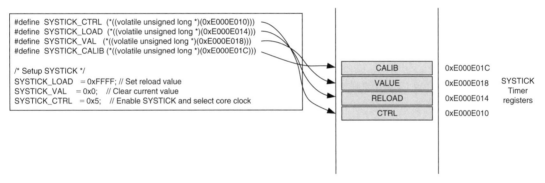

FIGURE 10.2

Accessing Peripheral Registers as Pointers.

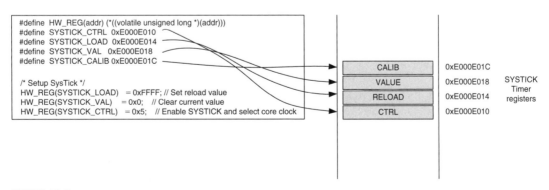

FIGURE 10.3

Alternative Way of Accessing Peripheral Registers as Pointers.

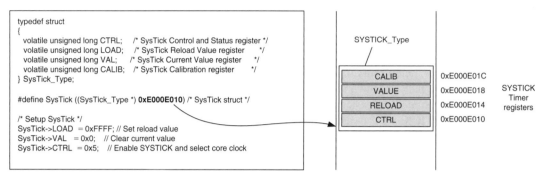

```
typedef struct
{
   volatile unsigned long CTRL;    /* SysTick Control and Status register */
   volatile unsigned long LOAD;    /* SysTick Reload Value register    */
   volatile unsigned long VAL;     /* SysTick Current Value register   */
   volatile unsigned long CALIB;   /* SysTick Calibration register    */
} SysTick_Type;

#define SysTick ((SysTick_Type *) 0xE000E010) /* SysTick struct */

/* Setup SysTick */
SysTick->LOAD  = 0xFFFF; // Set reload value
SysTick->VAL   = 0x0;   // Clear current value
SysTick->CTRL  = 0x5;   // Enable SYSTICK and select core clock
```

FIGURE 10.4

Accessing Peripheral Registers as Pointers to Elements in a Data Structure.

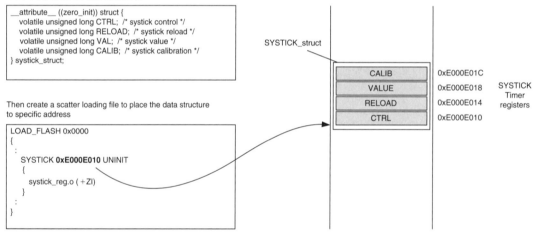

In the C file, define the data structure as

```
__attribute__ ((zero_init)) struct {
   volatile unsigned long CTRL; /* systick control */
   volatile unsigned long RELOAD; /* systick reload */
   volatile unsigned long VAL; /* systick value */
   volatile unsigned long CALIB; /* systick calibration */
} systick_struct;
```

Then create a scatter loading file to place the data structure
to specific address

```
LOAD_FLASH 0x0000
{
   :
   SYSTICK 0xE000E010 UNINIT
   {
      systick_reg.o ( +ZI)
   }
   :
}
```

FIGURE 10.5

Defining Peripheral-Based Address Using Scatter Loading File.

In this case (method is shown in Figure 10.5), the program code using the peripheral has to define the peripheral as a C pointer in an external object. The code for accessing the register is the same as in the second method.

Method 1 (shown in Figures 10.2 and 10.3) is the simplest, however, it can result in less efficient code compared with the others as the address value for the registers are stored separately as constant. As a result, the code size can be larger and might be slower as it requires more accesses to the program memory to set up the address values. However, for peripheral control code that only access to one register, the efficiency of method 1 is identical to others.

Method 2 (using data structure and a pointer defined in the C-code) is possibly the most commonly used. It allows the registers in a peripheral to share just one constant for base address value. The immediate offset address mode can be used for access of each register. This is the method used in CMSIS, which will be covered later in this chapter.

Method 3 (using scatter loading file or linker script, as shown in figure 10.5) has the same efficiency as method 2, but it is less portable due to the use of a scatter loading file (scatter loading file syntax is tool chain specific). Method 3 is required when you are developing a device driver library for a peripheral that is used in multiple devices, and the base address of the peripheral is not known until in the linking stage.

10.3.4 Intrinsic Functions

Use of the C language can often speed up application development, but in some cases, we need to use some instructions that cannot be generated using normal C-code. Some C compilers provide intrinsic functions for accessing these special instructions. Intrinsic functions are used just like normal C functions. For example, ARM compilers (including RealView C Compilers and Keil MDK-ARM) provide the intrinsic functions listed in Table 10.1 for commonly used instructions.

10.3.5 Embedded Assembler and Inline Assembler

As an alternative to using intrinsic functions, we can also directly access assembly instructions in C-code. This is often necessary in low-level system control or when you need to implement a timing critical routine and decide to implement it in assembly for the best performance. Most ARM C compilers allow you to include assembly code in form of *inline assembler*.

Table 10.1 Intrinsic Functions Provided in ARM Compilers

Assembly Instructions	ARM Compiler Intrinsic Functions
CLZ	unsigned char __clz(unsigned int val)
CLREX	void __clrex(void)
CPSID I	void __disable_irq(void)
CPSIE I	void __enable_irq(void)
CPSID F	void __disable_fiq(void)
CPSIE F	void __enable_fiq(void)
LDREX/LDREXB/LDREXH	unsigned int __ldrex(volatile void *ptr)
LDRT/LDRBT/LDRSBT/LDRHT/LDRSHT	unsigned int __ldrt(const volatile void *ptr)
NOP	void __nop(void)
RBIT	unsigned int __rbit(unsigned int val)
REV	unsigned int __rev(unsigned int val)
ROR	unsigned int __ror(unsigned int val, unsigned int shift)
SSAT	int __ssat(int val, unsigned int sat)
SEV	void __sev(void)
STREX/STREXB/STREXH	int __strex(unsigned int val, volatile void *ptr)
STRT/STRBT/STRHT	void int __strt(unsigned int val, const volatile void *ptr)
USAT	int __usat(unsigned int val, unsigned int sat)
WFE	void __wfe(void)
WFI	void __wfi(void)
BKPT	void __breakpoint(int val)

In the ARM compiler, you can add assembly code inside the C program. Traditionally, inline assembler is used, but the inline assembler in RealView C Compiler does not support instructions in Thumb-2 technology. Starting with RealView C Compiler version 3.0, a new feature called the Embedded Assembler is included, and it supports the instruction set in Thumb-2. For example, you can insert assembly functions in your C programs this way:

```
__asm void SetFaultMask(unsigned int new_value)
{
  // Assembly code here
  MSR FAULTMASK, new_value // Write new value to FAULTMASK
  BX LR                    // Return to calling program
}
```

Detailed descriptions of Embedded Assembler in RealView C Compiler can be found in the *RVCT 4.0 Compilation Tools Compiler Guide* [Ref. 6].

For the Cortex-M3, Embedded Assembler is useful for tasks, such as direct manipulation of the stacks and timing critical processing task (codec software).

10.4 CMSIS

10.4.1 Background of CMSIS

The Cortex-M3 microcontrollers are gaining momentum in the embedded application market, as more and more products based on the Cortex-M3 processor and software that support the Cortex-M3 processor are emerging. At the end of 2008, there were more than five C compiler vendors, and more than 15 embedded Operating Systems (OS) supporting the Cortex-M3 processor. There are also a number of companies providing embedded software solutions, including codecs, data processing libraries, and various software and debug solutions. The CMSIS was developed by ARM to allow users of the Cortex-M3 microcontrollers to gain the most benefit from all these software solutions and to allow them to develop their embedded application quickly and reliably (see Figure 10.6).

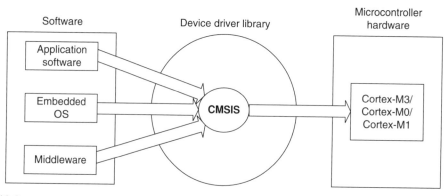

FIGURE 10.6

CMSIS Provides a Standardized Access Interface for Embedded Software Products.

The CMSIS was started in 2008 to improve software usability and inter-operability of ARM micro-controller software. It is integrated into the driver libraries provided by silicon vendors, providing a standardized software interface for the Cortex-M3 processor features, as well as a number of common system and I/O functions. The library is also supported by software companies including embedded OS vendors and compiler vendors.

The aims of CMSIS are to:

- improve software portability and reusability
- enable software solution suppliers to develop products that can work seamlessly with device libraries from various silicon vendors
- allow embedded developers to develop software quicker with an easy-to-use and standardized software interface
- allow embedded software to be used on multiple compiler products
- avoid device driver compatibility issues when using software solutions from multiple sources

The first release of CMSIS was available from fourth quarter of 2008 and has already become part of the device driver library from microcontroller vendors. The CMSIS is also available for Cortex-M0.

10.4.2 Areas of Standardization

The scope of CMSIS involves standardization in the following areas:

- *Hardware Abstraction Layer (HAL) for Cortex-M processor registers*: This includes standardized register definitions for NVIC, System Control Block registers, SYSTICK register, MPU registers, and a number of NVIC and core feature access functions.

- *Standardized system exception names*: This allows OS and middleware to use system exceptions easily without compatibility issues.

- *Standardized method of header file organization*: This makes it easier for users to learn new Cortex microcontroller products and improve software portability.

- *Common method for system initialization*: Each Microcontroller Unit (MCU) vendor provides a *SystemInit()* function in their device driver library for essential setup and configuration, such as initialization of clocks. Again, this helps new users to start to use Cortex-M microcontrollers and aids software portability.

- *Standardized intrinsic functions*: Intrinsic functions are normally used to produce instructions that cannot be generated by IEC/ISO C.* By having standardized intrinsic functions, software reusability and portability are considerably improved.

- *Common access functions for communication*: This provides a set of software interface functions for common communication interfaces including universal asynchronous receiver/transmitter (UART), Ethernet, and Serial Peripheral Interface (SPI). By having these common access functions in the device driver library, reusability and portability of embedded software are improved. At the time of writing this book, it is still under development.

- *Standardized way for embedded software to determine system clock frequency*: A software variable called *SystemFrequency* is defined in device driver code. This allows embedded OS to set up the SYSTICK unit based on the system clock frequency.

---

*C/C++ features are specified in a standard document "ISO/IEC 14882" prepared by the International Organization for Standards (ISO) and the International Electrotechnical Commission (IEC).

The CMSIS defines the basic requirements to achieve software reusability and portability. MCU vendors can include additional functions for each peripheral to enrich the features of their software solution. So using CMSIS does not limit the capability of the embedded products.

10.4.3 Organization of CMSIS

The CMSIS is divided into multiple layers as follows:

Core Peripheral Access Layer
• Name definitions, address definitions, and helper functions to access core registers and core peripherals

Middleware Access Layer
• Common method to access peripherals for the software industry (work in progress)
• Targeted communication interfaces include Ethernet, UART, and SPI.
• Allows portable software to perform communication tasks on any Cortex microcontrollers that support the required communication interface

Device Peripheral Access Layer (MCU specific)
• Name definitions, address definitions, and driver code to access peripherals

Access Functions for Peripherals (MCU specific)
• Optional additional helper functions for peripherals

The role of these layers is summarized in Figure 10.7.

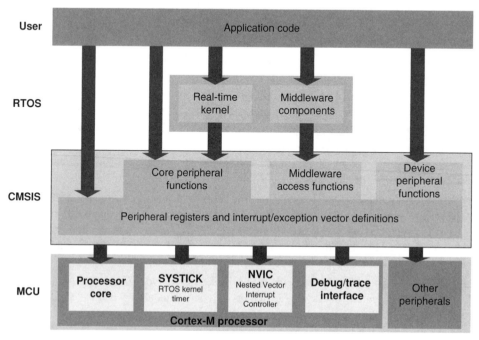

FIGURE 10.7

CMSIS Structure.

10.4.4 **Using CMSIS**

Since the CMSIS is incorporated inside the device driver library, there is no special setup requirement for using CMSIS in projects. For each MCU device, the MCU vendor provides a header file, which pulls in additional header files required by the device driver library, including the Core Peripheral Access Layer defined by ARM (see Figure 10.8).

The file *core_cm3.h* contains the peripheral register definitions and access functions for the Cortex-M3 processor peripherals like NVIC, System Control Block registers, and SYSTICK registers. The *core_cm3.h* file also contains declaration of CMSIS intrinsic functions to allow C applications to access instructions that cannot be generated using IEC/ISO C language. In addition, this file also contains a function for outputting a debug message via the Instrumentation Trace Module (ITM).

Note that in some cases, the intrinsic functions in CMSIS could have similar names compared with the intrinsic functions provided in the C compilers, whereas the CMSIS intrinsic functions are compiler independent.

The file *core_cm3.c* contains implementation of CMSIS intrinsic functions that cannot be implemented in core_cm3.h using simple definitions.

The *system_<device>.h* file contains microcontroller specific interrupt number definitions, and peripheral register definitions. The *system_<device>.c* file contains a microcontroller specific function called *SystemInit* for system initialization.

In addition, CMSIS compliant device drivers also contain start-up code (which contains the vector table) for various supported compilers, and CMSIS version of intrinsic functions to allow embedded software access to all processor core features on different C compiler products.

Examples of using CMSIS can be found on the microcontroller vendor's web site. You might also find examples in the device driver libraries itself. Alternatively, you can download the ARM CMSIS

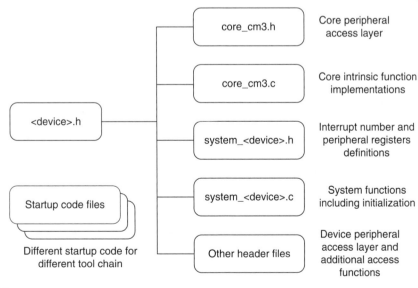

FIGURE 10.8

CMSIS Files.

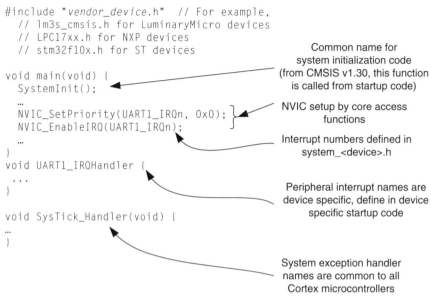

```
#include "vendor_device.h"  // For example,
   // lm3s_cmsis.h for LuminaryMicro devices
   // LPC17xx.h for NXP devices
   // stm32f10x.h for ST devices

void main(void) {
   SystemInit();
   ...
   NVIC_SetPriority(UART1_IRQn, 0x0);
   NVIC_EnableIRQ(UART1_IRQn);
   ...
}
void UART1_IRQHandler {
   ...
}

void SysTick_Handler(void) {
   ...
}
```

Common name for system initialization code (from CMSIS v1.30, this function is called from startup code)

NVIC setup by core access functions

Interrupt numbers defined in system_<device>.h

Peripheral interrupt names are device specific, define in device specific startup code

System exception handler names are common to all Cortex microcontrollers

FIGURE 10.9

CMSIS Example.

package from *www.onarm.com*, which contains examples and documentation. Documentation of the common functions can also be found in this package.

A simple example of using CMSIS in your application development is shown in Figure 10.9. To use the CMSIS to set up interrupts and exceptions, you need to use the exception/interrupt constants defined in the *system_<device>.h*. These exception and interrupt constants are different from the exception number used in the core internal registers (e.g., Interrupt Program Status Register [IPSR]). For CMSIS, negative numbers are for system exceptions and positive numbers are for peripheral interrupts.

For development of portable code, you should use the core access functions to access core functionalities and middleware access functions to access peripheral. This allows the porting of software to be minimized between different Cortex microcontrollers.

Details of common CMSIS access functions and intrinsic functions can be found in Appendix G.

10.4.5 Benefits of CMSIS

So what does CMSIS mean to end users?

The main advantage is much better software portability and reusability. Besides easy migration between different Cortex-M3 microcontrollers, it also allows software to be quickly ported between Cortex-M3 and other Cortex-M processors, reducing time to market.

For embedded OS vendors and middleware providers, the advantages of the CMSIS are significant. By using the CMSIS, their software products can become compatible with device drivers from multiple microcontroller vendors, including future microcontroller products that are yet to be released (see Figure 10.10). Without the CMSIS, the software vendors either have to include a small library for

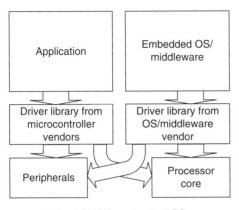 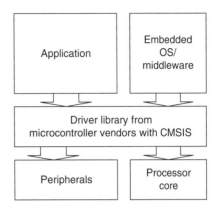

Without CMSIS, embedded OS or middleware needs to include processor core access functions and might need to include a few peripheral drivers

With CMSIS, embedded OS or middleware can use standardized core access functions in the driver library

FIGURE 10.10

CMSIS Avoids Overlapping Driver Code.

Cortex-M3 core functions or develop multiple configurations of their product so that it can work with device libraries from different microcontroller vendors.

The CMSIS has a small memory footprint (less than 1 KB for all core access functions and a few bytes of RAM). It also avoids overlapping of core peripheral driver code when reusing software code from other projects.

Since CMSIS is supported by multiple compiler vendors, embedded software can compile and run with different compilers. As a result, embedded OS and middleware can be MCU vendor independent and compiler tool vendor independent. Before availability of CMSIS, intrinsic functions were generally compiler specific and could cause problems in retargetting the software in a different compiler.

Since all CMSIS compliant device driver libraries have a similar structure, learning to use different Cortex-M3 microcontrollers is even easier as the software interface has similar look and feel (no need to relearn a new application programming interface).

CMSIS is tested by multiple parties and is Motor Industry Software Reliability Association (MISRA) compliant, thus reducing the validation effort required for developing your own NVIC or core feature access functions.

10.5 USING ASSEMBLY

For small projects, it is possible to develop the whole application in assembly language. However, this is often much harder for beginners. Using assembler, you might be able to get the best optimization you want, though it might increase your development time, and it could be easy to make mistakes. In addition, handling complex data structures or function library management can be extremely difficult in assembler. Yet even when the C language is used in a project, in some situations part of the program is implemented in assembly language as follows:

- Functions that cannot be implemented in C, such as direct manipulation of stack data or special instructions that cannot be generated by the C compiler in normal C-code
- Timing-critical routines
- Tight memory requirements, causing part of the program to be written in assembly to get the smallest memory size

10.5.1 The Interface between Assembly and C

In various situations, assembly code and the C program interact. For example,

- When embedded assembly (or inline assembler, in the case of the GNU tool chain) is used in C program code
- When C program code calls a function or subroutine implemented in assembler in a separate file
- When an assembly program calls a C function or subroutine

In these cases, it is important to understand how parameters and return results are passed between the calling program and the function being called. The mechanisms of these interactions are specified in the *ARM Architecture Procedure Call Standard [AAPCS, [Ref. 5]].*

For simple cases, when a calling program needs to pass parameters to a subroutine or function, it will use registers R0–R3, where R0 is the first parameter, R1 is the second, and so on. Similarly, R0 is used for returning a value at the end of a function. R0–R3 and R12 can be changed by a function or subroutine whereas the contents of R4–R11 should be restored to the previous state before entering the function, usually handled by stack PUSH and stack POP.

To make them easier to understand, the examples in this book do not strictly follow AAPCS practices. If a C function is called by an assembly code, the effect of a possible register change to R0–R3 and R12 will need to be taken into account. If the contents of these registers are needed at a later stage, these registers might need to be saved on the stack and restored after the C function completes. Since the example codes mostly only call assembly functions or subroutines that affect a few registers or restore the register contents at the end, it's not necessary to save registers R0–R3 and R12.

10.5.2 The First Step in Assembly Programming

This chapter reviews a few examples in assembly language. In most cases, you will be programming in C, but by looking into some assembler examples, we can gain a better understanding of how to use the Cortex-M3 processor. The examples here are based on ARM assembler tools (armasm) in RVDS. For users of Keil MDK-ARM, the command line options are slightly different. For other assembler tools, the file format and instruction syntax will also need to be modified. In addition, some development tools will actually do the startup code for you, so you might not need to worry about creating your assembly startup code.

The first simple program can be something like this

```
STACK_TOP EQU 0x20002000; constant for SP starting value

    AREA |Header Code |, CODE
    DCD STACK_TOP ; Stack top
```

```
      DCD Start      ; Reset vector
      ENTRY          ; Indicate program execution start here
Start ; Start of main program
      ; initialize registers
      MOV r0, #10    ; Starting loop counter value
      MOV r1, #0     ; starting result
      ; Calculated 10+9+8+...+1
loop
      ADD r1, r0     ; R1 = R1 + R0
      SUBS r0, #1    ; Decrement R0, update flag ("S" suffix)
      BNE loop       ; If result not zero jump to loop
      ; Result is now in R1
deadloop
      B deadloop     ; Infinite loop
      END            ; End of file
```

This simple program contains the initial stack pointer (SP) value, the initial program counter (PC) value, and setup registers and then does the required calculation in a loop.

Assuming you are using ARM RealView compilation tools, this program can be assembled using

```
$> armasm --cpu cortex-m3 -o test1.o test1.s
```

The *-o* option specifies the output file name. The test1.o is an object file. We then need to use a linker to create an executable image (ELF). This can be done by

```
$> armlink --rw_base 0x20000000 --ro_base 0x0 --map -o test1.elf test1.o
```

Here, *--ro-base 0x0* specifies that the read-only region (program ROM) starts at address 0x0; *--rw-base* specifies that the read/write region (data memory) starts at address 0x20000000. (In this example test1.s, we did not have any RAM data defined.) The *--map* option creates an image map, which is useful for understanding the memory layout of the compiled image.

Finally, we need to create the binary image

```
$> fromelf --bin --output test1.bin test1.elf
```

For checking that the image looks like what we wanted, we can also generate a disassembled code list file by

```
$> fromelf -c --output test1.list test1.elf
```

If everything works fine, you can then load your ELF image or binary image into your hardware or instruction set simulator for testing.

10.5.3 Producing Outputs

It is always more fun when you can connect your microcontroller to the outside world. The simplest way to do that is to turn on/off the LEDs. However, this practice is quite limiting because it can only represent very limited information. One of the most common output methods is to send text messages to a console. In embedded product development, this task is often handled by a UART interface connecting

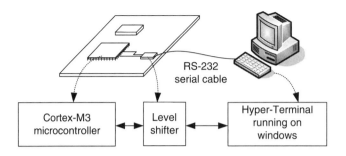

FIGURE 10.11

A Low-Cost Test Environment for Outputting Text Messages.

to a personal computer. For example, a computer running a Windows[1] system with the Hyper-Terminal program acting as a console can be a handy way to produce outputs (see Figure 10.11).

The Cortex-M3 processor does not contain a UART interface, but most Cortex-M3 microcontrollers come with UART provided by the chip manufacturers. The specification of the UART can differ among various devices, so we won't attempt to cover the topic in this book. Our next example assumes that a UART is available and has a status flag to indicate whether the transmit buffer is ready for sending out new data. A level shifter is needed in the connection because RS-232 has a different voltage level than the microcontroller I/O pins.

UART is not the only solution to output text messages. A number of features are implemented on the Cortex-M3 processor to help output debugging messages:

- *Semihosting*: Depending on the debugger and code library support, *semihosting* (outputting *printf* messages via a debug probe device) can be done via debug register in the NVIC. More information on this topic is covered in Chapter 15. In these cases, you can use *printf* within your C program, and the output will be displayed on the console/standard output (STDOUT) of the debugger software.

- *Instrumentation trace*: If the Cortex-M3 microcontroller provides a trace port and an external Trace Port Analyzer (TPA) is available, instead of using UART to output messages, we can use the ITM. The trace port works much faster than UART and can offer more data channels.

- *Instrumentation trace via Serial-Wire Viewer (SWV)*: Alternatively, the Cortex-M3 processor (revision 1 and later) also provides an SWV operation mode on the Trace Port Interface Unit (TPIU). This interface allows outputs from ITM to be captured using low-cost hardware instead of a TPA. However, the bandwidth provided with the SWV mode is limited, so it is not ideal for large amounts of data (e.g., instruction trace operation).

10.5.4 The "Hello World" Example

Before we try to write a "Hello world" program, we should figure out how to send one character through the UART. The code used to send a character can be implemented as a subroutine, which can

---

[1]Windows and Hyper-Terminal are trademarks of Microsoft Corporation.

be called by other message output codes. If the output device changes, we only need to change this subroutine and all the text messages can be output by a different device. This modification is usually called retargetting.

A simple routine to output a character could be something like this

```
UART0_BASE    EQU    0x4000C000
UART0_FLAG    EQU    UART0_BASE+0x018
UART0_DATA    EQU    UART0_BASE+0x000

Putc          ; Subroutine to send a character via UART
              ; Input R0 = character to send
              PUSH {R1,R2, LR}   ; Save registers
              LDR R1,=UART0_FLAG

PutcWaitLoop
              LDR R2,[R1]         ; Get status flag
              TST R2, #0x20       ; Check transmit buffer full flag
                                  ; bit
              BNE PutcWaitLoop    ; If busy then loop
              LDR R1,=UART0_DATA  ; otherwise
              STRB R0, [R1]       ; Output data to transmit buffer
              POP {R1,R2, PC}     ; Return
```

The register addresses and bit definitions here are just examples; you might need to change the value for your device. In addition, some UART might require a more complex status-checking process before the character is output to the transmit buffer. Furthermore, another subroutine call (*Uart0Initialize* in the following example) is required to initialize the UART, but this depends on the UART specification and will not be covered in this chapter. An example of UART initialization in C for Luminary Micro LM3S811 devices is covered in Chapter 20.

Now, we can use this subroutine to build a number of functions to display messages:

```
Puts          ; Subroutine to send string to UART
              ; Input R0 = starting address of string.
              ; The string should be null terminated
              PUSH {R0 ,R1, LR}   ; Save registers
              MOV R1, R0          ; Copy address to R1, because R0 will
                                  ; be used
PutsLoop                          ; as input for Putc
              LDRB R0,[R1],#1      ; Read one character and increment
                                  ; address
              CBZ  R0, PutsLoopExit ; if character is null, goto end
              BL   Putc           ; Output character to UART
              B    PutsLoop       ; Next character
PutsLoopExit
              POP {R0, R1, PC}    ; Return
```

With this subroutine, we are ready for our first "Hello world" program:

```
STACK_TOP     EQU    0x20002000; constant for SP starting value
UART0_BASE    EQU    0x4000C000
UART0_FLAG    EQU    UART0_BASE+0x018
UART0_DATA    EQU    UART0_BASE+0x000
```

```
                AREA | Header Code|, CODE
                DCD STACK_TOP ; Stack Pointer initial value
                DCD Start ; Reset vector
                ENTRY
Start           ; Start of main program
                MOV r0, #0          ; initialize registers
                MOV r1, #0
                MOV r2, #0
                MOV r3, #0
                MOV r4, #0
                BL  Uart0Initialize ; Initialize the UART0
                LDR r0,=HELLO_TXT   ; Set R0 to starting address of string
                BL  Puts
deadend
                B deadend ; Infinite loop
                ;--------------------------------
                ; subroutines
                ;--------------------------------
Puts            ; Subroutine to send string to UART
                ;Input R0 = starting address of string.
                ; The string should be null terminated
                PUSH {R0 ,R1, LR}  ; Save registers
                MOV R1, R0         ; Copy address to R1, because R0 will
                                   ; be used
PutsLoop                           ; as input for Putc
                LDRB R0,[R1],#1    ; Read one character and increment
                                   ; address
                CBZ  R0, PutsLoopExit ; if character is null, goto end
                BL   Putc            ; Output character to UART
                B    PutsLoop        ; Next character
PutsLoopExit
                POP {R0, R1, PC}   ; Return
                ;--------------------------------
Putc            ; Subroutine to send a character via UART
                ; Input R0 = character to send
                PUSH {R1,R2, LR}      ; Save registers
                LDR   R1,=UART0_FLAG
PutcWaitLoop
                LDR   R2,[R1]         ; Get status flag
                TST   R2, #0x20       ; Check transmit buffer full flag bit
                BNE   PutcWaitLoop    ; If busy then loop
                LDR   R1,=UART0_DATA  ; otherwise
                STR   R0, [R1]        ; Output data to transmit buffer
                POP  {R1,R2, PC}      ; Return
                ;--------------------------------
Uart0Initialize
                ; Device specific, not shown here
                BX   LR ; Return
                ;--------------------------------
HELLO_TXT
                DCB "Hello world\n",0 ; Null terminated Hello
                                      ; world string
                END    ; End of file
```

The only thing you need to add to this code is the details for the *Uart0Initialize* subroutine and modify the UART register address constants at the top of the file.

It will also be useful to have subroutines that output register values as well. To make things easier, they can all be based on *Putc* and *Puts* subroutines we have already done. The first subroutine is to display hexadecimal values.

```
PutHex ; Output register value in hexadecimal format
       ; Input R0 = value to be displayed
    PUSH {R0-R3,LR}
    MOV R3, R0    ; Save register value to R3 because R0 is used
                  ; for passing input parameter
    MOV R0,#'0' ; Starting the display with "0x"
    BL  Putc
    MOV R0,#'x'
    BL  Putc
    MOV R1, #8        ; Set loop counter
    MOV R2, #28       ; Rotate offset
PutHexLoop
    ROR R3,  R2       ; Rotate data value left by 4 bits
                      ; (right 28)
    AND R0,  R3,#0xF ; Extract the lowest 4 bit
    CMP R0,  #0xA     ; Convert to ASCII
    ITE GE
    ADDGE R0, #55     ; If larger or equal 10, then convert
                      ; to A-F
    ADDLT R0, #48     ; otherwise convert to 0-9
    BL Putc           ; Output 1 hex character
    SUBS R1,  #1      ; decrement loop counter
    BNE PutHexLoop    ; if all 8 hexadecimal character been
                      ; display then
    POP {R0-R3,PC}    ; return, otherwise process next 4-bit
```

This subroutine is useful for outputting register values. However, sometimes we also want to output register values in decimal. This sounds like a rather complex operation, but in the Cortex-M3 it is easy because of the hardware multiply and divide instructions. One of the other main problems is that during calculation, we will get output characters in reverse order, so we need to put the output results in a text buffer first, wait until the whole text is ready to display, and then use the *Puts* function to display the whole result. In this example, a part of the stack memory is used as the text buffer:

```
PutDec  ; Subroutine to display register value in decimal
        ; Input R0 = value to be displayed.
        ; Since it is 32 bit, the maximum number of character
        ; in decimal format, including null termination is 11
        PUSH {R0-R5, LR}  ; Save register values
        MOV R3, SP        ; Copy current Stack Pointer to R3
        SUB SP, SP, #12   ; Reserved 12 bytes as text buffer
        MOV R1, #0        ; Null character
        STRB R1,[R3, #-1]!; Put null character at end of text
                          ; buffer,pre-indexed
        MOV R5, #10       ; Set divide value
PutDecLoop
        UDIV R4, R0, R5 ; R4 = R0 / 10
```

```
            MUL R1, R4, R5      ; R1 = R4 * 10
            SUB R2, R0, R1      ; R2 = R0 - (R4 * 10) = remainder
            ADD R2, #48         ; convert to ASCII (R2 can only be 0-9)
            STRB R2,[R3, #-1]!  ; Put ascii character in text
                                ; buffer, pre-indexed
            MOVS R0, R4         ; Set R0 = Divide result and set Z flag
            ; if R4=0
            BNE PutDecLoop      ; If R0(R4) is already 0, then there
                                ; is no more digit
            MOV R0, R3          ; Put R0 to starting location of text
                                ; buffer
            BL Puts             ; Display the result using Puts
            ADD SP, SP, #12     ; Restore stack location
            POP {R0-R5, PC}     ; Return
```

With various features in the Cortex-M3 instruction set, the processing to convert values into decimal format display can be implemented in a very short subroutine.

10.5.5 **Using Data Memory**

Back to our first example: When we were doing the linking stage, we specified the read/write memory region. How do we put data there? The method is to define a data region in your assembly file. Using the same example from the beginning, we can store the data in the data memory at 0x20000000 (the SRAM region). The location of the data section is controlled by a command-line option when you run the linker:

```
STACK_TOP   EQU 0x20002000   ; constant for SP starting value
            AREA | Header Code|, CODE
            DCD STACK_TOP    ; SP initial value
            DCD Start        ; Reset vector
            ENTRY
Start       ; Start of main program
            ; initialize registers
            MOV r0, #10 ; Starting loop counter value
            MOV r1, #0  ; starting result
            ; Calculated 10+9+8+...+1
loop
            ADD r1, r0     ; R1 = R1 + R0
            SUBS r0, #1    ; Decrement R0, update flag ("S"
                           ; suffix)
            BNE loop       ; If result not zero jump to loop
            ; Result is now in R1
            LDR r0,=MyData1 ; Put address of MyData1 into R0
            STR r1,[r0]    ; Store the result in MyData1
deadloop
            B deadloop     ; Infinite loop
            AREA | Header Data|, DATA
            ALIGN 4
MyData1     DCD 0          ; Destination of calculation result
MyData2     DCD 0
            END            ; End of file
```

During the linking stage, the linker will put the DATA region into read/write memory, so the address for *MyData1* will be 0x20000000 in this case.

10.6 USING EXCLUSIVE ACCESS FOR SEMAPHORES

Exclusive access instructions are used for semaphore operations—for example, a MUTEX (Mutual Exclusion) to make sure that a resource is used by only one task. For instance, let's say that a data variable DeviceALocked in memory can be used to indicate that Device A is being used. If a task wants to use Device A, it should check the status by reading the variable DeviceALocked. If it is zero, it can write a 1 to DeviceALocked to lock the device. After it's finished using the device, it can then clear the DeviceALocked to zero so that other tasks can use it.

What will happen if two tasks try to access Device A at the same time? In that case, possibly both tasks will read the variable DeviceALocked, and both will get zero. Then both of them will try writing back 1 to the variable DeviceALocked to lock the device, and we'll end up with both tasks believing that they have exclusive access to Device A. That is where exclusive accesses are used. The STREX instruction has a return status, which indicates whether the exclusive store has been successful. If two tasks try to lock a device at the same time, the return status will be 1 (exclusive failed) and the task can then know that it needs to retry the lock.

Chapter 5 provided some background on the use of exclusive accesses. The flowchart in that earlier discussion is shown in Figure 10.12.

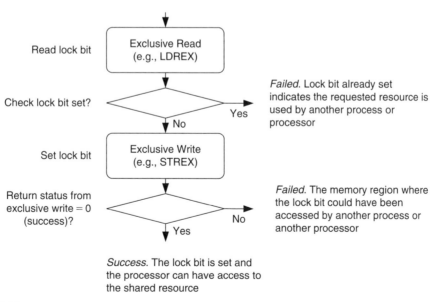

FIGURE 10.12

Using Exclusive Access for Semaphore Operations.

The operation can be carried out by the following C-code using intrinsic functions from CMSIS. Note that the data write operation of STREX will not be carried out if the exclusive monitor returns a fail status, preventing a lock bit being set when the exclusive access fails:

```
volatile unsigned int DeviceALocked; // lock variable

int LockDeviceA(void){
  unsigned int status; // variable to hold STREX status
  // Get the lock status and see if it is already locked
if (__LDREXW(&DeviceALocked) = 0) {
  // if not locked, try set lock to 1
  status = __STREXW(1, &DeviceALocked);
  if (status!=0) return (1); // return fail status
  else return(0); // return success status
} else {
  return(1); // return fail status
  }
}
```

The same operation can also be carried out by the following assembly code:

```
LockDeviceA
          ; A simple function to try to lock Device A
          ; Output R0 : 0 = Success, 1 = failed
          If successful, value of 1 will be written to variable
          ; DeviceALocked
          PUSH {R1, R2, LR}
TryToLockDeviceA
          LDR   R1,=DeviceALocked   ; Get the lock status
          LDREX R2,[R1]
          CMP   R2,#0               ; Check if it is locked
          BNE   LockDeviceAFailed
DeviceAIsNotLocked
          MOV   R0,#1               ; Try to write 1 to
                                    ; DeviceALocked
          STREX R2,R0,[R1]          ; Exclusive write
          CMP   R2, #0
          BNE LockDeviceAFailed     ; STREX Failed
LockDeviceASucceed
          MOV   R0,#0               ; Return success status
          POP {R1, R2, PC}          ; Return
LockDeviceAFailed
          MOV R0,#1                 ; Return fail status
          POP {R1, R2, PC}          ; Return
```

If the return status of this function is 1 (exclusive failed), the application tasks should wait a bit and retry later. In single-processor systems, the common cause of an exclusive access failing is an Interrupt occurring between the exclusive load and the exclusive store. If the code is run in privileged mode, this situation can be prevented by setting an Interrupt Mask register, such as PRIMASK, for a short time to increase the chance of getting the resource locked successfully.

In multiprocessor systems, aside from interrupts, the exclusive store could also fail if another processor has accessed the same memory region. To detect memory accesses from different processors, the bus infrastructure requires exclusive access monitor hardware to detect whether there is an access from a different bus master to a memory between the two exclusive accesses. However, in most low-cost Cortex-M3 microcontrollers, there is only one processor, so this monitor hardware is not required.

With this mechanism, we can be sure that only one task can have access to certain resources. If the application cannot gain the lock to the resource after a number of times, it might need to quit with a timeout error. For example, a task that locked a resource might have crashed and the lock remained set. In these situations, the OS should check which task is using the resource. If the task has completed or terminated without clearing the lock, the OS might need to unlock the resource.

If the process has started an exclusive access using LDREX and then found that the exclusive access is no longer needed, it can use the CLREX instruction to clear the local record in the exclusive access monitor. This can be done with CMSIS function:

```
void __CLREX(void);
```

If assembly language is used, the CLREX instruction can be used:

```
CLREX
```

or

```
CLREX.W
```

For the Cortex-M3 processor, all exclusive memory transfers must be carried out sequentially. However, if the exclusive access control code has to be reused on other ARM Cortex processors, the Data Memory Barrier (DMB) instruction might need to be inserted between exclusive transfers to ensure correct ordering of the memory accesses. Example code of using barrier instructions with exclusive accesses can be found in Section 14.3, Multiprocessor Communication.

10.7 USING BIT BAND FOR SEMAPHORES

It is possible to use the bit-band feature to carry semaphore operations, provided that the memory system supports locked transfers or only one bus master is present on the memory bus. With bit band, it is possible to carry out the semaphore in normal C-code, but the operation is different from using exclusive access. To use bit band as a resource allocation control, a memory location (such as word data) with a bit-band memory region is used, and each bit of this variable indicates that the resource is used by a certain task.

Since the bit-band alias writes are locked READ-MODIFY-WRITE transfers (the bus master cannot be switched to another one between the transfers), provided that all tasks only change the lock bit representing themselves, the lock bits of other tasks will not be lost, even if two tasks try to write to the same memory location at the same time. Unlike using exclusive accesses, it is possible for a resource to be "locked" simultaneously by two tasks for a short period of time until one of them detects the conflict and releases the lock (see Figure 10.13).

Using bit band for semaphores can work only if all the tasks in the system change only the lock bit they are assigned to using the bit-band alias. If any of the tasks change the lock variable using a normal write, the semaphore can fail because another task sets a lock bit just before the write to the lock variable, the previous lock bit set by the other task will be lost.

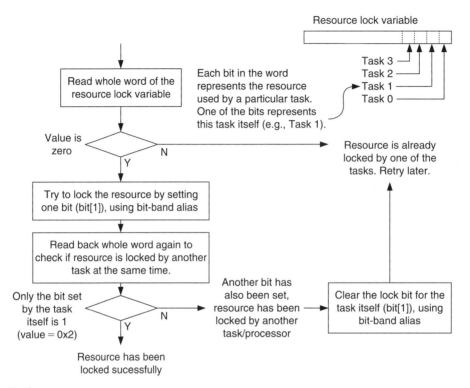

FIGURE 10.13

Mutex Implemented Using Bit Band as a Semaphore Control.

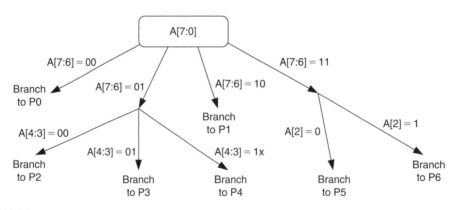

FIGURE 10.14

Bit Field Decoder: Example Use of UBFX and TBB Instructions.

10.8 **WORKING WITH BIT FIELD EXTRACT AND TABLE BRANCH**

We examined the unsigned bit field extract (UBFX) and Table Branch (TBB/TBH) instructions in Chapter 4. These two instructions can work together to form a very powerful branching tree. This capability is very useful in data communication applications where the data sequence can have different meanings with different headers. For example, let's say that the following decision tree based on Input A is to be coded in assembler (see Figure 10.14).

```
DecodeA
    LDR   R0,=A          ; Get the value of A from memory
    LDR   R0,[R0]
    UBFX  R1, R0, #6, #2  ; Extract bit[7:6] into R1
    TBB   [PC, R1]
BrTable1
    DCB   ((P0       -BrTable1)/2) ; Branch to P0       if A[7:6] = 00
    DCB   ((DecodeA1-BrTable1)/2) ; Branch to DecodeA1 if A[7:6] = 01
    DCB   ((P1       -BrTable1)/2) ; Branch to P1       if A[7:6] = 10
    DCB   ((DecodeA2-BrTable1)/2) ; Branch to DecodeA1 if A[7:6] = 11
DecodeA1
    UBFX  R1, R0, #3, #2 ; Extract bit[4:3] into R1
    TBB   [PC, R1]
BrTable2
    DCB   ((P2 -BrTable2)/2) ; Branch to P2   if A[4:3] = 00
    DCB   ((P3 -BrTable2)/2) ; Branch to P3   if A[4:3] = 01
    DCB   ((P4 -BrTable2)/2) ; Branch to P4   if A[4:3] = 10
    DCB   ((P4 -BrTable2)/2) ; Branch to P4   if A[4:3] = 11
DecodeA2
    TST   R0, #4 ; Only 1 bit is tested, so no need to use UBFX
    BEQ   P5
    B     P6
P0 ... ; Process 0
P1 ... ; Process 1
P2 ... ; Process 2
P3 ... ; Process 3
P4 ... ; Process 4
P5 ... ; Process 5
P6 ... ; Process 6
```

This code completes the decision tree in a short assembler code sequence. If the branch target addresses are at a larger offset, some of the TBB instructions would have to be replaced by TBH instructions.

Exception Programming

IN THIS CHAPTER

Using Interrupts 183
Exception/Interrupt Handlers 188
Software Interrupts 189
Example of Vector Table Relocation 190
Using SVC 193
SVC Example: Use for Text Message Output Functions 194
Using SVC with C 197

11.1 USING INTERRUPTS

Interrupts are used in almost all embedded applications. In the Cortex™-M3 processor, the interrupt controller Nested Vectored Interrupt Controller (NVIC) handles a number of processing tasks for you, including priority checking and stacking/unstacking of registers. However, a number of tasks have to be prepared before interrupts can be used:

- Stack setup
- Vector table setup
- Interrupt priority setup
- Enable the interrupt

11.1.1 Stack Setup

For simple application development, you can use the Main Stack Pointer (MSP) for the whole program. That way you need to reserve memory that's just large enough and set the MSP to the top of the stack. When determining the stack size required, besides checking the stack level that could be used by the software, you also need to check how many levels of nested interrupts can occur.

For each level of nested interrupts, you need at least eight words of stack. The processing inside interrupt handlers might need extra stack space as well.

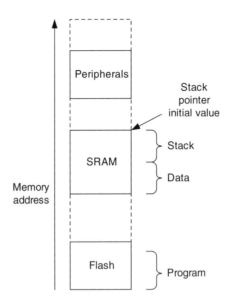

FIGURE 11.1

A Simple Memory Usage Example.

Because the stack operation in the Cortex-M3 is full descending, it is common to put the stack initial value at the end of the static memory so that the free space in the Static Random Access Memory (SRAM) is not fragmented (see Figure 11.1).

For applications that use separate stacks for user code and kernel code, the main stack should have enough memory for the nested interrupt handlers as well as the stack memory used by the kernel code. The process stack should have enough memory for the user application code plus one level of stacking space (eight words). This is because stacking from the user thread to the first level of the interrupt handler uses the process stack.

11.1.2 Vector Table Setup

For simple applications that have fixed interrupt handlers, the vector table can be coded in Flash or ROM. In this case, there is no need to set up the vector table during run time. However, in some applications, it is necessary to change the interrupt handlers for different situations. Then, you will need to relocate the vector table to writable memory.

Before the vector table is relocated, you might need to copy the existing vector table content to the new vector table location. This includes vector addresses for fault handlers, the nonmaskable interrupt (NMI), system calls, and so on. Otherwise, invalid vector addresses will be fetched by the processor if these exceptions take place after the vector table relocation.

After the necessary vector table items are set up and the vector table is relocated, we can add new vectors to the vector table. For users of Cortex Microcontroller Software Interface Standard (CMSIS) compliant driver libraries, the vector table offset register can be accessed by "*SCB->VTOR*" in the core peripheral definition.

```
void SetVector(unsigned int ExcpType, unsigned int VectorAddress)
{ // Calculate vector location = VTOR + (Exception_Type * 4)
```

```
    *((volatile unsigned int *) (SCB->VTOR + (ExcpType << 2))) =
      VectorAddress | 0x1;
     // LSB of vector set to 1 to indicate Thumb
    return;
  }
```

For users who prefer programming in assembly, this can be done by the following code example:

```
        ; Subroutine for setting vector of an exception based on
        ; exception type
        ; (For IRQs add 16 : IRQ #0 = exception type 16)
SetVector
        ; Input R0 = exception type
        ; Input R1 = vector address value
        PUSH {R2, LR}
        LDR R2,=0xE000ED08        ; Vector table offset register
        LDR R2, [R2]
        ORR R1, R1, #1            ; Set LSB of vector to indicate Thumb
        STR R1, [R2, R0, LSL #2]  ; Write vector to VectTblOffset+
                                  ; ExcpType*4
        POP {R2, PC}             ; Return
```

The setting of least significant bit (LSB) to 1 in the vector is not necessary in most case, as the compiler or assembler should recognize the address as a Thumb® instruction address and set it automatically.

11.1.3 Interrupt Priority Setup

By default, after a reset, all exceptions with programmable priority are in priority level 0. For hard fault exceptions and NMI, the priority levels are −1 and −2, respectively. For users of CMSIS compliant device driver libraries, you can use the CMSIS function to set priority level value. For example, to set the priority of interrupt request (IRQ) #4 to 0xC0, you can use

```
NVIC_SetPriority(IRQ4_IRQn, 0xC); // This function
// automatically shifts the priority value to implemented bits
// in the priority level registers
```

The constant IRQ4_IRQn above is just an example of an interrupt identifier. When using CMSIS interrupt control functions, it is recommended to use the interrupt identifiers defined in the header file (*device.h* as shown in Figure 10.8) to help readability and portability.

You can use the *NVIC_SetPriority* function with another CMSIS function that calculates the priority level value based on the preempt priority, subpriority and priority group setting:

```
NVIC_SetPriority(IRQ4_IRQn, NVIC_EncodePriority(PriorityGroup,
    PreemptPriority, SubPriority));
```

Additional details on these functions are described in Appendix G.

If you are programming in assembly language, to program priority-level registers, we can take advantage of the fact that the registers are byte addressable, making the coding easier. For example:

```
        ; Setting IRQ #4 priority to 0xC0
        LDR R0, =0xE000E400 ; External Interrupt Priority Reg starting
                            ; address
```

```
LDR   R1, =0xC0        ; Priority level
STRB  R1, [R0, #4]     ; Set IRQ #4 priority (Byte write)
```

In the Cortex-M3, the width of the interrupt priority configuration registers is specified by chip manufacturers. The minimum width is 3 bits, and the maximum width is 8 bits. In a CMSIS compliant device driver, the width of a priority level register is specified by __NVIC_PRIO_BITS. You can determine the implemented width by writing 0xFF to one of the priority configuration registers and reading it back. For example, you can do it in assembly with the following code:

```
; Determine the implemented priority width
LDR   R0,=0xE000E400 ; Priority Configuration register for
                     ; external interrupt #0
LDR   R1,=0xFF
STRB  R1,[R0]        ; Write 0xFF (note : byte size write)
LDRB  R1,[R0]        ; Read back (e.g. 0xE0 for 3-bits)
RBIT  R2, R1         ; Bit reverse R2 (e.g. 0x07000000 for
                     ; 3-bits)
CLZ   R1, R2         ; Count leading zeros (e.g. 0x5 for 3-bits)
MOV   R2, #8
SUB   R2, R2, R1     ; Get implemented width of priority
                     ; (e.g. 8-5=3 for 3-bits)
MOV   R1, #0x0
STRB  R1,[R0]        ; Restore to reset value (0x0)
```

If your application needs to be portable, it is best to use priority levels 0x00, 0x20, 0x40, 0x60, 0x80, 0xA0, 0xC0, and 0xE0 only. This is because all Cortex-M3 devices have these priority levels.

Do not forget to set up the priority for system exceptions and fault handler exceptions as well. If it is necessary for some of the important interrupts to have higher priority than other system exceptions or fault handlers, you will need to reduce the priority level of these system exceptions and fault handlers so that the important interrupts can preempt these handlers.

11.1.4 Enable the Interrupt

After the vector table and interrupt priority are set up, it's time to enable the interrupt. However, two steps might be required before you actually enable the interrupt:

1. If the vector table is located in a memory region that is write buffered, a Data Synchronization Barrier (DSB) instruction might be needed to ensure that the vector table memory is updated. In

---

ACCESSING NVIC INTERRUPT REGISTERS

For best software compatibility, CMSIS core peripheral access functions should be used for accessing the NVIC registers including interrupt configurations. Details of the CMSIS core peripheral access function are covered in Appendix G.

You can also develop your own NVIC interrupt control function if necessary; selecting the right transfer size can make your program development easier. For the Cortex-M3 processor, most registers in the NVIC can be accessed using word, half word, or byte transfers. For example, priority-level registers are best programmed with byte transfers. In this way, there is no need to worry about accidentally changing the priority of other exceptions. However, this method will not work with Cortex-M0 because the NVIC registers in Cortex-M0 only accept word size transfers.

most cases, the memory write should be completed within a few clock cycles. However, if your software needs to be portable between different ARM processors, this step ensures that the core will get the updated vector if the interrupt takes place immediately after being enabled.

2. An interrupt might already be pended or asserted beforehand, so it might be needed to clear the pending status. For example, signal glitches during power-up might have accidentally triggered some interrupt generation logic. In addition, in some peripherals such as a universal asynchronous receiver/transmitter (UART), noise from the UART receiver before connection might be mistaken as data and can cause an interrupt to be pended. Therefore, it can be safer to check and clear the pending status of an interrupt before enabling it. Depending on the peripheral design, the peripheral might also need some reinitialization if the pending status was set already.

Inside the NVIC, two separate register addresses are used for enabling and disabling interrupts. This duality ensures that each interrupt can be enabled or disabled without affecting or losing the other interrupt enable status. Otherwise, through software-based READ-MODIFY-WRITE, changes in enable register status carried out by interrupt handlers could be lost. To set an enable, the software needs to compute the correct bit location in the SETEN registers in the NVIC and write 1 to it. Similarly, to clear an interrupt, the software needs to write a 1 to the corresponding bit in the CLREN registers:

For users of CMSIS compliant driver libraries, the interrupt enable/disable feature can be accessed by the "*NVIC_EnableIRQ*" and "*NVIC_DisableIRQ*" functions. For example:

```
NVIC_EnableIRQ(UART1_IRQn);   // Enable UART#1 interrupt
                              // UART1_IRQn is MCU specific and is defined
                              // in the device driver library

NVIC_DisableIRQ(UART1_IRQn); // Disable UART#1 interrupt
```

Details of these functions are described in Appendix G.

Assembly language users can create an assembly function to carry out the same operation:

```
      ; A subroutine to enable an IRQ based on IRQ number
EnableIRQ
      ; Input R0 = IRQ number
      PUSH    {R0-R2, LR}
      AND.W   R1, R0, #0x1F  ; Generate enable bit pattern for
                             ; the IRQ
      MOV     R2, #1
      LSL     R2, R2, R1     ; Bit pattern = (0x1 << (N & 0x1F))
      AND.W   R1, R0, #0xE0  ; Generate address offset if IRQ number
                             ; is above 31
      LSR     R1, R1, #3     ; Address offset = (N/32)*4 (Each word
                             ; has 32 IRQ enable)
      LDR     R0,=0xE000E100 ; SETEN register for external interrupt
                             ; #31-#0
      STR R2, [R0, R1]       ; Write bit pattern to SETEN register
      POP     {R0-R2, PC}    ; Restore registers and Return
```

Likewise, we can write another subroutine for disabling IRQ:

```
      ; A subroutine to disable an IRQ based on IRQ number
DisableIRQ
      ; Input R0 = IRQ number
```

```
    PUSH  {R0-R2, LR}
    AND.W R1, R0, #0x1F   ; Generate Disable bit pattern for
                          ; the IRQ
    MOV   R2, #1
    LSL   R2, R2, R1      ; Bit pattern = (0x1 << (N & 0x1F))
    AND.  W R1, R0, #0xE0 ; Generate address offset if IRQ number
                          ; is above 31
    LSR   R1, R1, #3      ; Address offset = (N/32)*4 (Each word
                          ; has 32 IRQ enable)
    LDR   R0,=0xE000E180  ; CLREN register for external interrupt
                          ; #31-#0
    STR   R2, [R0, R1]    ; Write bit pattern to CLREN register
    POP   {R0-R2, PC}     ; Restore registers and Return
```

Similar subroutines can be developed for setting and clearing IRQ pending status registers.

11.2 EXCEPTION/INTERRUPT HANDLERS

In the Cortex-M3, interrupt handlers can be programmed completely in C, whereas in ARM7, an assembly handler is commonly used to ensure that all registers are saved, and in cases of systems with nested interrupt support, the processor needs to switch to a different mode to prevent losing information. These steps are not required in the Cortex-M3, making programming much easier.

In C language, an interrupt handler could be like

```
void UART1_Handler(void) {
   ... // processing task for the peripheral
   return;
}
```

For users of the CMSIS compliant device driver library, the interrupt handler name should match the interrupt handler name defined by the Microcontroller Unit (MCU) vendor to ensure that the vector is set up in the vector table correctly. You can find the handler function name in the vector table inside the startup codes. For example, for a Keil Microcontroller Development Kit user, the file is *startup_<device>.s*.

For users of ARM RealView Compilers or the Keil Microcontroller Development Kit, for clarity, you can add the optional __irq keyword. For example:

```
__irq void UART1_Handler(void) {
   ... // process IRQ request for the peripheral
   ... // Deassert IRQ request in peripheral
   return;
}
```

In assembler, a simple exception handler might look like this:

```
irq1_handler
    ; Process IRQ request
    ...
    ; Deassert IRQ request in peripheral
    ...
    ; Interrupt return
    BX  LR
```

The deassertion of an IRQ inside the interrupt service routine depends on the peripheral design. If the peripheral generates IRQs in the form of pulses, this step is not required. With the Cortex-M3, if a peripheral generates IRQs in the form of pulses, the NVIC can store the request as a pending request status. Once the processor enters the exception handler, the pending status is cleared automatically. This is different from traditional ARM processors that a peripheral has to maintain its IRQ until it is served because the interrupt controllers designed for previous ARM cores like ARM7TDMI do not have the pending memory.

In some cases, where the peripheral can generate multiple IRQs in a short period, the deassertion of the IRQ in the peripheral might have to be done conditionally to ensure that no requests are missed.

In many cases, the interrupt handler requires more than R0–R3 and R12 to process the interrupt, so we might need to save some other registers as well. For C language users, there is no need to worry about this, as the C function saves additional registers automatically if required. For assembly language users, their interrupt handlers have to perform stack PUSH and POP to ensure the values of R4–R11 are preserved.

The following example saves all registers that are not saved during the stacking process, but if some of the registers are not used by the exception handler, they can be omitted from the saved register list:

```
irq1_handler
    PUSH {R4-R11, LR} ; Save all registers that are not saved
                      ; during stacking
    ; Process IRQ request
    ...
    ; Deassert IRQ request in peripheral (optional)
    ...
    POP {R4-R11, PC}  ; Restore registers and Interrupt return
```

Because POP is one of the instructions that can start interrupt returns, we can combine the register restore and interrupt return in the same instruction.

Depending on the design of a peripheral, it might be necessary for an exception handler to program the peripheral to deassert the exception request. If the exception request from the peripheral to the NVIC is a pulse signal, then there is no need for the exception handler to clear the exception request. Otherwise, the exception handler needs to clear the exception request so that it won't be pending again immediately after exception exits. In traditional ARM processors, a peripheral has to maintain its IRQ until it is served because the interrupt controllers designed for previous ARM cores do not have the pending memory.

With the Cortex-M3, if a peripheral generates IRQs in the form of pulses, the NVIC can store the request as a pending request status. Once the processor enters the exception handler, the pending status is cleared automatically. In this way, the exception handler does not have to program the peripheral to clear the IRQ.

11.3 SOFTWARE INTERRUPTS

There are various ways to trigger an interrupt:

- External interrupt input
- Setting an interrupt pending register in the NVIC (see Chapter 8)
- Via the Software Trigger Interrupt register (STIR) in the NVIC (see Chapter 8)

In most cases, some of the interrupts are unused and can be used as software interrupts. Software interrupts can work similar to supervisor call (SVC), allowing accesses to system services. However, by default, user programs cannot access the NVIC; they can only access the NVIC's STIR if the USERSETMPEND bit in the NVIC Configuration Control register is set (see Table D.18 in Appendix D).

Unlike the SVC, software interrupts are not precise. In other words, the interrupt preemption does not necessarily happen immediately, even when there is no blocking from Interrupt Mask registers or other interrupt service routines. As a result, if the instruction immediately following the write to the NVIC STIR depends on the result of the software interrupt, the operation could fail because the software interrupt could invoke after the instruction is executed.

To solve this problem, use the DSB instruction. For example, users of CMSIS compliant device driver libraries can use the following code:

```
NVIC_SetPendingIRQ(SOFTWARE_INTERRUPT_NUMBER);
__DSB();
```

For assembly language users:

```
MOV   R0, #SOFTWARE_INTERRUPT_NUMBER
LDR   R1,=0xE000EF00  ; NVIC Software Interrupt Trigger
                      ; Register address
STR   R0, [R1]        ; Trigger software interrupt
DSB                   ; Data synchronization barrier
...
```

However, there is still another possible problem. If the Interrupt Mask register is set or if the program code generating the software interrupt is an exception handler itself, there could be a chance that the software interrupt cannot execute. Therefore, the program code generating the software interrupt should check to see whether the software interrupt has been executed. This can be done by having a software flag set by the software interrupt handler.

Finally, setting USERSETMPEND can lead to another problem. After this is set, user programs can trigger any software interrupt except system exceptions. As a result, if the USERSETMPEND is used and the system contains untrusted user programs, exception handlers need to check whether the exception is allowed because it could have been triggered from user programs. Ideally, if a system contains untrusted user programs, it is best to provide system services only via SVC.

11.4 EXAMPLE OF VECTOR TABLE RELOCATION

In Chapter 7, we mentioned that the starting vector table should contain a reset vector, an NMI vector, and a hard fault vector because the NMI and hard fault handler can take place without any exception enabling. After the program starts, we can then relocate the vector table to a different place in the SRAM if necessary. In most simple applications, there is no need to relocate the vector table.

If it is necessary to relocate the vector table, then the following steps would be required:

• *Reserve a memory space for the new vector table*: You might need to use linker scripts to reserve the memory space. The vector table address should be aligned to the vector table size, extended to the next larger power of 2.

- *Copy the existing vector table to the new vector table*: Before relocating the vector table, you need to ensure that the new vector table contains valid vector entries for all required exceptions including NMI, hard fault, and all enabled exceptions.
- *Write the new exception vector into the new vector table and write to Vector Table Offset Register to relocate the vector table.*

An example of relocating the vector table is covered in Chapter 8. In the following assembly example, we demonstrate reservation of memory space for the vector table in the beginning of SRAM and then the other data variables following it:

```
STACK_TOP       EQU 0x20002000      ; constant for the SP starting value
NVIC_SETEN      EQU 0xE000E100      ; NVIC Interrupt Set Enable Registers
                                    ; base address
NVIC_VECTTBL    EQU 0xE000ED08      ; Vector Table Offset Register
NVIC_AIRCR      EQU 0xE000ED0C      ; Application Interrupt and Reset
                                    ; Control Register
NVIC_IRQPRI     EQU 0xE000E400      ; Interrupt Priority Level register

        AREA  | Header Code   |, CODE
        DCD   STACK_TOP        ; SP initial value
        DCD   Start            ; Reset vector
        DCD   Nmi_Handler      ; NMI handler
        DCD   Hf_Handler       ; Hard fault handler
        ENTRY
Start ; Start of main program
        ; initialize registers
        MOV r0, #0     ; initialize registers
        MOV r1, #0
        ...

        ; Copy old vector table to new vector table
        LDR   r0,=0
        LDR   r1,=VectorTableBase
        LDMIA r0!,{r2-r5}      ; Copy 4 words
        STMIA r1!,{r2-r5}

        DSB    ; Data synchronization barrier.

        ; Set vector table offset register
        LDR   r0,=NVIC_VECTTBL
        LDR   r1,=VectorTableBase
        STR   r1,[r0]

        ...
        ; Setup Priority group register
        LDR   r0,=NVIC_AIRCR
        LDR   r1,=0x05FA0500 ; Priority group 5
        STR   R1,[r0]

        ; Setup IRQ 0 vector
        MOV   r0, #0              ; IRQ#0
        LDR   r1, =Irq0_Handler
        BL    SetupIrqHandler
```

```
    ; Setup priority
    LDR    r0,=NVIC_IRQPRI
    LDR    r1,=0xC0        ; IRQ#0 priority
    STRB   r1,[r0,#0]      ; Set IRQ0 priority at offset=0.
        ; Note : Byte store
        ;(IRQ#1 will have offset = 1)
    DSB ; Data synchronization barrier. Make sure
        ; everything ready before enabling interrupt
    MOV  r0, #0      ; select IRQ#0
    BL   EnableIRQ


    ...
    ;------------------------
    ; functions
SetupIrqHandler
    ; Input R0 = IRQ number
    ; R1 = IRQ handler
    PUSH   {R0, R2, LR}
    LDR    R2,=NVIC_VECTTBL ; Get vector table offset
    LDR    R2,[R2]
    ADD    R0, #16          ; Exception number = IRQ number + 16
    LSL    R0, R0, #2       ; Times 4 (each vector is 4 bytes)
    ADD    R2, R0           ; Find vector address
    STR    R1,[R2]          ; store vector handler
    POP    {R0, R2, PC}     ; Return
EnableIRQ
    ; Input R0 = IRQ number
    PUSH   {R0 - R3, LR}
    AND    R1, R0, #0x1F ; Get lower 5 bit to find bit pattern
    MOV    R2, #1
    LSL    R2, R2, R1     ; Bit pattern in R2
    BIC    R0, #0x1F
    LSR    R0, #3         ; word offset. (IRQ number can be
                          ; higher than 32)
    LDR    R1, =NVIC_SETEN
    STR    R2,[R1, R0]    ; Set enable bit
    POP    {R0 - R3, PC} ; Return
    ;------------------------
    ; Exception handlers
Hf_Handler
    ...      ; insert your code here
    BX   LR ; Return
Nmi_Handler
    ...      ; insert your code here
    BX   LR ; Return
Irq0_Handler
    ...      ; insert your code here
    BX   LR ; Return
    ;------------------------
    AREA | Header Data|, DATA
    ALIGN 4
    ; Relocated vector table
VectorTableBase SPACE 256 ; Number of bytes
```

```
VectorTableEnd          ; (256 / 4 = up to 64 exceptions)
MyData1    DCD 0         ; Variables
MyData2    DCD 0

           END           ; End of file
```

This is a slightly long example. Let's start from the end, the data region, first.

In the data memory region (almost the end of the program), we define a space of 256 bytes as a vector table (SPACE 256). This allows up to 64 exception vectors to be stored here. You might want to change the size if you want less or more space for the vector table. The other software variables follow the vector table space, so the variable MyData1 is now in address 0x20000100.

At the beginning of the code, we defined a number of address constants for the rest of the program. So, instead of using numbers, we can use these constant names to make the program easier to understand.

The initial vector table now contains the reset vector, the NMI vector, and the hard fault handler vector. The preceding example code illustrates how to set up the exception vectors and does not contain actual NMI, hard fault, or IRQ handlers. Depending on the actual application, these handlers will have to be developed. The example uses branch with exchange state (BX) Link register (LR) as the exception return, but that could be replaced by other valid exception return instructions.

After the initialization of registers, we copy the vector handlers to the new vector table in the SRAM. This is done by one multiple load and one multiple store instruction. If more vectors need to be copied, we can simply add extra load/store multiple instructions or increase the number of words to be copied for each pair of load and store instructions.

After the vector table is ready, we can relocate the vector table to the new one in the SRAM. However, to ensure that the transfer of the vector handler is complete, the DSB instruction is used.

We then need to set up the rest of the interrupt setting. The first one is the priority group setup. This needs to be done only once. In the example, two subroutines called SetupIrqHandler and EnableIRQ have been developed to make it easier to set up interrupts. Using the same code and simply changing the NVIC_SETEN to NVIC_CLREN, we can also add a similar function called DisableIRQ. After the handler and priority level have been set up, the IRQ can then be enabled.

11.5 USING SVC

SVC is a common way to allow user applications to access the application programming interface (API) in an OS. This is because the user applications only need to know what parameters to pass to the OS; they don't need to know the memory address of API functions.

SVC instructions contain a parameter, which is 8-bit immediate data inside the instruction. The value is required for using the SVC instruction. For example:

```
SVC #3 ; Call system service number 3
```

The alternative syntax can also be used (without the "#"):

```
SVC 3 ; Call system service number 3
```

Inside the SVC handler, the parameter can be extracted back from the instruction by locating the executed SVC instruction from the stacked PC. To do this, the procedures illustrated in Figure 11.2 can be used.

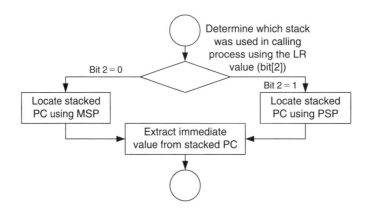

FIGURE 11.2

One Way to Extract the SVC Parameter.

Here's some simple assembly code to do this:

```
svc_handler
    TST     LR, #0x4        ; Test EXC_RETURN number in LR bit 2
    ITE     EQ              ; if zero (equal) then
    MRSEQ   R0, MSP         ; Main Stack was used, put MSP in R0
    MRSNE   R0, PSP         ; else, Process Stack was used, put PSP
                            ; in   R0
    LDR     R1,[R0,#24]     ; Get stacked PC from stack
    LDRB    R0,[R1,#-2]     ; Get the immediate data from the
                            ; instruction
    ; Now the immediate data is in R0
    ...
    BX  LR                  ; Return to calling function
```

Once the calling parameter of the SVC is determined, the corresponding SVC function can be executed. An efficient way to branch into the correct SVC service code is to use table branch instructions such as Table Branch Byte (TBB) and Table Branch Halfword (TBH). However, if the table branch instruction is used, unless it is certain that the SVC calling parameter contains a correct value, you should do a value check on the parameter to prevent invalid SVC calling from crashing the system.

Note that passing of parameters to the SVC handler and the return value from the SVC handler has to be carried out via stack frame. The reason for this is covered in the next section.

Because an SVC cannot request another SVC service via the exception mechanism, the SVC handler should directly call another SVC function (for example, BL).

11.6 SVC EXAMPLE: USE FOR TEXT MESSAGE OUTPUT FUNCTIONS

Previously we developed various subroutines for output functions. Sometimes it is not good enough to use BL to call the subroutines—for example, when the software code is running in nonprivileged access level and the text output I/O need privileged accesses. In these cases, we might want to use

SVC to act as an entry point for the output functions. For example, a user program can use SVC with different parameters to access different services:

```
LDR    R0,=HELLO_TXT
SVC    #0    ; Display string pointed to by R0
MOV    R0,#'A'
SVC    #1    ; Display character in R0
LDR    R0,=0xC123456
SVC    #2    ; Display hexadecimal value in R0
MOV    R0,#1234
SVC    #3    ; Display decimal value in R0
```

To use SVC, we might need to set up the SVC handler if the vector table is relocated to SRAM. We can modify the function that we have created to handle the interrupt (SetupIrqHandler function in previous section). The only difference is that this function takes an exception type as input (SVC is exception type 11). In addition, this time we have further optimized the code to use the 32-bit Thumb instruction features:

```
SetupExcpHandler    ; Setup vector in relocated vector table in SRAM
    ;Input R0 = Exception number
    ;       R1 = Exception handler
    PUSH    {R0, R2, LR}
    LDR     R2,=NVIC_VECTTBL ; Get vector table offset
    LDR     R2,[R2]
    STR.W   R1,[R2, R0, LSL #2] ; store vector handler in [R2+R0<<2]
    POP     {R0, R2, PC} ; Return
```

For *svc_handler*, the SVC calling number can be extracted as in the previous example, and the parameter passed to the SVC can be accessed by reading from the stack. In addition, the decision branches to reach various functions are added:

```
svc_handler
    TST     LR, #0x4        ; Test EXC_RETURN number in LR bit 2
    ITTEE   EQ              ; if zero (equal) then
    MRSEQ   R1, MSP         ; Main Stack was used, put MSP in R1
    MRSNE   R1, PSP         ; else, Process Stack was used, put PSP
                            ; in R1
    LDR     R0,[R1,#0]      ; Get stacked R0 from stack
    LDR     R1,[R1,#24]     ; Get stacked PC from stack
    LDRB    R1,[R1,#-2]     ; Get the immediate data from the
                            ; instruction
    ; Now the immediate data is in R1, input parameter is in R0
    PUSH    {LR}            ; Store LR to stack
    CBNZ    R1,svc_handler_1
    BL      Puts            ; Branch to Puts
    B       svc_handler_end
svc_handler_1
    CMP     R1,#1
    BNE     svc_handler_2
    BL      Putc            ; Branch to Putc
    B       svc_handler_end
```

```
svc_handler_2
      CMP     R1,#2
      BNE     svc_handler_3
      BL      PutHex          ; Branch to PutHex
      B       svc_handler_end
svc_handler_3
      CMP     R1,#3
      BNE     svc_handler_4
      BL      PutDec          ; Branch to PutDec
      B       svc_handler_end
svc_handler_4
      B       error           ; input not known
      . . .
svc_handler_end
      POP     {PC}            ; Return
```

The *svc_handler* code should be put close together with the outputting functions so that we can ensure that they are within the allowed branch range.

Notice that instead of the current contents of the register bank, the stacked register contents are used for parameter passing. This is because if a higher-priority interrupt takes place when the SVC is executed, the SVC starts immediately after other interrupt handlers (tail chaining), and the contents of R0–R3 and R12 might be changed by the executed interrupt handler. This is caused by the characteristic that unstacking is not carried out if there is tail chaining of interrupts. For example:

1. A parameter is put in R0.
2. SVC is executed at the same time as a higher-priority interrupt takes place.
3. Stacking is carried out, and R0–R3, R12, LR, PC, and xPSR are saved to the stack.
4. The interrupt handler is executed. R0–R3 and R12 can be changed by the handler. This is acceptable because these registers will be restored by hardware unstacking.
5. The SVC handler tail chains the interrupt handler. When SVC is entered, the contents in R0–R3 and R12 can be different from the value when SVC is called. However, the correct parameter is stored in the stack and can be accessed by the SVC handler.

MAKE THE MOST OF THE ADDRESSING MODES

From the code examples of the *SetupIrqHandler* and *SetupExcpHandler* routines, we find that the code can be shortened a lot if we use the addressing mode feature in the Cortex-M3. In *SetupIrqHandler*, the destination address of the IRQ vector is calculated, and then, the store is carried out:

```
SetupIrqHandler /* R0 = IRQ number, R1 = handler address */
  PUSH {R0, R2, LR}
  LDR  R2,=NVIC_VECTTBL ; Get vector table offset        ; Step 1
  LDR  R2,[R2]                                            ; Step 2
  ADD  R0, #16 ; Exception number = IRQ number + 16       ; Step 3
  LSL  R0, R0, #2    ; Times 4 (each vector is 4 bytes) : Step 4
  ADD  R2, R0        ; Find vector address                ; Step 5
  STR  R1,[R2]       ; store vector handler               ; Step 6
  POP  {R0, R2, PC}  ; Return
```

```
      In SetupExcpHandler, the operation Steps 4–6 are reduced to just one step:

SetupExcpHandler /* R0 = exception number, R1 = handler address */
        PUSH {R0, R2, LR}
        LDR   R2,=NVIC_VECTTBL      ; Get vector table offset
        LDR   R2,[R2]
        STR.W R1,[R2, R0, LSL #2] ; store vector handler in
                                  ; [R2+R0<<2]
        POP {R0, R2, PC}   ; Return
```

In general, we can reduce the number of instructions required if the data address is like one of these:

- Rn + (2^N) × Rm
- Rn +/– immediate_offset

For the *SetupIrqHandler* routine, the shortest code we can get is this:

```
SetupIrqHandler
        PUSH {R0, R2, LR}
        LDR   R2,=NVIC_VECTTBL      ; Get vector table offset   ; Step 1
        LDR   R2,[R2]                                           ; Step 2
        ADD   R2, #(16*4)           ; Get IRQ vector start      ; Step 3
        STR.W R1,[R2, R0, LSL #2] ; Store vector handler        ; Step 4
        POP   {R0, R2, PC}         ; Return
```

11.7 USING SVC WITH C

In most cases, an assembler handler code is needed for parameter passing to SVC functions. This is because the parameters should be passed by the stack, not by registers, as explained earlier. If the SVC handler is to be developed in C, a simple assembly wrapper code can be used to obtain the stacked register location and pass it on to the SVC handler. The SVC handler can then extract the SVC number and parameters from the stack pointer information. Assuming that the RealView Development Suite (RVDS) or Keil Microcontroller Development Kit for ARM (MDK-ARM) is used, the assembler wrapper can be implemented with an Embedded Assembler:

```
// Assembler wrapper for extracting stack frame starting location.
// Starting address of stack frame is put into R0 and then branch
// to the actual SVC handler.
__asm void svc_handler_wrapper(void)
{
 TST   LR, #4
 ITE   EQ
 MRSEQ R0, MSP
 MRSNE R0, PSP
 B __cpp(svc_handler)
} // No need to add return (BX LR) at the end of this wrapper
  // because return of svc_handler will return execution to where
  // SVC is called from
```

The rest of the SVC handler can then be implemented in C using R0 as input (stack frame starting location), which is used to extract the SVC number and passing parameters (R0–R3):

```
// SVC handler in C, with stack frame location as an input parameter
// used as a memory pointer to an array of arguments.
// svc_args[0] = R0 , svc_args[1] = R1
// svc_args[2] = R2 , svc_args[3] = R3
// svc_args[4] = R12, svc_args[5] = LR
// svc_args[6] = Return address (Stacked PC)
// svc_args[7] = xPSR
void svc_handler(unsigned int * svc_args)
{
  unsigned int svc_number;
  unsigned int svc_r0;
  unsigned int svc_r1;
  unsigned int svc_r2;
  unsigned int svc_r3;

  svc_number = ((char *) svc_args[6])[-2]; // Memory[(Stacked PC)-2]
  svc_r0 = ((unsigned long) svc_args[0]);
  svc_r1 = ((unsigned long) svc_args[1]);
  svc_r2 = ((unsigned long) svc_args[2]);
  svc_r3 = ((unsigned long) svc_args[3]);
  printf ("SVC number = %xn", svc_number);
  printf ("SVC parameter 0 = %x\n", svc_r0);
  printf ("SVC parameter 1 = %x\n", svc_r1);
  printf ("SVC parameter 2 = %x\n", svc_r2);
  printf ("SVC parameter 3 = %x\n", svc_r3);
  return;
}
```

Note that SVC cannot return results to the calling program in the same way as in normal C functions. Normal C functions return values by defining the function with a data type such as *unsigned int func( )* and use *return* to pass the return value, which actually puts the value in register R0. If an SVC handler put return values in register R0–R3 when exiting the handler, the register values would be overwritten by the unstacking sequence. Therefore, if an SVC has to return results to a calling program, it must directly modify the stack frame so that the value can be loaded into the register during unstacking.

To call an SVC inside a C program for ARM RVDS or Keil MDK-ARM, we can use the _ _*svc compiler* keyword. For example, if four variables are to be passed to an SVC function number 3, an SVC named *call_svc_3* can be declared as

```
void __svc(0x03) call_svc_3(unsigned long svc_r0, unsigned long
svc_r1, unsigned long svc_r2, unsigned long svc_r3);
```

This will then allow the C program code to call the SVC function by

```
int main(void)
{
  unsigned long p0, p1, p2, p3; // parameters to pass to SVC handler
  ...
  call_svc_3(p0, p1, p2, p3); // call SVC number 3, with parameters
                              // p0, p1, p2, p3 pass to the SVC
  ...
  return;
}
```

Detailed information on using the _ _*svc* keyword in RVDS or RealView C Compiler can be found in the *RVCT 4.0 Compilation Tools Compiler Reference Guide* [Ref. 8].

For users of the Gnu's Not Unix (GNU) tool chain, because there is no _ _*svc* keyword in GNU C Compiler (GCC), the SVC has to be accessed by an inline assembler. For example, if the SVC call number 3 is needed with one input variable and it returns one variable via register R0 (according to the *AAPCS* [Ref. 5], the first passing variable uses register R0), the following inline assembler code can be used to call the SVC:

```
int MyDataIn = 0x123;
__asm __volatile ("mov R0, %0\n"
                  "svc 3 \n" : "" : ""r" (MyDataIn) );
```

This inline assembler code can be broken down into the following parts, with input data specified by *r* (*MyDataIn*) and no output field (indicated as " " in the preceding code):

```
__asm ( assembler_code : output_list : input_list )
```

More examples using inline assembler in the GNU tool chain can be found in Chapter 19. For complete details on passing parameters to or from inline assembler, refer to the GNU tool chain documentation.

Advanced Programming Features and System Behavior

12

IN THIS CHAPTER

Running a System with Two Separate Stacks 201
Double-Word Stack Alignment 204
Nonbase Thread Enable 205
Performance Considerations 206
Lockup Situations 208
FAULTMASK 210

12.1 RUNNING A SYSTEM WITH TWO SEPARATE STACKS

One of the important features of ARMv7-M architecture is the capability to allow the user application stack to be separated from the privileged/kernel stack. If the optional Memory Protection Unit (MPU) is implemented, it could be used to block user applications from accessing kernel stack memory so that they cannot crash the kernel by memory corruption.

Typically, a robust system based on the Cortex™-M3 has the following properties:

- Exception handlers using Main Stack Pointer (MSP)
- Kernel code invoked by a System Tick (SYSTICK) exception at regular intervals, running in the privileged access level for task scheduling and system management
- User applications running as threads with the user access level (nonprivileged); these applications use Process Stack Pointer (PSP)
- Stack memory for kernel and exception handlers is pointed to by the MSP, and the stack memory is restricted to privileged accesses only, if the MPU is available
- Stack memory for user applications is pointed to by the PSP

Assume that the system memory has a Static Random Access Memory (SRAM) memory and a Memory Protection Unit (MPU), we could set up the MPU so that the SRAM is divided into two regions for user and privileged access (see Figure 12.1). Each region is used by application data, as well as by stack memory space. Since stack operation in the Cortex-M3 is full descending, the initial value of stack pointers needs to be pointed to the top of the regions.

201

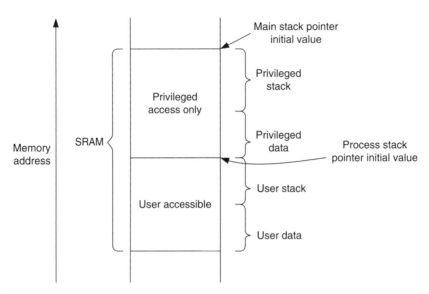

FIGURE 12.1

Example Memory Use with Privileged Data and User Application Data.

After power-up, only the MSP is initialized (by fetching address 0x0 in the power-up sequence). Additional steps are required to set up a completely robust two-stack system. For applications in assembly code, it can simply be

```
; Start at privileged level (this code locates in user
; accessible memory)
BL      MpuSetup           ; Setup MPU regions and enable memory
                           ; protection
LDR     R0,=PSP_TOP        ; Setup Process SP to top of process stack
MSR     PSP, R0
BL      SystickSetup       ; Setup Systick and systick exception to
                           ; invoke OS kernel at regular intervals
MOV     R0, #0x3           ; Setup CONTROL register so that user
                           ; program use PSP,
MSR     CONTROL, R0  ; and switch current access level to user
ISB                        ; Instruction Synchronization Barrier
B       UserApplicationStart   ; Now we are in user access
                           ; level. Start user code
```

This arrangement is fine for assembler, but for C programs, switching stack pointers in the middle of a C function can cause loss of local variables (because in C functions or subroutines, local variables may be put onto stack memory). *The Cortex-M3 Technical Reference Manual (TRM)* [Ref. 1] suggests that we use an interrupt service routine (ISR) like Supervisor Call (SVC) to invoke the kernel, and then change the stack pointer by modifying the EXC_RETURN value (see Figure 12.2).

In most cases, EXC_RETURN modification and stack switching are included in the operating system (OS). After the user application starts, the SYSTICK exception can be used regularly to invoke the OS for system management and possibly arrange context switching, if needed (see Figure 12.3).

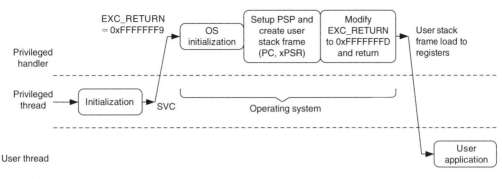

FIGURE 12.2

Initialization of Multiple Stacks in a Simple OS.

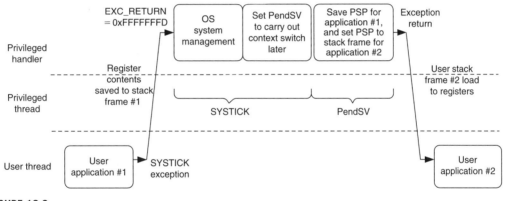

FIGURE 12.3

Context Switching in a Simple OS.

Note that context switching is carried out in PendSV (a low-priority exception) to prevent context switching at the middle of an interrupt handler.

However, many applications do not require an OS, but it is still helpful to use separate stacks for different sections of application code as a way to improve reliability. One possible way to handle this is to start Cortex-M3 with the MSP pointed to a process stack region. This way the initialization is done with the process stack region but using MSP. Before starting the user application, the following code is executed:

```
; Start at privileged level, MSP point to User stack
MpuSetup();              // Setup MPU regions and enable memory protection
SystickSetup();         // Setup Systick and systick exception for routine
                        // system management code
SwitchStackPointer();   // Call an assembly subroutine to switch SP
    /*; ------Inside SwitchStackPointer -----
    PUSH {R0, R1, LR}
```

```
        MRS R0, MSP            ; Save current stack pointer
        LDR R1, =MSP_TOP       ; Change MSP to new location
        MSR MSP, R1
        MSR PSP, R0            ; Store current stack pointer in PSP
        MOV R0, #0x3
        MSR CONTROL, R0        ; Switch to user mode, and use PSP as
                               ; current stack
        POP {R0, R1, PC}       ; Return
        ; ------ Back to C program -----*/
    ; Now we are in User mode, using PSP and the local variables
    ; still here
    UserApplicationStart();    // Start application code in user mode
```

12.2 DOUBLE-WORD STACK ALIGNMENT

In applications that conform to AAPCS[1], it is necessary to ensure that the stack pointer value at function entry should be aligned to the double word address. To achieve this requirement, the stacking address of registers at exception handling is adjusted accordingly. This is a configurable option on the Cortex-M3 processor. To enable this feature, the STKALIGN bit in the Nested Vectored Interrupt Controller (NVIC) Configuration Control register needs to be set (see Table D.18 in Appendix D). For example, if CMSIS compliant device driver is used in C language project

```
    SCB->CCR = SCB->CCR | 0x200;
```

If the project is in C but CMSIS is not used,

```
    #define NVIC_CCR *((volatile unsigned long *) (0xE000ED14))
    NVIC_CCR = NVIC_CCR | 0x200; /* Set STKALIGN in NVIC */
```

This can also be done in assembly language

```
    LDR     R0,=0xE000ED14       ; Set R0 to be address of NVIC CCR
    LDR     R1, [R0]
    ORR.W   R1, R1, #0x200       ; Set STKALIGN bit
    STR     R1, [R0]             ; Write to NVIC CCR
```

When the STKALIGN bit is set during exception stacking, bit 9 of the stacked xPSR (combined Program Status Register) is used to indicate whether a stack pointer adjustment has been made to align the stacking. When unstacking, the stack pointer (SP) adjustment checks bit 9 of the stacked xPSR and adjusts the SP accordingly.

To prevent stack data corruption, the STKALIGN bit must not be changed within an exception handler. This can cause a mismatch of stack pointer location before and after the exception.

This feature is available from Cortex-M3 revision 1 onward. Early Cortex-M3 products based on revision 0 do not have this feature. In Cortex-M3 revision 2, this feature is enabled by default whereas in revision 1, this needs to be turned on by software.

This feature should be used if the AAPCS conformation is required.

---

[1]Procedure Call Standard for the ARM Architecture (AAPCS) [Ref. 5]. An advisory note has been published on the ARM web site regarding SP alignment and AAPCS; see http://infocenter.arm.com/help/topic/com.arm.doc.ihi0046a/IHI0046A_ABI_Advisory_1.pdf.

12.3 **NONBASE THREAD ENABLE**

In the Cortex-M3, it is possible to switch a running interrupt handler from privileged level to user access level. This is needed when the interrupt handler code is part of a user application and should not be allowed to have privileged access. This feature is enabled by the Nonbase Thread Enable (NONBASETHRDENA) bit in the NVIC Configuration Control register.

> ### USE THIS FEATURE WITH CAUTION
> Because of the need to manually adjust the stack and modify the stacked data, this feature should be avoided in normal application programming. If it is necessary to use this feature, it must be done very carefully, and the system designer must ensure that the interrupt service routine is terminated correctly. Otherwise, it could cause some interrupts with the same or lower priority levels to be masked.

To use this feature, an exception handler redirection is involved. The vector in the vector table points to a handler running in privileged mode but located in user mode accessible memory

```
redirect_handler
    PUSH    {LR}
    SVC     0          ; A SVC function to change from privileged to
                       ; user mode
    BL User_IRQ_Handler
    SVC     1          ; A SVC function to change back from user to
                       ; privileged mode
    POP     {PC}       ; Return
```

The SVC handler is divided into three parts as follows:

- Determine the parameter when calling SVC.
- SVC service #0 enables the NONBASETHRDENA, adjusts the user stack and EXC_RETURN value, and returns to the redirect handler in user mode, using the process stack.
- SVC service #1 disables the NONBASETHRDENA, restores the user stack pointer position, and returns to the redirect handler in privileged mode, using the main stack.

```
svc_handler
    TST     LR, #0x4        ; Test EXC_RETURN bit 2
    ITE     EQ              ; if zero then
    MRSEQ   R0, MSP         ; Get correct stack pointer to R0
    MRSNE   R0, PSP
    LDR     R1,[R0, #24]    ; Get stacked PC
    LDRB    R0,[R1, #-2]    ; Get parameter at stacked PC - 2
    CBZ     r0, svc_service_0
    CMP     r0, #1
    BEQ     svc_service_1
    B.W     Unknown_SVC_Request

svc_service_0           ; Service to switch handler from privileged mode to
                        ; user mode
    MRS     R0, PSP         ; Adjust PSP
```

```
        SUB     R0,  R0, #0x20     ; PSP = PSP - 0x20
        MSR     PSP,  R0
        MOV     R1, #0x20                  ; Copy stack frame from main stack to
                                           ; process stack

svc_service_0_copy_loop
        SUBS    R1, R1, #4
        LDR     R2,[SP, R1]
        STR     R2,[R0, R1]
        CMP     R1, #0
        BNE     svc_service_0_copy_loop
        STRB    R1,[R0, #0x1C]     ; Clear stacked IPSR of user stack to 0
        LDR     R0, =0xE000ED14    ; Set Non-base thread enable in CCR
        LDR     r1,[r0]
        ORR     r1, #1
        STR     r1,[r0]
        ORR     LR, #0xC           ; Change LR to return to thread, using PSP
        BX      LR

svc_service_1                      ; Service to switch handler back from user mode to
                                   ; privileged mode
        MRS     R0, PSP            ; Update stacked PC in privileged
                                   ; stack so that it
        LDR     R1,[R0, #0x18]     ; return to the instruction after 2nd
                                   ; SVC in redirect
        STR     R1,[SP, #0x18]     ; handler
        MRS     R0, PSP            ; Adjust PSP back to what it was
                                   ; before 1st SVC
        ADD     R0, R0, #0x20
        MSR     PSP, R0
        LDR     R0, =0xE000ED14    ; Clear Non-base thread enable in CCR
        LDR     r1,[r0]
        BIC     r1, #1
        STR     r1,[r0]
        BIC     LR, #0xC           ; Return to handler mode, using main
                                   ; stack
        BX      LR
```

The SVC services are used because the only way you can change the Interrupt Status register (IPSR) is via an exception return. Other exceptions, such as software-triggered interrupts, could be used, but they are not recommended because they are imprecise and could be masked, which means that there is a possibility that the required stack copying and switch operation is not carried out immediately. The sequence of the code is illustrated in Figure 12.4, which shows the stack pointer changes and the current exception priority.

In this figure, the manual adjustment of the PSP inside the SVC services is highlighted by circles indicated by dotted lines.

12.4 PERFORMANCE CONSIDERATIONS

To get the best out of the Cortex-M3, a few aspects need to be considered. First, we need to avoid memory wait states. During the design stage of the microcontroller or SoC, the designer should

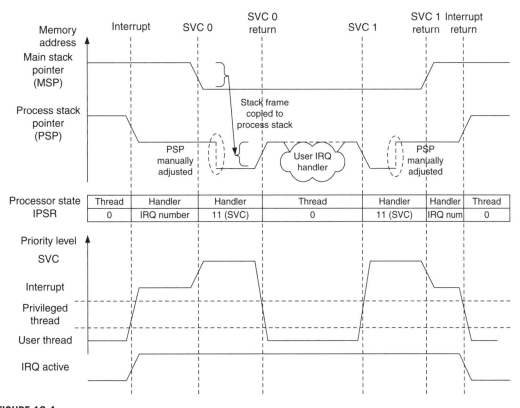

FIGURE 12.4

Operation of NONBASETHRDENA.

optimize the memory system design to allow instruction and data accesses to be carried out at the same time, and use 32-bit memories, if possible. For developers, the memory map should be arranged so that program code is executed from the code region and the majority of data accesses is done via the system bus. This way data accesses can be carried out at the same time as instruction fetches.

Second, the interrupt vector table should also be put into the code region, if possible. Thus, vector fetch and stacking can be carried out at the same time. If the vector table is located in the SRAM, extra clock cycles might result in interrupt latency because both vector fetch and stacking could share the same system bus (unless the stack is located in the code region, which uses a D-Code bus).

If possible, avoid using unaligned transfers. An unaligned transfer might take two or more Advanced High-Performance Bus (AHB) transfers to complete and will slow program performance, so plan your data structure carefully. In assembly language with ARM tools, you can use the ALIGN directive to ensure that a data location is aligned.

Most of you might be using C language for development, but for those who are using assembly, you can use a few tricks to speed up parts of the program.

1. Use memory access instruction with offset. When multiple memory locations in a small region are to be accessed, instead of writing

```
LDR     R0, =0xE000E400      ; Set interrupt priority #3,#2,#1,#0
LDR     R1, =0xE0C02000      ; priority levels
STR     R1,[R0]
LDR     R0, =0xE000E404      ; Set interrupt priority #7,#6,#5,#4
LDR     R1, =0xE0E0E0E0      ; priority levels
STR     R1,[R0]
```

you can reduce the program code to the following:

```
LDR     R0, =0xE000E400      ; Set interrupt priority #3,#2,#1,#0
LDR     R1, =0xE0C02000      ; priority levels
STR     R1,[R0]
LDR     R1,=0xE0E0E0E0       ; priority levels
STR     R1,[R0,#4]           ; Set interrupt priority #7,#6,#5,#4
```

The second store uses an offset of the first address and hence reduces the number of instructions.

2. Combine multiple memory accesses into Load/Store Multiple instructions (LDM/STM). The preceding example can be further reduced by using STM instruction as follows:

```
LDR     R0,=0xE000E400       ; Set interrupt priority base
LDR     R1,=0xE0C02000       ; priority levels #3,#2,#1,#0
LDR     R2,=0xE0E0E0E0       ; priority levels #7,#6,#5,#4
STMIA   R0, {R1, R2}
```

3. Use IF-THEN (IT) instruction blocks to replace small conditional branches. Since the Cortex-M3 is a pipelined processor, a branch penalty happens when a branch operation is taken. If the conditional branch operation is used to skip a few instructions, this can be replaced by the IT instruction block, which might save a few clock cycles.

4. If an operation can be carried out by either two Thumb® instructions or a single Thumb-2 instruction, the Thumb-2 instruction method should be used because it gives a shorter execution time, despite the fact that the memory size is the same.

12.5 LOCKUP SITUATIONS

When an error condition occurs, the corresponding fault handler will be triggered. If another fault takes place inside the usage fault/bus fault/memory management fault handler, the hard fault handler will be triggered. However, what if we get another fault inside the hard fault handler? In this case, a lockup situation will take place (see Figure 12.5).

12.5.1 What Happens During Lockup?

During lockup, the program counter will be forced to 0xFFFFFFFX and will keep fetching from that address. In addition, an output signal called LOCKUP from the Cortex-M3 will be inserted to indicate the situation. Chip designers might use this signal to trigger a reset at the system reset generator.

Lockup can take place when

- Faults occur inside the hard fault handler (double fault)
- Faults occur inside the nonmaskable interrupt (NMI) handler
- Bus faults occur during the reset sequence (initial SP or program counter (PC) fetch)

For double-fault situations, it is still possible for the core to respond to an NMI and execute the NMI handler. But after the handler completes, it will return to the lockup state, with the program counter restored to 0xFFFFFFFX. In this case, the system locks up and the current priority level is held at −1. If an NMI occurs, the processor will still preempt and execute the NMI handler because the NMI has a higher priority (−2) than the current priority level (−1). When the NMI is complete and returns to the lockup state, the current exception priority is returned to −1.

Normally, the best way to exit a lockup is to perform a reset. Alternatively, for a system with a debugger attached, it is possible to halt the core, change the PC to a different value, and start the program execution from there. In most cases this might not be a good idea, since a number of registers, including the interrupt system, might need reinitialization before the system can be returned to normal operation.

You might wonder why we do not simply reset the core when a lockup takes place. You might want to do that in a live system, but during software development, we should first try to find out the cause of the problem. If we reset the core immediately, we might not be able to analyze what went wrong because registers will be reset and hardware status will be changed. In most Cortex-M3 microcontrollers, a watchdog timer can be used to reset the core if it enters the lockup state.

Note that a bus fault that occurs during stack when entering a hard fault handler or NMI handler does not cause lockup, but the bus fault handler will be pended.

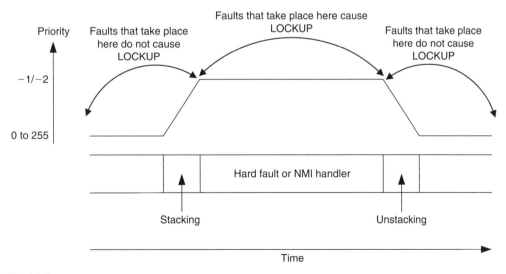

FIGURE 12.5

Only a Fault Occurring During a Hard Fault or NMI Handler Will Cause Lockup.

12.5.2 Avoiding Lockup

It is important to take extra care to prevent lockup problems when you're developing an NMI or hard fault handler. For example, we can avoid unnecessary stack accesses in a hard fault handler unless we know that the memory is functioning correctly and the stack pointer is still valid. In developing complex systems, one of the possible causes of a bus fault or memory fault is stack pointer corruption. If we start the hard fault handler with something like this

```
hard_fault_handler
    PUSH    {R4-R7,LR}      ; Bad idea unless you are sure that the
                            ; stack is safe to use!
    . . .
```

and if the fault was caused by a stack error, we could enter lockup in our hard fault handler straight away. In general, when programming hard fault, bus fault, and memory management fault handlers, it might be worth checking whether the stack pointer is in a valid range before we carry out more stack operations. For coding NMI handlers, we can try to reduce risk caused by stack operation by using R0–R3 and R12 only, since they are already stacked.

One approach for developing hard fault and NMI handlers is to carry out only the essential tasks inside the handlers, and the rest of the tasks, such as error reporting, can be pended using a separate exception, such as PendSV or a software interrupt. This helps to ensure that the hard fault handler or NMI is small and robust.

Furthermore, we should ensure that the NMI and hard fault handler code will not try to use SVC instructions. Since SVC always has lower priority than hard fault and NMI, using SVC in these handlers will cause lockup. This might look simple, but when your application is complex and you call functions from different files in your NMI and hard fault handler, you might accidentally call a function that contains an SVC instruction. Therefore, before you develop your software, you need to carefully plan the SVC implementation.

12.6 FAULTMASK

FAULTMASK is used to escalate a configurable fault handler (bus fault, usage fault or memory management fault) to hard fault level without the need to invoke hard fault by a real fault. This allows the configurable fault handler to pretend to be the hard fault handler. By doing this, the fault handler can have the ability to

1. Mask bus fault by setting HFHFNMIGN in Configuration Control register. It can be used to probe the bus system without causing lockup. For example, for checking if a bus bridge is working correctly.
2. Bypass the MPU. This allows the fault handler to access an MPU protected memory location without reprogramming the MPU just to carry out a few transfers to fix faults.

The FAULTMASK usage is different from PRIMASK. PRIMASK is generally used in timing critical code, but it doesn't have the ability to mask bus fault or bypass MPU. With PRIMASK set, all configurable faults will be escalated to hard fault handler. FAULTMASK is used to allow a configurable fault handler to solve memory-related problems by using features normally only available for a hard fault handler. However, when FAULTMASK is set, faults such as incorrect undefined instruction, or using SVC in the wrong priority level, can still cause lockup.

The Memory Protection Unit

IN THIS CHAPTER

Overview 211
MPU Registers... 212
Setting Up the MPU 218
Typical Setup 225

13.1 OVERVIEW

The Cortex™-M3 design includes an optional Memory Protection Unit (MPU). Including the MPU in the microcontrollers or system-on-chip (SoC) products provides memory protection features, which can make the developed products more robust. The MPU needs to be programmed and enabled before use. If the MPU is not enabled, the memory system behavior is the same as though no MPU is present.

The MPU can improve the reliability of an embedded system by

- Preventing user applications from corrupting data used by the operating system
- Separating data between processing tasks by blocking tasks from accessing others' data
- Allowing memory regions to be defined as read-only so that vital data can be protected
- Detecting unexpected memory accesses (for example, stack corruption)

In addition, the MPU can also be used to define memory access characteristics such as caching and buffering behaviors for different regions.

The MPU sets up the protection by defining the memory map as a number of regions. Up to eight regions can be defined, but it is also possible to define a default background memory map for privileged accesses. Accesses to memory locations that are not defined in the MPU regions or not permitted by the region settings will cause the memory management fault exception to take place.

MPU regions can be overlapped. If a memory location falls on two regions, the memory access attributes and permission will be based on the highest-numbered region. For example, if a transfer address is within the address range defined for region 1 and region 4, the region 4 settings will be used.

13.2 MPU REGISTERS

The MPU contains a number of registers. The first one is the MPU Type register. The MPU Type register can be used to determine whether the MPU is fitted. If the DREGION field is read as 0, the MPU is not implemented (see Table 13.1).

The MPU is controlled by a number of registers. The first one is the MPU Control register (see Table 13.2). This register has three control bits. After reset, the reset value of this register is zero, which disables the MPU. To enable the MPU, the software should set up the settings for each MPU regions, and then, set the ENABLE bit in the MPU Control register.

By using PRIVDEFENA and if no other regions are set up, privileged programs will be able to access all memory locations, and only user programs will be blocked. However, if other MPU regions are programmed and enabled, they can override the background region. For example, for two systems with similar region setups but only one with PRIVDEFENA set to 1 (the right-hand side in Figure 13.1), the one with PRIVDEFENA set to 1 will allow privileged access to background regions.

Setting the enable bit in the MPU Control register is usually the last step in the MPU setup code. Otherwise, the MPU might generate faults by accident before the region configuration is done. In some situations, it might be worth clearing the MPU Enable at the start of the MPU configuration routine to make sure that the MPU faults won't be triggered by accident during setup of MPU regions.

Table 13.1 MPU Type Register (0xE000ED90)

Bits	Name	Type	Reset Value	Description
23:16	IREGION	R	0	Number of instruction regions supported by this MPU; because ARMv7-M architecture uses a unified MPU, this is always 0
15:8	DREGION	R	0 or 8	Number of regions supported by this MPU; in the Cortex-M3, this is either 0 (MPU not present) or 8 (MPU present)
0	SEPARATE	R	0	This is always 0, as the MPU is unified

Table 13.2 MPU Control Register (0xE000ED94)

Bits	Name	Type	Reset Value	Description
2	PRIVDEFENA	R/W	0	Privileged default memory map enable; when set to 1 and if the MPU is enabled, the default memory map will be used for privileged accesses as a background region. If this bit is not set, the background region is disabled and any access not covered by any enabled region will cause a fault.
1	HFNMIENA	R/W	0	If set to 1, it enables the MPU during the hard fault handler and nonmaskable interrupt (NMI) handler; otherwise, the MPU is not enabled (bypassed) for the hard fault handler and NMI.
0	ENABLE	R/W	0	It enables the MPU if set to 1.

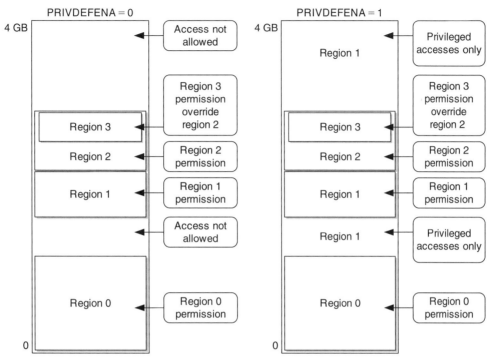

FIGURE 13.1

The Effect of PRIVDEFENA.

Table 13.3 MPU Region Number Register (0xE000ED98)

Bits	Name	Type	Reset Value	Description
7:0	REGION	R/W	—	Select the region that is being programmed. Because eight regions are supported in the Cortex-M3 MPU, only bit [2:0] of this register is implemented.

The next MPU control register is the MPU Region Number register (see Table 13.3), before each region is set up, write to this register to select the region to be programmed.

The starting address of each region is defined by the MPU Region Base Address register (see Table 13.4). Using the VALID and REGION fields in this register, we can skip the step of programming the MPU Region Number register. This might reduce the complexity of the program code, especially if the whole MPU setup is defined in a lookup table.

We also need to define the properties of each region. This is controlled by the MPU Region Base Attribute and Size register (see Table 13.5).

The REGION SIZE field (5 bits) in the MPU Region Base Attribute and Size register determines the size of the region (see Table 13.6).

Table 13.4 MPU Region Base Address Register (0xE000ED9C)

Bits	Name	Type	Reset Value	Description
31:N	ADDR	R/W	—	Base address of the region; N is dependent on the region size—for example, a 64 KB size region will have a base address field of [31:16].
4	VALID	R/W	—	If this is 1, the REGION defined in bit [3:0] will be used in this programming step; otherwise, the region selected by the MPU Region Number register is used.
3:0	REGION	R/W	—	This field overrides the MPU Region Number register if VALID is 1; otherwise, it is ignored. Because eight regions are supported in the Cortex-M3 MPU, the region number override is ignored if the value of the REGION field is larger than 7.

Table 13.5 MPU Region Base Attribute and Size Register (0xE000EDA0)

Bits	Name	Type	Reset Value	Description
31:29	Reserved	—	—	—
28	XN	R/W	—	Instruction Access Disable (1 = disable instruction fetch from this region; an attempt to do so will result in a memory management fault)
27	Reserved	—	—	—
26:24	AP	R/W	—	Data Access Permission field
23:22	Reserved	—	—	—
21:19	TEX	R/W	—	Type Extension field
18	S	R/W	—	Shareable
17	C	R/W	—	Cacheable
16	B	R/W	—	Bufferable
15:8	SRD	R/W	—	Subregion disable
7:6	Reserved	—	—	—
5:1	REGION SIZE	R/W	—	MPU Protection Region size
0	ENABLE	R/W	—	Region enable

Table 13.6 Encoding of REGION Field for Different Memory Region Sizes

REGION Size	Size
b00000	Reserved
b00001	Reserved
b00010	Reserved
b00011	Reserved

Table 13.6 Encoding of REGION Field for Different Memory Region Sizes
Continued

REGION Size	Size
b00100	32 bytes
b00101	64 bytes
b00110	128 bytes
b00111	256 bytes
b01000	512 bytes
b01001	1 KB
b01010	2 KB
b01011	4 KB
b01100	8 KB
b01101	16 KB
b01110	32 KB
b01111	64 KB
b10000	128 KB
b10001	256 KB
b10010	512 KB
b10011	1 MB
b10100	2 MB
b10101	4 MB
b10110	8 MB
b10111	16 MB
b11000	32 MB
b11001	64 MB
b11010	128 MB
b11011	256 MB
b11100	512 MB
b11101	1 GB
b11110	2 GB
b11111	4 GB

The subregion disable field (bit [15:8] of the MPU Region Base Attribute and Size register) is used to divide a region into eight equal subregions and then to define each as enabled or disabled. If a subregion is disabled and overlaps another region, the access rules for the other region are applied. If the subregion is disabled and does not overlap any other region, access to this memory range will result in a memory management fault. Subregions cannot be used if the region size is 128 bytes or less. The data Access Permission (AP) field (bit [26:24]) defines the AP of the region (see Table 13.7).

The XN (Execute Never) field (bit [28]) decides whether an instruction fetch from this region is allowed. When this field is set to 1, all instructions fetched from this region will generate a memory management fault when they enter the execution stage.

Table 13.7 Encoding of AP Field for Various Access Permission Configurations

AP Value	Privileged Access	User Access	Description
000	No access	No access	No access
001	Read/write	No access	Privileged access only
010	Read/write	Read only	Write in a user program generates a fault
011	Read/write	Read/write	Full access
100	Unpredictable	Unpredictable	Unpredictable
101	Read only	No access	Privileged read only
110	Read only	Read only	Read only
111	Read only	Read only	Read only

Table 13.8 ARMv7-M Memory Attributes

TEX	C	B	Description	Region Shareability
b000	0	0	Strongly ordered (transfers carry out and complete in programmed order)	Shareable
b000	0	1	Shared device (write can be buffered)	Shareable
b000	1	0	Outer and inner write-through; no write allocate	[S]
b000	1	1	Outer and inner write-back; no write allocate	[S]
b001	0	0	Outer and inner non cacheable	[S]
b001	0	1	Reserved	Reserved
b001	1	0	Implementation defined	–
b001	1	1	Outer and inner write-back; write and read allocate	[S]
b010	0	0	Nonshared device	Not shared
b010	0	1	Reserved	Reserved
b010	1	X	Reserved	Reserved
b1BB	A	A	Cached memory; BB = outer policy, AA = inner policy	[S]

Note: [S] indicates that shareability is determined by the S bit field (shared by multiple processors).

The TEX, S, B, and C fields (bit [21:16]) are more complex. Despite that the Cortex-M3 processor does not have cache, its implementation follows ARMv7-M architecture, which can support external cache and more advanced memory systems. Therefore, the region access properties can be programmed to support different types of memory management models.

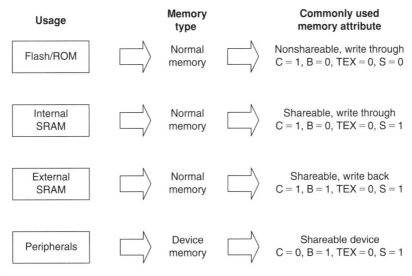

Usage		Memory type		Commonly used memory attribute
Flash/ROM		Normal memory		Nonshareable, write through C = 1, B = 0, TEX = 0, S = 0
Internal SRAM		Normal memory		Shareable, write through C = 1, B = 0, TEX = 0, S = 1
External SRAM		Normal memory		Shareable, write back C = 1, B = 1, TEX = 0, S = 1
Peripherals		Device memory		Shareable device C = 0, B = 1, TEX = 0, S = 1

FIGURE 13.2

Commonly Used Memory Attributes in Microcontrollers.

In v6 and v7 architecture, the memory system can have two cache levels: inner cache and outer cache. They can have different caching policies. Because the Cortex-M3 processor itself does not have a cache controller, the cache policy only affects write buffering in the internal BusMatrix and possibly the memory controller (see Table 13.8). For most microcontrollers, the usage of memory attributes can be simplified to just a few memory types (see Figure 13.2).

If you are using a microcontroller with cache memory, then you should program the MPU according to the cache policy you want to use (e.g., cache disable/write through cache/write back cache). When TEX[2] is 1, the cache policy for outer cache and inner cache is as shown in Table 13.9.

For more information on cache behavior and cache policy, refer to the *ARM Architecture Application Level Reference Manual* [Ref. 2].

Table 13.9 Encoding of Inner and Outer Cache Policy When Most Significant Bit of TEX Is Set to 1

Memory Attribute Encoding (AA and BB)	Cache Policy
00	Noncacheable
01	Write back, write, and read allocate
10	Write through, no write allocate
11	Write back, no write allocate

13.3 SETTING UP THE MPU

The MPU register might look complicated, but as long as you have a clear idea of the memory regions that are required for your application, it should not be difficult. Typically, you need to have the following memory regions:

- Program code for privileged programs (for example, OS kernel and exception handlers)
- Program code for user programs
- Data memory for privileged and user programs in various memory regions (e.g., data and stack of the application situated in the SRAM (Static Random Access Memory) memory region-- 0x20000000 to 0x3FFFFFFF)
- Other peripherals

It is not necessary to set up a region for the memory in the private peripheral bus range. The MPU automatically recognizes the private peripheral bus memory addresses and allows privileged software to perform data accesses in this region.

For Cortex-M3 products, most memory regions can be set up with TEX = b000, C = 1, B = 1. System devices such as the Nested Vectored Interrupt Controller (NVIC) should be strongly ordered, and peripheral regions can be programmed as shared devices (TEX = b000, C = 0, B = 1). However, if you want to make sure that any bus faults occurring in the region are precise bus faults, you should use a strongly ordered memory attribute (TEX = b000, C = 0, B = 0) so that write buffering is disabled. However, doing so can reduce system performance.

For users of a Cortex Microcontroller Software Interface Standard (CMSIS) compliant device driver, the MPU registers can be accessed using the following register names as shown in Table 13.10. A simple flow for an MPU setup routine is shown in Figure 13.3 on page 220.

Before the MPU is enabled and if the vector table is relocated to RAM, remember to set up the fault handler for the memory management fault in the vector table, and enable the memory management fault in the System Handler Control and State register. They are needed to allow the memory management fault handler to be executed if an MPU violation takes place.

Table 13.10 MPU Register Names in CMSIS

Register Names	MPU Register	Address
MPU->TYPE	MPU Type register	0xE000ED90
MPU->CTRL	MPU Control register	0xE000ED94
MPU->RNR	MPU Region Number register	0xE000ED98
MPU->RBAR	MPU Region Base Address register	0xE000ED9C
MPU->RASR	MPU Region Attribute and Size register	0xE000EDA0
MPU->RBAR_A1	MPU Alias 1 Region Base Address register	0xE000EDA4
MPU->RBAR_A2	MPU Alias 2 Region Base Address register	0xE000EDAC
MPU->RBAR_A3	MPU Alias 3 Region Base Address register	0xE000EDB4
MPU->RASR_A1	MPU Alias 1 Region Attribute and Size register	0xE000EDA8
MPU->RASR_A2	MPU Alias 2 Region Attribute and Size register	0xE000EDB0
MPU->RASR_A3	MPU Alias 3 Region Attribute and Size register	0xE000EDB8

For a simple case of only four required regions, the MPU setup code (without the region checking and enabling) looks like this:

```
MPU->RNR  = 0;          // MPU Region Number Register
                        // select region 0
MPU->RBAR = 0x00000000; // MPU Region Base Address Register
                        // Base Address = 0x00000000
MPU->RASR = 0x0307002F; // Region Attribute and Size Register
                        // R/W, TEX=0,S=1,C=1,B=1, 16MB, Enable=1
MPU->RNR  = 1;          // select region 1
MPU->RBAR = 0x20000000; // Base Address = 0x20000000
MPU->RASR = 0x03070033; // R/W, TEX=0,S=1,C=1,B=1, 64MB, Enable=1
MPU->RNR  = 2;          // select region 2
MPU->RBAR = 0x40000000; // Base Address = 0x40000000
MPU->RASR = 0x03050033; // R/W, TEX=0,S=1,C=0,B=1, 64MB, Enable=1
MPU->RNR  = 3;          // select region 3
MPU->RBAR = 0xA0000000; // Base Address = 0xA0000000
MPU->RASR = 0x01040027; // Privileged R/W, TEX=0,S=1,C=0,B=0,
                        // 1MB, Enable=1
MPU->CTRL = 1;          // MPU Control register - Enable MPU
```

This can also be coded in assembly language:

```
    LDR   R0,=0xE000ED98  ; Region number register
    MOV   R1,#0           ; Select region 0
    STR   R1, [R0]
    LDR   R1,=0x00000000  ; Base Address = 0x00000000
    STR   R1, [R0, #4]    ; MPU Region Base Address Register
    LDR   R1,=0x0307002F  ; R/W, TEX=0,S=1,C=1,B=1, 16MB, Enable=1
    STR   R1, [R0, #8]    ; MPU Region Attribute and Size Register
    MOV   R1,#1           ; Select region 1
    STR   R1, [R0]
    LDR   R1,=0x20000000  ; Base Address = 0x20000000
    STR   R1, [R0, #4]    ; MPU Region Base Address Register
    LDR   R1,=0x03070033  ; R/W, TEX=0,S=1,C=1,B=1, 64MB, Enable=1
    STR   R1, [R0, #8]    ; MPU Region Attribute and Size Register
    MOV   R1,#2           ; Select region 2
    STR   R1, [R0]
    LDR   R1,=0x40000000  ; Base Address = 0x40000000
    STR   R1, [R0, #4]    ; MPU Region Base Address Register
    LDR   R1,=0x03050033  ; R/W, TEX=0,S=1,C=0,B=1, 64MB, Enable=1
    STR   R1, [R0, #8]    ; MPU Region Attribute and Size Register
    MOV   R1,#3           ; Select region 3
    STR   R1, [R0]
    LDR   R1,=0xA0000000  ; Base Address = 0xA0000000
    STR   R1, [R0, #4]    ; MPU Region Base Address Register
    LDR   R1,=0x01040027  ; Privileged R/W, TEX=0,S=1,C=0,B=0, 1MB,
                          ; Enable=1
    STR   R1, [R0, #8]    ; MPU Region Attribute and Size Register
    MOV   R1,#1           ; Enable MPU
    STR   R1, [R0,#-4]    ; MPU Control register
                          ; (0xE000ED98-4=0xE000ED94)
```

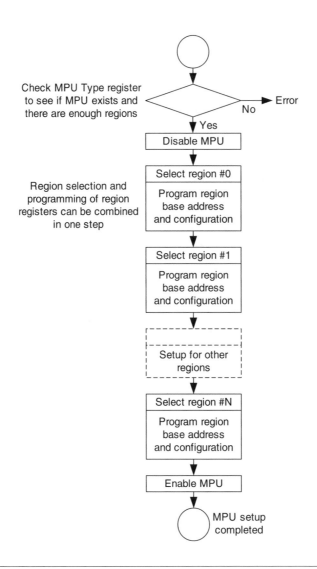

FIGURE 13.3

Example Steps to Set Up the MPU.

This provides four regions:

- *Code*: 0x00000000–0x00FFFFFF (16 MB), full access, cacheable
- *Data*: 0x20000000–0x02003FFFF (64 MB), full access, cacheable
- *Peripheral*: 0x40000000–0x5FFFFFFF (64 MB), full access, shared device
- *External device*: 0xA0000000–0xA00FFFFF (1 MB), privileged access, strongly ordered, XN

By combining region selection and writing to the base address register, we can shorten the code to this:

```
MPU->RBAR = 0x00000010;  // MPU Region Base Address Register
                         // Base Address = 0x00000000, valid, region 0
MPU->RASR = 0x0307002F;  // Region Attribute and Size Register
                         // R/W, TEX=0,S=1,C=1,B=1, 16MB, Enable=1
MPU->RBAR = 0x20000011;  // Base Address = 0x20000000, valid, region 1
MPU->RASR = 0x03070033;  // R/W, TEX=0,S=1,C=1,B=1, 64MB, Enable=1
MPU->RBAR = 0x40000012;  // Base Address = 0x40000000, valid, region 2
MPU->RASR = 0x03050033;  // R/W, TEX=0,S=1,C=0,B=1, 64MB, Enable=1
MPU->RBAR = 0xA0000013;  // Base Address = 0xA0000000, valid, region 3
MPU->RASR = 0x01040027;  // Privileged R/W, TEX=0,S=1,C=0,B=0,
                         // 1MB, Enable=1
MPU->CTRL = 1;           // MPU Control register - Enable MPU
```

Or, in assembly:

```
LDR     R0,=0xE000ED9C      ; Region Base Address register
LDR     R1,=0x00000010      ; Base Address = 0x00000000, region 0,
                            ; valid=1
STR     R1, [R0, #0]        ; MPU Region Base Address Register
LDR     R1,=0x0307002F      ; R/W, TEX=0,S=1,C=1,B=1, 16MB, Enable=1
STR     R1, [R0, #4]        ; MPU Region Attribute and Size Register
LDR     R1,=0x20000011      ; Base Address = 0x20000000, region 1,
                            ; valid=1
STR     R1, [R0, #0]        ; MPU Region Base Address Register
LDR     R1,=0x03070033      ; R/W, TEX=0,S=1,C=1,B=1, 64MB, Enable=1
STR     R1, [R0, #4]        ; MPU Region Attribute and Size Register
LDR     R1,=0x40000012      ; Base Address = 0x40000000, region 2,
                            ; valid=1
STR     R1, [R0, #0]        ; MPU Region Base Address Register
LDR     R1,=0x03050033      ; R/W, TEX=0,S=1,C=0,B=1, 64MB, Enable=1
STR     R1, [R0, #4]        ; MPU Region Attribute and Size Register
LDR     R1,=0xA0000013      ; Base Address = 0xA0000000, region 3,
                            ; valid=1
STR     R1, [R0, #0]        ; MPU Region Base Address Register
LDR     R1,=0x01040027      ; R/W, TEX=0,S=1,C=0,B=0, 1MB, Enable=1
STR     R1, [R0, #4]        ; MPU Region Attribute and Size Register
MOV     R1,#1               ; Enable MPU
STR     R1, [R0,#-8]        ; MPU Control register
                            ; (0xE000ED9C-8=0xE000ED94)
```

We've shortened the code quite a bit. However, you can make further enhancements to create even faster setup code. This is done using MPU aliased register addresses (see Table D.34 in Appendix D). The aliased register addresses follow the MPU Region Attribute and Size registers and are aliased to the MPU Base Address register and the MPU Region Attribute and Size register. They produce a continuous address of eight words, making it possible to use Load/Store Multiple (LDM and STM) instructions:

```
LDR     R0,=0xE000ED9C      ; Region Base Address register
LDR     R1,=MPUconfig       ; Table of predefined MPU setup variables
```

```
        LDMIA   R1!, {R2, R3, R4, R5}; Read 4 words from table
        STMIA   R0!, {R2, R3, R4, R5}; write 4 words to MPU
        LDMIA   R1!, {R2, R3, R4, R5}; Read next 4 words from table
        STMIA   R0!, {R2, R3, R4, R5}; write next 4 words to MPU
        B       MPUconfigEnd
        ALIGN 4    ; This is needed to make sure the following table
                   ; is word aligned
MPUconfig          ; so that we can use load multiple instruction
        DCD     0x00000010  ; Base Address = 0x00000000, region 0,
                            ; valid=1
        DCD     0x0307002F   ; R/W, TEX=0,S=1,C=1,B=1, 16MB, Enable=1
        DCD     0x20000011   ; Base Address = 0x08000000, region 1,
                            ; valid=1
        DCD     0x03070033   ; R/W, TEX=0,S=1,C=1,B=1, 64MB, Enable=1
        DCD     0x40000012   ; Base Address = 0x40000000, region 2,
                            ; valid=1
        DCD     0x03050033   ; R/W, TEX=0,S=1,C=0,B=1, 64MB, Enable=1
        DCD     0xA0000013   ; Base Address = 0xA0000000, region 3,
                            ; valid=1
        DCD     0x01040027   ; R/W, TEX=0,S=1,C=0,B=0, 1MB, Enable=1
MPUconfigEnd
        LDR     R0,=0xE000ED94 ; MPU Control register
        MOV     R1,#1        ; Enable MPU
        STR     R1, [R0]
```

This solution, of course, can be used only if all the required information is known beforehand. Otherwise, a more generic approach has to be used. One way to handle this is to use a subroutine (*MpuRegionSetup*) that can set up a region based on a number of input parameters and then call it several times to set up different regions:

```
void MpuRegionSetup(unsigned int addr, unsigned int region,
  unsigned int size, unsigned int ap, unsigned int MemAttrib,
  unsigned int srd, unsigned int XN, unsigned int enable)
{ // Setup procedure for each region
  MPU->RBAR = (addr & 0xFFFFFFE0) | (region & 0xF) | 0x10;
  MPU->RASR = ((XN & 0x1)<<28) | ((ap & 0x7)<<24) |
            ((MemAttrib & 0x3F)<<16) | ((srd&0xFF)<<8) |
            ((size & 0x1F)<<1)| (enable & 0x1);
  return;
}
void MpuRegionDisable(unsigned int region)
{ // Function to disable an unused region
  MPU->RBAR = (region & 0xF) | 0x10;
  MPU->RASR = 0; // disable
  return;
}
void MpuSetup(void)
{ // Setup the whole MPU
  MPU->CTRL = 0; // Disable MPU first
  MpuRegionSetup(0x00000000, 0, 0x17, 3, 7, 0, 0, 1); // Region 0,16M
  MpuRegionSetup(0x20000000, 1, 0x19, 3, 7, 0, 0, 1); // Region 1,64M
  MpuRegionSetup(0x40000000, 2, 0x19, 3, 5, 0, 0, 1); // Region 2,64M
```

```
       MpuRegionSetup(0xA0000000, 3, 0x13, 1, 4, 0, 0, 1); // Region 3, 1M
       MpuRegionDisable(4); // Disable unused region 4
       MpuRegionDisable(5); // Disable unused region 5
       MpuRegionDisable(6); // Disable unused region 6
       MpuRegionDisable(7); // Disable unused region 7
       MPU->CTRL = 1; // Enable MPU
       return;
   }
```

In this example, we included a subroutine that is used to disable a region that is not used. This is necessary if you do not know whether a region has been programmed previously. If an unused region is previously programmed to be enabled, it needs to be disabled so that it doesn't affect the new configuration.

The MPU setup routines can be rewritten in assembly as

```
MpuSetup     ; A subroutine to setup the MPU by calling subroutines that
             ; setup regions
      PUSH   {R0-R6,LR}
      LDR    R0,=0xE000ED94      ; MPU Control Register
      MOV    R1,#0
      STR    R1,[R0]             ; Disable MPU
      ; ---  Region #0 ---
      LDR    R0,=0x00000000      ; Region 0: Base Address  = 0x00000000
      MOV    R1,#0x0             ; Region 0: Region number = 0
      MOV    R2,#0x17            ; Region 0: Size          = 0x17 (16MB)
      MOV    R3,#0x3             ; Region 0: AP            = 0x3 (full
                                                                  access)
      MOV    R4,#0x7             ; Region 0: MemAttrib     = 0x7
      MOV    R5,#0x0             ; Region 0: Sub R disable = 0
      MOV    R6,#0x1             ; Region 0: {XN, Enable}  = 0,1
      BL     MpuRegionSetup
      ; ---  Region #1 ---
      LDR    R0,=0x20000000      ; Region 1: Base Address  = 0x20000000
      MOV    R1,#0x1             ; Region 1: Region number = 1
      MOV    R2,#0x19            ; Region 1: Size          = 0x19 (64MB)
      MOV    R3,#0x3             ; Region 1: AP            = 0x3 (full
                                                                  access)
      MOV    R4,#0x7             ; Region 1: MemAttrib     = 0x7
      MOV    R5,#0x0             ; Region 1: Sub R disable = 0
      MOV    R6,#0x1             ; Region 1: {XN, Enable}  = 0,1
      BL     MpuRegionSetup
      ...                        ; setup for region #2 and #3
      ; ---  Region #4-#7 Disable ---
      MOV    R0,#4
      BL     MpuRegionDisable
      MOV    R0,#5
      BL     MpuRegionDisable
      MOV    R0,#6
      BL     MpuRegionDisable
      MOV    R0,#7
      BL     MpuRegionDisable
      LDR    R0,=0xE000ED94      ; MPU Control Register
```

```
        MOV    R1,#1
        STR    R1,[R0]              ; Enable MPU
        POP    {R0-R6,PC}           ; Return

MpuRegionSetup
        ; MPU region setup subroutine
        ; Input R0 : Base Address
        ;       R1 : Region number
        ;       R2 : Size
        ;       R3 : AP (access permission)
        ;       R4 : MemAttrib ({TEX[2:0], S, C, B})
        ;       R5 : Sub region disable
        ;       R6 : {XN,Enable}
        PUSH   {R0-R1, LR}
        BIC    R0, R0, #0x1F        ; Clear unused bits in address
        BFI    R0, R1, #0, #4       ; Insert region number to R0[3:0]
        ORR    R0, R0, #0x10        ; Set valid bit
        LDR    R1,=0xE000ED9C       ; MPU Region Base Address Register
        STR    R0,[R1]              ; Set base address reg
        AND    R0, R6, #0x01        ; Get Enable bit
        UBFX   R1, R6, #1, #1       ; Get XN bit
        BFI    R0, R1, #28, #1      ; Insert XN to R0[28]
        BFI    R0, R2, #1 , #5      ; Insert Region Size field (R2[4:0]) to
                                    ; R0[5:1]
        BFI    R0, R3, #24, #3      ; Insert AP fields (R3[2:0]) to R0[26:24]
        BFI    R0, R4, #16, #6      ; Insert memattrib field (R4[5:0]) to
                                    ; R0[21:16]
        BFI    R0, R5, #8, #8       ; Insert subregion disable (SRD) fields
                                    ; to R0[15:8]
        LDR    R1,=0xE000EDA0       ; MPU Region Base Size and Attribute
                                    ; Register
        STR    R0,[R1]              ; Set base attribute and size reg
        POP    {R0-R1, PC}          ; Return

MpuRegionDisable
        ; Subroutine to disable unused region
        ; Input R0 : Region number
        PUSH   {R1, LR}
        AND    R0, R0, #0xF         ; Clear unused bits in Region Number
        ORR    R0, R0, #0x10        ; Set valid bit
        LDR    R1,=0xE000ED9C       ; MPU Region Base Address Register
        STR    R0,[R1]
        MOV    R0, #0
        LDR    R1,=0xE000EDA0       ; MPU Region Base Size and Attribute
                                    ; Register
        STR    R0,[R1]              ; Set base attribute and size reg to 0
                                    ; (disabled)
        POP    {R1, PC}             ; Return
```

The example shows the application of the Bit Field Insert (BFI) instruction in the Cortex-M3. This can greatly simplify bit-field merging operations.

13.4 TYPICAL SETUP

In typical applications, the MPU is used when there is a need to prevent user programs from accessing privileged process data and program regions. Usually, this is done by the embedded OS. Between each context switching, the MPU is reprogrammed by the OS to allow user applications to access their application code and data and any other resources they are entitled to access. When developing the setup routine for the MPU, you need to consider a number of regions:

1. Code region:
   a. Privileged code, including a starting vector table
   b. User code

2. SRAM region:
   a. Privileged data, including the main stack
   b. User data, including the process stack
   c. Privileged bit-band alias region
   d. User bit-band alias region

3. Peripherals:
   a. Privileged peripherals
   b. User peripherals
   c. Privileged peripheral bit-band alias region
   d. User peripheral bit-band alias region

From this list, we have identified 10 regions; more than the eight regions supported by the Cortex-M3 MPU. However, we can define the privileged regions by means of a background region (PRIVDEFENA set to 1), so there are only five user regions to set up, leaving three spare MPU regions. The unused regions might still be used for setting up additional regions in external memory, to protect read-only data or to completely block some part of the memory if necessary. Alternatively, some of the regions could be merged together to reduce the number of regions required.

13.4.1 Example Use of the Subregion Disable

In some cases, we might have some peripherals accessible by user programs, and a few should be protected to be privileged accesses only, resulting in fragmentation of user-accessible peripheral memory space. In this kind of scenario, we could do one of these things:

- Define multiple user regions
- Define privileged regions inside the user peripheral region
- Use subregion disable within the user region

The first two methods can use up available regions very easily. With the third solution, using the subregion disable feature, we can easily set up AP to separate peripheral blocks without using extra regions. For example, see Figure 13.4.

The same techniques can be applied to memory regions as well. However, it is more likely that peripherals will have a fragmented privilege setup.

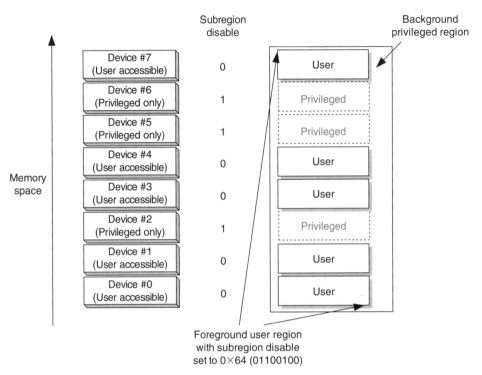

FIGURE 13.4

Using Subregion Disable to Control Access Rights to Separated Peripherals.

Table 13.11 Memory Region Arrangement for MPU Setup Example Code

Address	Description	Size	Type	Memory Attributes (C, B, A, S, XN)	MPU Region
0x00000000–0x00003FFF	Privileged program	16 KB	Read only	C, –, A, –, –	Background
0x00004000–0x00007FFF	User program	16 KB	Read only	C, –, A, –, –	Region #0
0x20000000–0x20000FFF	User data	4 KB	Full access	C, B, A, –, –	Region #1
0x20001000–0x20001FFF	Privileged data	4 KB	Privileged accesses	C, B, A, –, –	Background
0x22000000–0x2201FFFF	User data bit-band alias	128 KB	Full access	C, B, A, –, –	Region #2
0x22020000–0x2203FFFF	Privileged data bit-band alias	128 KB	Full access	C, B, A, –, –	Background

Table 13.11 Memory Region Arrangement for MPU Setup Example Code *Continued*

Address	Description	Size	Type	Memory Attributes (C, B, A, S, XN)	MPU Region
0x40000000–0x400FFFFF	User peripherals	1 MB	Full access	–, B, –, –, XN	Region #3
0x40040000–0x4005FFFF	Privileged peripherals within user peripheral region	128 KB	Privileged accesses	–, B, –, –, XN	Disabled subregions in Region #3
0x42000000–0x43FFFFFF	User peripherals bit-band alias	32 MB	Full access	–, B, –, –, XN	Region #4
0x42800000–0x42BFFFFF	Privileged peripherals bit-band alias within user region	4 MB	Privileged accesses	–, B, –, –, XN	Disabled subregion in Region #4
0x60000000–0x60FFFFFF	External RAM	16 MB	Full access	C, B, A, –, –	Region #5
0xE0000000–0xF00FFFFF	NVIC, debug, and private peripheral bus	1 MB	Privileged accesses	–, –, –, –, XN	Background

Note: A in memory attribute refers to cache allocate.

Let's assume that the memory regions in Table 13.11 will be used. After the required regions are defined, we can create the MPU setup code. To make the code easier to understand and modify, we used the function that we created earlier to develop the completed MPU setup example:

```
void MpuSetup(void)
{ // Setup the whole MPU
MPU->CTRL = 0; // Disable MPU first

// Parameters:  Address  Region Size AP Mem SRD XN Enable
MpuRegionSetup(0x00004000, 0, 0x0D, 3, 0x2, 0, 0, 1); // Region 0
// 0x00004000-0x00007FFF: user program , 16kB, full access,
// MemAttrib = 0x2 (TEX=0,S=0,C=1,B=0), Subregion disable = 0, XN=0

MpuRegionSetup(0x20000000, 1, 0x0B, 3, 0xB, 0, 0, 1); // Region 1
// 0x20000000-0x20000FFF: user data, 4kB, full access,
// MemAttrib = 0xB (TEX=1,S=0,C=1,B=1), Subregion disable = 0, XN=0

MpuRegionSetup(0x22000000, 2, 0x10, 3, 0xB, 0, 0, 1); // Region 2
// 0x22000000-0x2201FFFF: user bit band, 128kB, full access,
// MemAttrib = 0xB (TEX=1,S=0,C=1,B=1), Subregion disable = 0, XN=0

MpuRegionSetup(0x40000000, 3, 0x13, 3, 0x1,0x64,0,1); // Region 3
// 0x40000000-0x400FFFFF: user peripherals, 1MB, full access,
// MemAttrib = 0x1 (TEX=0,S=0,C=0,B=1), Subregion disable=0x64, XN=0
// Note: Sub-region disable = 0x64 based on figure 13.4

MpuRegionSetup(0x42000000, 4, 0x18, 3, 0x1,0x64,0,1); // Region 4
// 0x42000000-0x43FFFFFF: user peripheral bit band, 32MB, full access,
// MemAttrib = 0x1 (TEX=0,S=0,C=0,B=1), Subregion disable=0x64, XN=0
```

```
// Note: Sub-region disable = 0x64 based on figure 13.4

MpuRegionSetup(0x60000000, 5, 0x17, 3, 0x3, 0, 0, 1); // Region 5
// 0x60000000-0x60FFFFFF: external RAM, 16MB, full access,
// MemAttrib = 0x3 (TEX=0,S=0,C=1,B=1), Subregion disable = 0, XN=0

MpuRegionDisable(6); // Disable unused region 6
MpuRegionDisable(7); // Disable unused region 7
MPU->CTRL = 5; // Enable MPU with Default memory map enabled
               // for privileged accesses
return;
}
```

Other Cortex-M3 Features

14

IN THIS CHAPTER

The SYSTICK Timer 229
Power Management 232
Multiprocessor Communication 236
Self-Reset Control 241

14.1 THE SYSTICK TIMER

The SYSTICK register in the Nested Vectored Interrupt Controller (NVIC) was covered briefly in Chapter 8. As we saw, the SYSTICK timer is a 24-bit down counter. Once it reaches zero, the counter loads the reload value from the RELOAD register. It does not stop until the enable bit in the SYSTICK Control and Status register is cleared (see Figure 14.1).

The Cortex™-M3 processor allows two different clock sources for the SYSTICK counter. The first one is the core free-running clock (not from the system clock HCLK, so it does not stop when the system clock is stopped). The second one is an external reference clock. This clock signal must be at least two times slower than the free-running clock because this signal is sampled by the free-running clock. Because a chip designer might decide to omit this external reference clock in the design, it might not be available. To determine whether the external clock source is available, you should check bit 31 of the SYSTICK Calibration register. The chip designer should connect this pin to an appropriate value based on the design.

When the SYSTICK timer changes from 1 to 0, it will set the COUNTFLAG bit in the SYSTICK Control and Status register. The COUNTFLAG can be cleared by one of the following:

- Read of the SYSTICK Control and Status register by the processor
- Clear of the SYSTICK counter value by writing any value to the SYSTICK Current Value register

The SYSTICK counter can be used to generate SYSTICK exceptions at regular intervals. This is often necessary for the OS, for task and resources management. To enable SYSTICK exception

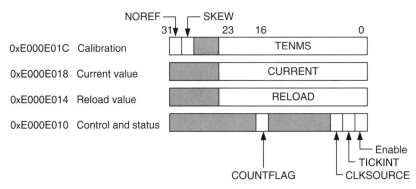

FIGURE 14.1

SYSTICK Registers in the NVIC.

generation, the TICKINT bit should be set. In addition, if the vector table has been relocated to Static Random Access Memory (SRAM), it would be necessary to set up the SYSTICK exception handler in the vector table. For example:

```
*((volatile unsigned int *)(SCB->VTOR+(15<<2))) = (unsigned int) SysTick_Handler;
```

This can be written in assembly language as

```
; Setup SYSTICK exception handler (only needed if vector table
; is located in RAM)
MOV R0, #0xF              ; Exception type 15
LDR R1, =SysTick_handler ; address of exception handler
LDR R2, =0xE000ED08      ; Vector table offset register
LDR R2, [R2]
STR R1, [R2, R0, LSL #2]  ; Write vector to
                         ; VectTblOffset+ExcpType*4
```

For users of Cortex Microcontroller Software Interface Standard (CMSIS) compliant device driver, a function call "SysTick_Config" is available for configuration of the SYSTICK Timer. Please refer to Appendix G for information on this function. You can also access the SYSTICK registers directly via the following register names:

- SysTick->CTRL (Control and Status register)
- SysTick->LOAD (Reload Value register)
- SysTick->VAL (Current Value register)
- SysTick->CALIB (Calibration Value register)

For example, to generate SYSTICK exception every 1024 processor clock cycle, you can use the following C code:

```
SysTick->LOAD = 1023;// Count down from 1023 to 0
SysTick->VAL  = 0;   // Clear current value to 0
SysTick->CTRL = 0x7; // Enable SysTick, enable SysTick
                     // exception and use processor clock
```

The same operation can be written in assembly language as follows:

```
; Enable SYSTICK timer operation and enable SYSTICK interrupt
LDR   R0, =0xE000E010  ; SYSTICK control and status register
MOV   R1, #0
STR   R1, [R0]           ; Stop counter to prevent interrupt
                         ; triggered accidentally
LDR   R1, =1023          ; Trigger every 1024 cycles (since counter
                         ; decrement from 1023 to 0, total of 1024
                         ; cycles, reload value is set to 1023)
STR   R1, [R0,#4]        ; Write reload value to reload register
                         ; address
STR   R1, [R0,#8]        ; Write any value to current value
                         ; register to clear current value to 0 and
                         ; clear COUNTFLAG
MOV   R1, #0x7           ; Clock source = core clock, Enable
                         ; Interrupt, Enable
                         ; SYSTICK counter
STR   R1, [R0]           ; Start counter
```

The SYSTICK counter provides a simple way to allow timing calibration information to be accessed. The top level of the Cortex-M3 processor has a 24-bit input to which a chip designer can input a reload value that can be used to generate a 10-ms time interval. This value can be accessed by the SYSTICK Calibration register. However, this option is not necessarily available, so you'll need to check the device's datasheet to see if you can use this feature.

The SYSTICK counter can also be used as an alarm timer that starts a certain task after a number of clock cycles. For example, if a task has to be started to execute after 300 clock cycles, we could set up the task at the SYSTICK exception handler and program the SYSTICK timer so that the task will be executed when the 300 cycle count is reached:

```
volatile int SysTickFired; // A global software flag to
                           // indicate SysTickAlarm executed
...
// Optional:Setup SYSTICK Handler, only needed if vector table
// relocated to SRAM
*((volatile unsigned int *)(SCB->VTOR+(15<<2))) = (unsigned int) SysTickAlarm;

SysTick->CTRL = 0x0;       // Disable SysTick
SysTick->LOAD = (300-12);  // Set Reload value
                           // Minus 12 because of exception latency
SysTick->VAL  = 0;    // Clear current value to 0
SysTickFired  = 0;    // Setup software flag to zero
SysTick->CTRL = 0x7;  // Enable SysTick, enable SysTick
                      // exception and use processor clock
while (SysTickFired == 0); // Wait until software flag is set by
                           // SYSTICK handler
```

The exception handler can be written as follows:

```
void SysTickAlarm(void) // SYSTICK exception handler
{
```

```
SysTick->CTRL = 0x0;              // Disable SysTick
                                  // Execute required processing task
SCB->ICSR = SCB->ICSR & (0xFDFFFFFF); // Clear SYSTICK pend bit
                                  // in case it has been pended again
SysTickFired++;                   // Update software flag so that the
                                  // main program know that SysTick alarm
                                  // task has been carried out
return;
}
```

The counter starts with an initial value of zero because it was manually cleared from the main program. It then immediately reloads to 288 (300 – 12). We subtract 12 from the count because this is the number of clock cycles for minimum exception latency. However, if another exception with the same or a higher priority is running when the SYSTICK counter reaches zero, the start of the exception could be delayed.

Note that the subtraction of 12 cycles from the reload value in this example is required for only one-shot alarm timer usage. For periodic counting usage, the reload value should be the number of clock cycles per period minus 1.

Because the SYSTICK counter does not stop automatically, we need to stop it within the SYSTICK handler (*SysTickAlarm*). Furthermore, there's a chance that the SYSTICK exception could have been pended again if it was delayed by processing of other exceptions, so the pending status of a SYSTICK exception needs to be cleared if the SYSTICK exception uses a one-off processing.

In the final step of the SYSTICK exception handler, we set a software variable called *SysTickFired* so that the main program knows the required task has been carried out.

14.2 POWER MANAGEMENT

14.2.1 Sleep Modes

The Cortex-M3 provides sleep modes as a power management feature. During sleep mode, the system clock can be stopped, but the free-running clock input could still be running to allow the processor to be woken by an interrupt. The two sleep modes are as follows:

- Sleep: Indicated by the SLEEPING signal from the Cortex-M3 processor
- Deep sleep: Indicated by the SLEEPDEEP signal from the Cortex-M3 processor

To decide which sleep mode will be used, the NVIC System Control register has a bit field called SLEEPDEEP (see Table 14.1). The actions of SLEEPING and SLEEPDEEP depend on the particular Microcontroller Unit (MCU) implementation. In some implementations, the action will be the same in both cases.

The sleep modes are invoked by Wait-For-Interrupt (WFI) or Wait-For-Event (WFE) instructions. Events can be interrupts, a previously triggered interrupt, or an external event signal pulse via the Receive Event (RXEV) signal. Inside the processor, there is a latch for events, so a past event can wake up a processor from WFE (see Figure 14.2).

For users of a CMSIS compliant device driver, WFI and WFE instructions can be accessed by __WFI( ) and __WFE( ) intrinsic functions. The System Control register can be accessed using the "SCB->SCR" register name.

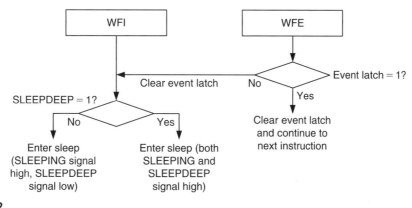

FIGURE 14.2

Sleep Operations.

Bits	Name	Type	Reset Value	Description
Table 14.1 System Control Register (0xE000ED10)				
4	SEVONPEND	R/W	0	Send Event on Pending; wakes up from WFE if a new interrupt is pended, regardless of whether the interrupt has priority higher than the current level
3	Reserved	—	—	—
2	SLEEPDEEP	R/W	0	Enable SLEEPDEEP output signal when entering sleep mode
1	SLEEPONEXIT	R/W	0	Enable Sleep-On-Exit feature
0	Reserved	—	—	—

What exactly happens when the processor enters sleep mode depends on the chip design. The common case is that some of the clock signals can be stopped to reduce power consumption. However, the chip can also be designed to shut down part of the chip to further reduce power, or it is also possible that a design can shut down the chip completely, and all the clock signals will be stopped. In a case where the chip is shut down completely, the only way to wake the system from sleep is via a system reset.

To wake the processor from WFI sleep, the interrupt will have to be higher priority than the current priority level (if it is an executing interrupt) and higher than the level set by the BASEPRI register or mask registers (PRIMASK and FAULTMASK). If an interrupt is not going to be accepted due to priority level, it will not wake up a sleep caused by WFI.

The situation for WFE is slightly different. If the interrupt triggered during sleep has lower or equal priority than the mask registers or BASEPRI registers and if the SEVONPEND is set, it could still wake the processor from sleep. The rules of waking the Cortex-M3 processor from sleep modes are summarized in Table 14.2.

Table 14.2 WFI and WFE Wakeup Behavior

WFI Behavior	Wake Up	IRQ Execution
IRQ with BASEPRI		
IRQ priority > BASEPRIv	Y	Y
IRQ priority =< BASEPRI	N	N
IRQ with BASEPRI and PRIMASK		
IRQ priority > BASEPRI	Y	N
IRQ priority =< BASEPRI	N	N
WFE Behavior		
IRQ with BASEPRI, SEVONPEND = 0		
IRQ priority > BASEPRI	Y	Y
IRQ priority =< BASEPRI	N	N
IRQ with BASEPRI, SEVONPEND = 1		
IRQ priority > BASEPRI	Y	Y
IRQ priority =< BASEPRI	Y	N
IRQ with BASEPRI and PRIMASK, SEVONPEND = 0		
IRQ priority > BASEPRI	N	N
IRQ priority =< BASEPRI	N	N
IRQ with BASEPRI & PRIMASK, SEVONPEND = 1		
IRQ priority > BASEPRI	Y	N
IRQ priority =< BASEPRI	Y	N

14.2.2 Sleep-On-Exit Feature

Another feature of sleep mode is that it can be programmed to go back to sleep automatically after the interrupt routine exit. In this way, we can make the core sleep all the time unless an interrupt needs to be served. To use this feature, we need to set the SLEEPONEXIT bit in the System Control register (see Figure 14.3).

Note that if the Sleep-On-Exit feature is enabled, the processor can enter sleep at any exception return to thread level, even if no WFE/WFI instruction is executed. To ensure that the processor only enter sleep when required, set the SLEEPONEXIT bit only when the system is ready for entering sleep.

14.2.3 Wakeup Interrupt Controller

Starting from revision 2 of Cortex-M3, additional low-power features have been added. A new unit called the Wakeup Interrupt Controller (WIC) is available as an optional component. This controller is coupled to the existing NVIC and is used to generate a wakeup request when an interrupt arrives.

From a software point of view, the WFI and WFE behaviors remain the same. There are no programmable registers in the WIC, as it gets all the required interrupt information via the interface between WIC and NVIC. By using the WIC, the clock signals going into the processor core can be completely stopped. When an interrupt request arrives, the WIC can send a wakeup request to the system controller or Power Management Unit (PMU) in the chip to restore the processor clock (figure 14.4).

The availability of the WIC also provides a new method for reducing power consumption during sleep mode. By using new technologies in digital logic design, it is now possible to power down most of

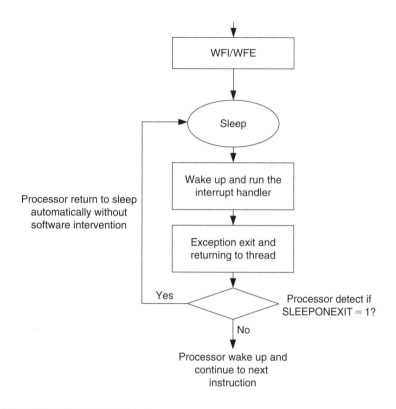

FIGURE 14.3

Example Use of the Sleep-On-Exit Feature.

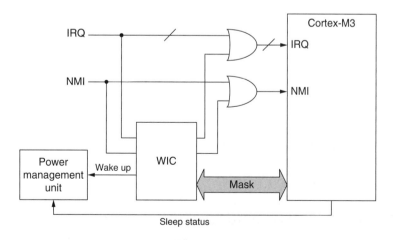

FIGURE 14.4

WIC Mirrors the Interrupt Detection Function when clock signals to Cortex-M3 stops.

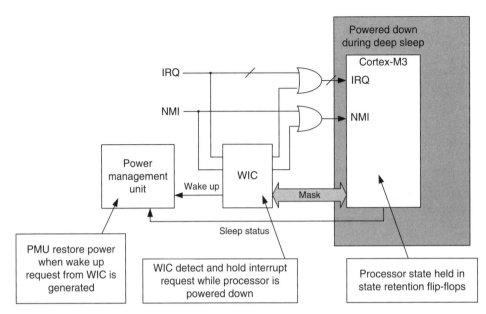

FIGURE 14.5

WIC Mirrors the Interrupt Detection Function when Cortex-M3 is in state retention.

the Cortex-M3 processor, leaving a small portion of the logic gates to retain the current state of the logic. This technology is called State Retention Power Gating (SRPG). By using SRPG and WIC together, most portions of the Cortex-M3 processor can be powered down during deep sleep, leaving a small amount of logic for state retention (see Figure 14.5). During this power down state, the WIC remains operational and generates a wakeup request to power up and restore the system state when an interrupt arrives. As a result, the processor can resume operation and service the interrupt request in a very short time. The maximum interrupt latency with such arrangement depends on the time required to power up the system. In most cases, it is in the range of 20 to 30 clock cycles. Normal sleep (SLEEPDEEP bit in the System Control register is zero) does not trigger the power down feature.

The new power down capability is optional and may not be included in some microcontroller products. It requires an on-chip PMU developed by silicon vendors to control the power up and power down sequences and might need to be programmed before the power down feature is used. Please refer to the silicon vendor's documentation for further information. A couple of points to be aware of: the power down feature stops the SYSTICK timer during deep sleep, and the power down feature is disabled when a debugger is attached (this is required because debugger needs to access the debug registers regularly to examine the status of the processor).

14.3 MULTIPROCESSOR COMMUNICATION

The Cortex-M3 comes with a simple multiprocessor communication interface for event communication. The processor has one output signal, called Transmit Event (TXEV), for sending out events, and an input signal, called RXEV, for receiving events. For a system with two processors, the event communication signal connection can be implemented as shown in Figure 14.6.

As mentioned in the previous section "Power Management," the processor can enter sleep when the WFE instruction is executed and can continue the instruction execution when an external event is received. If we use an instruction called Send Event (SEV), one processor can wake up another processor that is in sleep mode and make sure both processors start executing a task at the same time (see Figure 14.7).

For users of a CMSIS compliant device driver, SEV instruction can be accessed by the __SEV( ) intrinsic functions. Using this feature, we can make both processors start executing a task at the same time (possibly with small timing differences, depending on actual chip implementation and the software

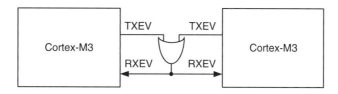

FIGURE 14.6

Event Communication Connection in a Two-Processor System.

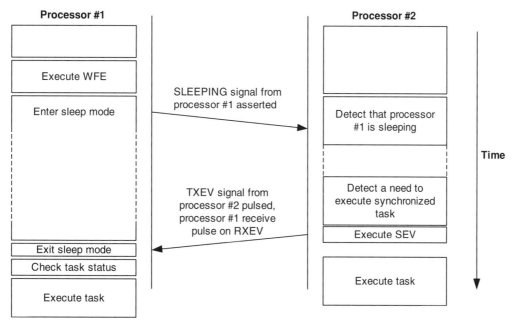

FIGURE 14.7

Using Event Signals to Synchronize Tasks.

code for checking task status). The number of processors invoked can be any number, but it requires that one processor acts as a master to generate the event pulse to other processors.

It is important to note that the processor could also be woken by other events, such as interrupt and debugging events. Therefore, before starting the required synchronized task, it is necessary to check whether the wakeup was caused by task synchronization. In most multitasking systems, an interprocessor messaging system like mailbox is still required to ensure that the tasks are synchronized correctly.

It is also important to note that execution of WFE does not always cause the processor to enter sleep mode. Therefore, WFE is normally used with looping (to reduce system power consumption) and status checking code to check if the required synchronized task should be carried out after the WFE, as shown in Figure 14.8.

When the WFE instruction is executed, it first checks the local event latch. If the latch is not set, the core enters sleep mode. If the latch is set, it will be cleared and the instruction execution continues without entering sleep mode. The local event latch can be set by previously occurring exceptions and by the SEV instruction. So, if you execute an SEV and then execute a WFE, the processor will not enter sleep and will simply continue on to the next instruction, with the event latch cleared by WFE.

An example of WFE usage is semaphore in a multiprocessor system. In a typical scenario like Mutual Exclusion (MUTEX), system-level exclusive-access monitor and exclusive-access instructions are used for spin locks for granting accesses to shared memory or a shared peripheral. A process requiring a resource would need to call a function to gain the "lock":

```
void get_lock(volatile int * Lock_Variable)
{ // __LDREXW and __STREXW are intrinsic functions in CMSIS
  // compliant device driver libraries
    int status = 0;
    do {
    while ( __LDREXW(&Lock_Variable) != 0); // Wait until lock
                                            // variable is free
    status = __STREXW(1, &Lock_Variable);   // Try set Lock_Variable
                                            // to 1 using STREX
    } while (status != 0); // retry until lock successfully
    __DMB();                                 // Data memory Barrier
    return;
}
```

The same process can be carried out in assembly code:

```
get_lock                            ; an assembly function to get the lock
    LDR      r0, =Lock_Variable
    MOVS     r2, #1                 ; use for locking STREX
get_lock_loop
    LDREX    r1, [r0]
    CMP      r1, #0
    BNE      get_lock_loop          ; It is locked, retry again
    STREX    r1, r2, [r0]           ; Try set Lock_Variable to 1 using STREX
    CMP      r1, #0                 ; Check return status of STREX
    BNE      get_lock_loop          ; STREX was not successful, retry
    DMB                            ; Data Memory Barrier
    BX       LR                     ; Return
```

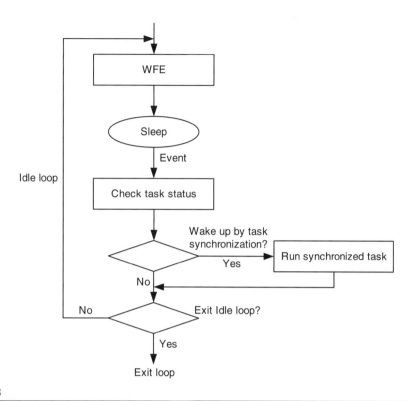

FIGURE 14.8

Example Use of the WFE Feature.

On the other hand, a process using the resource should unlock the resource when it is no longer required:

```
void free_lock(volatile int * Lock_Variable)
{
  __DMB();              // Data memory Barrier
  Lock_Variable = 0;    // Free the lock
  return;
}
```

The same operation can be written in assembly as follows:

```
free_lock                  ; an assembly function to free the lock
    LDR  r0, =Lock_Variable
    MOVS r1, #0
    DMB                    ; Data Memory Barrier
    STR  r1, [r0]         ; Clear lock
    BX   LR               ; Return
```

The spin lock can result in unnecessary power consumption when the processor is idle. As a result, we add WFE into these operations to reduce power consumption, while allowing the processor waiting for the lock to be woken up as soon as the resource is free.

```
void get_lock_with_WFE(volatile int * Lock_Variable)
{
  int status = 0;
  do {
  while ( __LDREXW(&Lock_Variable) != 0){ // Wait until lock
    __WFE();} // variable is free, if not, enter sleep until event
  status = __STREXW(1, &Lock_Variable);     // Try set Lock_Variable
                                            // to 1 using STREX
  } while (status != 0); // retry until lock successfully
  __DMB();                                  // Data memory Barrier
  return;
}
```

For the function to free the lock, the SEV instruction is used to wake up other processors that are waiting for the lock.

```
void free_lock(volatile int * Lock_Variable)
{
  __DMB();// Data memory Barrier
  Lock_Variable = 0;// Free the lock
  __DSB();// ensure the store is complete
  __SEV();// Send Event to wake up other processors
  return;
}
```

The same operation can be written in assembly as follows:

```
get_lock_with_WFE        ; an assembly function to get the lock
  LDR   r0, =Lock_Variable
  MOVS r2, #1            ; use for locking STREX
get_lock_loop
  LDREX  r1,[r0]
  CBNZ   r1, lock_is_set; If lock is set, sleep and retry later
  STREX  r1, r2, [r0]   ; Try set Lock_Variable to 1 using STREX
  CMP    r1, #0         ; Check return status of STREX
  BNE get_lock_loop     ; STREX was not successful, retry
  DMB                   ; Data Memory Barrier
  BX   LR               ; Return
lock_is_set
  WFE                   ; Wait for event
  B    get_lock_loop    ; woken up, retry again
```

And for the function that frees the lock, it can be written in assembly as follows:

```
free_lock_with_SEV       ; an assembly function to free the lock
  LDR   r0, =Lock_Variable
  MOVS r1, #0
  DMB                   ; Data Memory Barrier
  STR   r1, [r0]        ; Clear lock
  DSB                   ; ensure the store is complete
  SEV                   ; Send Event to wake up other processors
  BX    LR              ; Return
```

By combining event communication interface and necessary semaphore code, the power consumption during spin lock can be reduced. Similar techniques can be created for message passing and tasks synchronizations.

In most Cortex-M3-based products, there will be only one processor, and the RXEV input is likely tied to 0 or connected to peripherals that generate events.

14.4 SELF-RESET CONTROL

The Cortex-M3 provides two self-reset control features. The first reset feature is the SYSRESETREQ (System Reset Request) bit in the same NVIC register. It allows the Cortex-M3 processor to assert a reset request signal to the system's reset generator. Because the system reset generator is not part of Cortex-M3 design, the implementation of this reset feature depends on the chip design. Therefore, it is necessary to carefully check the chip's specification to determine which part of the chip is reset by this reset control bit.

For users of a CMSIS compliant device driver, the *NVIC_SystemReset()* function can be used to trigger the system reset using SYSRESETREQ. (A summary of this function can be found in Appendix G.) Users not using CMSIS can use:

```
*((volatile unsigned int *)(0xE000ED0C))= 0x05FA0004;
// Set SYSRESETREQ bit (05FA is a write access key)
while(1); // a deadloop is used to ensure no other
          // instructions follow the reset is executed
```

Assembly language users can generate the system reset request using the following example code:

```
        LDR  R0,=0xE000ED0C   ; NVIC AIRCR address
        LDR  R1,=0x05FA0004   ; Set SYSRESETREQ bit (05FA is a write
                              ; access key)
        STR  R1,[R0]
deadloop
        B deadloop            ; a deadloop is used to ensure no other
                              ; instructions follow the reset is executed
```

The second reset feature is the VECTRESET control bit in the NVIC Application Interrupt and Reset Control register (bit [0]). Writing 1 to this bit will reset the Cortex-M3 processor, excluding the debug logic. This does not reset any circuit outside the Cortex-M3 processor. For example, if the system-on-chip (SoC) contains a universal asynchronous receiver/transmitter (UART), writing to this bit does not reset the UART or any peripherals outside the Cortex-M3. This feature is mainly targeted for debug, or in some case, where the software needs to reset the processor only but not the rest of the system.

```
*((volatile unsigned int *)(0xE000ED0C))= 0x05FA0001;
// Set VECTRESET bit (05FA is a write access key)
while(1); // a deadloop is used to ensure no other
          // instructions follow the reset is executed
```

The same operation can be carried out in the following assembly code:

```
        LDR  R0,=0xE000ED0C   ; NVIC AIRCR address
        LDR  R1,=0x05FA0001   ; Set VECTRESET bit (05FA is a write
                              ; access key)
        STR  R1,[R0]
deadloop
        B deadloop            ; a deadloop is used to ensure no other
                              ; instructions following the reset is
                              ; executed
```

In general, software reset should be generated using SYSRESETREQ instead of VECTRESET. This ensures most parts of the system will be reset at the same time. Depending on the chip design, it might or might not reset all peripherals, the chip and the clocking control logic including Phase-Locked Loop (PLL). Please refer to the manufacturer datasheet for details.

Note that the delay from assertion of SYSRESETREQ to actual reset from the reset generator can be an issue in some cases. If there is a delay in the reset generator, you might find the processor still accepting interrupts after the reset request is set. If you want to stop the core from accepting interrupts before running this code, you can set the Interrupt Mask register (e.g., PRIMASK or FAULTMASK) before requesting the reset.

Debug Architecture

15

IN THIS CHAPTER

Debugging Features Overview..243
CoreSight Overview..244
Debug Modes 248
Debugging Events 250
Breakpoint in the Cortex-M3 251
Accessing Register Content in Debug..253
Other Core Debugging Features 254

15.1 DEBUGGING FEATURES OVERVIEW

The Cortex™-M3 processor provides a comprehensive debugging environment. Based on the nature of operations, the debugging features can be classified into two groups.

1. Invasive debugging:
- Program halt and stepping
- Hardware breakpoints
- Breakpoint instruction
- Data watchpoint on access to data address, address range, or data value
- Register value accesses (both read or write)
- Debug monitor exception
- ROM-based debugging (Flash Patch)

2. Noninvasive debugging:
- Memory accesses (memory contents can be accessed even when the core is running)
- Instruction trace (through the optional Embedded Trace Module)
- Data trace
- Software trace (through the Instrumentation Trace Module)
- Profiling (through the Data Watchpoint and Trace [DWT] Module)

DOI: 10.1016/B978-1-85617-963-8.00018-1

243

A number of debugging components are included in the Cortex-M3 processor. The debugging system is based on the CoreSight debug architecture, allowing a standardized solution to access debugging controls, gather trace information, and detect debugging system configuration.

15.2 CORESIGHT OVERVIEW

The CoreSight debug architecture covers a wide area, including the debugging interface protocol, debugging bus protocol, control of debugging components, security features, trace data interface, and more. The *CoreSight Technology System Design Guide* [Ref. 3] is a useful document for getting an overview of the architecture. In addition, a number of sections in the *Cortex-M3 Technical Reference Manual* [Ref. 1] are descriptions of the debugging components in Cortex-M3 design. These components are normally used only by debugger software, not by application code. However, it is still useful to briefly review these items so that we can have a better understanding of how the debugging system works.

15.2.1 Processor Debugging Interface

Unlike traditional ARM7 or ARM9, the debugging system of the Cortex-M3 processor is based on the CoreSight debug architecture. Traditionally, ARM processors provide a Joint Test Action Group (JTAG) interface, allowing registers to be accessed and memory interface to be controlled. In the Cortex-M3, the control to the debug logic on the processor is carried out through a bus interface called the Debug Access Port (DAP), which is similar to Advanced Peripheral Bus (APB) in Advanced Microcontroller Bus Architecture (AMBA). The DAP is controlled by another component that converts JTAG or Serial-Wire (SW) communication protocols into the DAP bus interface protocol.

Because the internal debug bus is similar to APB, a generic bus protocol, it is easy to connect multiple debugging components, resulting in a very scalable debugging system. In addition, by separating the debug interface and debug control hardware, the actual interface type used on the chip can become transparent; hence, the same debugging tasks can be carried out no matter what debugging interface you use.

The actual debugging functions in the Cortex-M3 processor core are controlled by the Nested Vectored Interrupt Controller (NVIC) and a number of other debugging components, such as the Flash Patch and Breakpoint (FPB), DWT, and Instrumentation Trace Macrocell (ITM). The NVIC contains a number of registers to control the core debugging actions, such as halt and stepping, whereas the other blocks support features such as watchpoints, breakpoints, and debug message outputs.

15.2.2 The Debug Host Interface

CoreSight technology supports a number of interface types for connection between the debug host and the system-on-chip (SoC). Traditionally, this has always been JTAG. Now, because the processor debugging interface has been changed to a generic bus interface, by putting a different interface module between the debug host and the debug interface of the processor, we can come up with different chips that have different debug host interfaces, without redesigning the debug interface on the processor.

Currently, Cortex-M3 systems support two types of debug host interface. The first one is the well-known JTAG interface, and the second one is a new interface protocol called *Serial-Wire*. The SW interface reduces the number of signals to two. Several types of debug host interface modules (called Debug Port, or DP) are available from ARM and they provide the support for the different interface protocols. The debugger hardware is connected to one side of a DP, and the other side is connected to the DAP, a generic bus interface on the processor.

WHY SERIAL-WIRE?

The Cortex-M3 is targeted at the low-cost microcontroller market in which some devices have very low pin counts. For example, some of the low-end versions are in 28-pin packages. Despite the fact that JTAG is a very popular protocol, using four pins to debug is a lot for a 28-pin device. Therefore, SW is an attractive solution because it can reduce the number of debug pins to two.

15.2.3 **DP Module, AP Module, and DAP**

The connection from external debugging hardware to the debug interface in the Cortex-M3 processor is divided into multiple stages (see Figure 15.1).

The DP interface module (normally either Serial-Wire JTAG Debug Port [SWJ-DP] or Serial-Wire Debug Port [SW-DP]) first converts the external signals into a generic 32-bit debug bus (a DAP bus in the diagram). SWJ-DP supports both JTAG and SW, and SW-DP supports SW only. In the ARM CoreSight product series, there is also a JTAG Debug Port, which only supports the JTAG protocol; chip manufacturers can choose to use one of these DP modules to suit their needs. The address of the DAP bus is 32-bit, with the upper 8 bits of the address bus used to select which device is being accessed. Up to 256 devices can be attached to the DAP bus. Inside the Cortex-M3 processor, only one of the device addresses is used, so you can attach 255 more access port (AP) devices to the DAP bus if needed. So, theoretically, you can have hundreds of processors in one chip sharing one JTAG or SW debug connection.

After passing through the DAP interface in the Cortex-M3 processor, an AP device called Advanced High-Performance Bus Access Port (AHB-AP) is connected. This acts as a bus bridge to convert commands into AHB transfers, which are inserted into the internal bus network inside the Cortex-M3. This allows the memory map of the Cortex-M3, including the debug control registers in the NVIC, to be accessed.

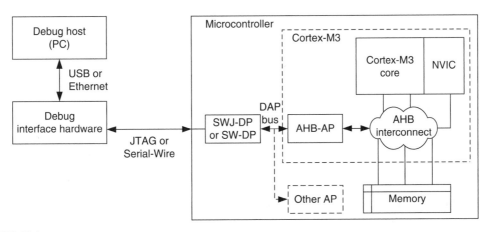

FIGURE 15.1

Connection from the Debug Host to the Cortex-M3.

In the CoreSight product series, several types of Access Port (AP) devices are available, including an Advanced Peripheral Bus Access Port (APB-AP) and a JTAG Access Port (JTAG-AP). The APB-AP can be used to generate APB transfers, and the JTAG-AP can be used to control traditional JTAG-based test interfaces such as the debug interface on ARM7.

15.2.4 Trace Interface

Another part of the CoreSight architecture concerns tracing. In the Cortex-M3, there can be three types of trace sources:

- Instruction trace: Generated by the Embedded Trace Macrocell (ETM)
- Data trace: Generated by the DWT unit
- Debug message: Generated by ITM, provides message output such as *printf* in the debugger graphical user interface

During tracing, the trace results, in the form of data packets, are output from the trace sources like ETM, using a trace data bus interface called *Advanced Trace Bus* (ATB). Based on the CoreSight architecture, if an SoC contains multiple trace sources (e.g., multiprocessors), the ATB data stream can be merged using ATB merger hardware (in the CoreSight architecture this hardware is called *ATB funnel*). The final data stream on the chip can then be connected to a Trace Port Interface Unit (TPIU) and exported to external trace hardware. Once the data reach the debug host (for example, a PC), the data stream can then be converted back into multiple data streams.

Despite the Cortex-M3 having multiple trace sources, its debugging components are designed to handle trace merging so that there is no need to add ATB funnel modules. The trace output interface can be connected directly to a special version of the TPIU designed for the Cortex-M3. The trace data are then captured by external hardware and collected by the debug host (e.g., a PC) for analysis.

15.2.5 CoreSight Characteristics

The CoreSight-based design has a number of advantages:

- The memory content and peripheral registers can be examined even when the processor is running.
- Multiple processor debug interfaces can be controlled with a single piece of debugger hardware. For example, if JTAG is used, only one Test Access Port (TAP) controller is required, even when there are multiple processors on the chip.
- Internal debugging interfaces are based on simple bus design, making it scalable and easy to develop additional test logic for other parts of the chip or SoC.
- It allows multiple trace data streams to be collected in one trace capture device and separated back into multiple streams on the debug host.

The debugging system used in the Cortex-M3 processor is slightly different from the standard CoreSight implementation.

- Trace components are specially designed in the Cortex-M3. Some of the ATB interface is 8 bits wide in the Cortex-M3, whereas in CoreSight the width is 32 bits.
- The debug implementation in the Cortex-M3 does not support TrustZone.[1]

---

[1]TrustZone is an ARM technology that provides security features to embedded products.

- The debug components are part of the system memory map, whereas in standard CoreSight systems, a separate bus (with a separate memory map) is used for controlling debug components. For example, the conceptual system connection in a CoreSight system can be like the one shown in Figure 15.2.

In the Cortex-M3, the debugging devices share the same system memory map (see Figure 15.3).

Additional information about the CoreSight debug architecture can be found in the *CoreSight Technology System Design Guide* [Ref. 3].

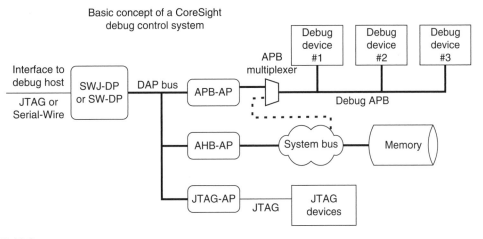

FIGURE 15.2

Design Concept of a CoreSight System.

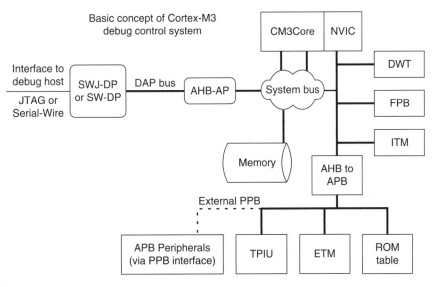

FIGURE 15.3

The Debug System in the Cortex-M3.

Although the debug components in Cortex-M3 are build differently from normal CoreSight systems, the communication interface and protocols in the Cortex-M3 are compliant to CoreSight architecture and can be directly attached to other CoreSight systems. For example, CoreSight debug components like CoreSight TPIU, DPs, and trace infrastructure blocks can be used with Cortex-M3 and allow it to be extended to a multicore debug system.

15.3 DEBUG MODES

There are two types of debug operation modes in the Cortex-M3. The first one is halt, whereby the processor stops program execution completely. The second one is the debug monitor exception, whereby the processor executes an exception handler to carry out the debugging tasks while still allowing higher-priority exceptions to take place. Debug monitor is exception type 12 and its priority is programmable. It can be invoked by means of debug events, as well as by manually setting the pending bit. In summary:

1. Halt mode:
   * Instruction execution is stopped
   * The System Tick Timer (SYSTICK) counter is stopped
   * Supports single-step operations
   * Interrupts can be pended and can be invoked during single stepping or be masked so that external interrupts are ignored during stepping

2. Debug monitor mode:
   * Processor executes exception handler type 12 (debug monitor)
   * SYSTICK counter continues to run
   * New arrive interrupts may or may not preempt, depending on the priority of the debug monitor and the priority of the new interrupt
   * If the debug event takes place when a higher-priority interrupt is running, the debug event will be missed
   * Supports single-step operations
   * Memory contents (for example, stack memory) could be changed by the debug monitor handler during stacking and handler execution

The reason for having a debug monitor is that in some electronic systems, stopping a processor for a debugging operation can be infeasible. For example, in automotive engine control or hard disk controller applications, the processor should continue to serve interrupt requests during debugging, to ensure safety of operations or to prevent damage to the device being tested. With a debug monitor, the debugger can stop and debug the thread level application and lower-priority interrupt handlers, whereas higher-priority interrupts and exceptions can still be executed.

To enter halt mode, the C_DEBUGEN bit in the NVIC Debug Halting Control and Status register (DHCSR) must be set. This bit can only be programmed through the DAP, so you cannot halt the Cortex-M3 processor without a debugger. After C_DEBUGEN is set, the core can be halted by setting the C_HALT bit in DHCSR. This bit can be set by either the debugger or by the software running on the processor itself.

The bit field definition of DHCSR differs between read operations and write operations. For write operations, a debug key value must be used on bit 31 to bit 16. For read operations, there is no debug key and the return value of the upper half word contains the status bits (see Table 15.1).

Table 15.1 Debug Halting Control and Status Register (0xE000EDF0)

Bits	Name	Type	Reset Value	Description
31:16	KEY	W	—	Debug key; value of 0xA05F must be written to this field to write to this register, otherwise the write will be ignored
25	S_RESET_ST	R	—	Core has been reset or being reset; this bit is clear on read
24	S_RETIRE_ST	R	—	Instruction is completed since last read; this bit is clear on read
19	S_LOCKUP	R	—	When this bit is 1, the core is in a locked-up state
18	S_SLEEP	R	—	When this bit is 1, the core is in sleep mode
17	S_HALT	R	—	When this bit is 1, the core is halted
16	S_REGRDY	R	—	Register read/write operation is completed
15:6	Reserved	—	—	Reserved
5	C_SNAPSTALL	R/W	0*	Use to break a stalled memory access
4	Reserved	—	—	Reserved
3	C_MASKINTS	R/W	0*	Mask interrupts while stepping; can only be modified when the processor is halted
2	C_STEP	R/W	0*	Single step the processor; valid only if C_DEBUGEN is set
1	C_HALT	R/W	0*	Halt the processor core; valid only if C_DEBUGEN is set
0	C_DEBUGEN	R/W	0*	Enable halt mode debug

* The control bit in DHCSR is reset by power on reset. System reset (for example, by the Application Interrupt and Reset Control register of NVIC) does not reset the debug controls.

In normal situations, the DHCSR is used only by the debugger. Application codes should not change DHCSR contents to avoid causing problems to debugger tools.

- The control bit in DHCSR is reset by power-on reset. System reset (for example, by the Application Interrupt and Reset Control register of NVIC) does not reset the debug controls. For debugging using the debug monitor, a different NVIC register, the NVIC's Debug Exception and Monitor Control register, is used to control the debug activities (see Table 15.2). Aside from the debug monitor control bits, the Debug Exception and Monitor Control register contains the trace system enable bit (TRCENA) and a number of vector catch (VC) control bits. The VC feature can be used only with halt mode debugging. When a fault (or core reset) takes place and the corresponding VC control bit is set, the halt request will be set and the core will halt as soon as the current instruction completes.
- The TRCENA control bit and VC control bits in Debug Exception and Monitor Control register (DEMCR) are reset by power-on reset. System reset does not reset these bits. The control bits for monitor mode debug, however, are reset by power-on reset as well as system reset.

Table 15.2 Debug Exception and Monitor Control Register (0xE000EDFC)

Bits	Name	Type	Reset Value	Description
24	TRCENA	R/W	0*	Trace system enable; to use DWT, ETM, ITM, and TPIU, this bit must be set to 1
23:20	Reserved	—	—	Reserved
19	MON_REQ	R/W	0	Indication that the debug monitor is caused by a manual pending request rather than hardware debug events
18	MON_STEP	R/W	0	Single step the processor; valid only if MON_EN is set
17	MON_PEND	R/W	0	Pend the monitor exception request; the core will enter monitor exceptions when priority allows
16	MON_EN	R/W	0	Enable the debug monitor exception
15:11	Reserved	—	—	Reserved
10	VC_HARDERR	R/W	0*	Debug trap on hard faults
9	VC_INTERR	R/W	0*	Debug trap on interrupt/exception service errors
8	VC_BUSERR	R/W	0*	Debug trap on bus faults
7	VC_STATERR	R/W	0*	Debug trap on usage fault state errors
6	VC_CHKERR	R/W	0*	Debug trap on usage fault-enabled checking errors (e.g., unaligned, divide by zero)
5	VC_NOCPERR	R/W	0*	Debug trap on usage fault, no coprocessor errors
4	VC_MMERR	R/W	0*	Debug trap on memory management fault
3:1	Reserved	—	—	Reserved
0	VC_CORERESET	R/W	0*	Debug trap on core reset

* The control bit in DHCSR is reset by power on reset. System reset (for example, by the Application Interrupt and Reset Control register of NVIC) does not reset the debug controls.

15.4 DEBUGGING EVENTS

The Cortex-M3 can enter debug mode (both halt or debug monitor exception) for a number of possible reasons. For halt mode debugging, the processor will enter halt mode if conditions resemble those shown in Figure 15.4.

The external debug request is from a signal called *EDBGREQ* on the Cortex-M3 processor. The actual connection of this signal depends on the microcontroller or SoC design. In some cases, this signal could be tied low and never occur. However, this can be connected to accept debug events from additional debug components (chip manufacturers can add extra debug components to the SoC) or, if the design is a multiprocessor system, it could be linked to debug events from another processor.

After debugging is completed, the program execution can be returned to normal by clearing the C_HALT bit. Similarly, for debugging with the debug monitor exceptions, a number of debug events can cause a debug monitor to take place (see Figure 15.5).

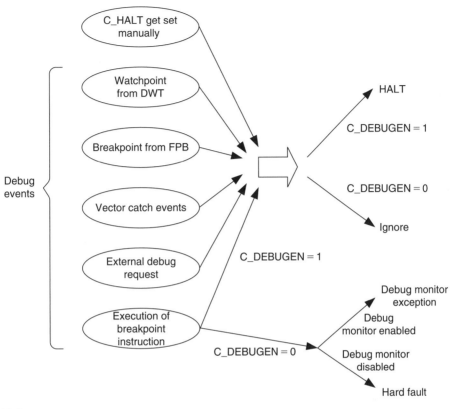

FIGURE 15.4

Debugging Events for Halt Mode Debugging.

For debug monitor, the behavior is a bit different from halt mode debugging. This is because the debug monitor exception is just one type of exception and can be affected by the current priority of the processor if it is running another exception handler.

After debugging is completed, the program execution can be returned to normal by carrying out an exception return.

15.5 BREAKPOINT IN THE CORTEX-M3

One of the most commonly used debug features in most microcontrollers is the breakpoint feature. In the Cortex-M3, the following two types of breakpoint mechanisms are supported:

- Breakpoint instruction
- Breakpoint using address comparators in the FPB

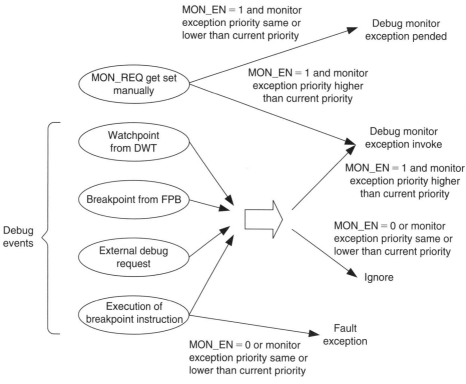

MON_EN = 1 and monitor
exception priority same or
lower than current priority

Debug monitor
exception pended

MON_REQ get set
manually

MON_EN = 1 and monitor
exception priority higher
than current priority

Watchpoint
from DWT

Debug monitor
exception invoke

Breakpoint from FPB

MON_EN = 1 and monitor
exception priority higher
than current priority

Debug
events

External debug
request

MON_EN = 0 or monitor
exception priority same or
lower than current priority

Ignore

Execution of
breakpoint instruction

Fault
exception

MON_EN = 0 or monitor
exception priority same or
lower than current priority

FIGURE 15.5

Debugging Events for Debug Monitor Exceptions.

The breakpoint instruction (*BKPT #immed8*) is a 16-bit Thumb® instruction with encoding 0xBExx. The lower 8 bits depend on the immediate data given following the instruction. When this instruction is executed, it generates a debug event and can be used to halt the processor core if C_DBGEN is set, or if the debug monitor is enabled, it can be used to trigger the debug monitor exception. Because the debug monitor is one type of exception with programmable priority, it can only be used in thread or exception handlers with priority lower than itself. As a result, if debug monitor is used for debugging, the BKPT instructions should not be used in exception handlers such as nonmaskable interrupt or hard fault, and the debug monitor can only be pended and executed after the exception handler is completed.

When the debug monitor exception returns, it is returned to the address of the BKPT instruction, not the address after the BKPT instruction. This is because in normal use of breakpoint instructions, the BKPT is used to replace a normal instruction, and when the breakpoint is hit and the debug action is carried out, the instruction memory is restored to the original instruction, and the rest of the instruction memory is unaffected.

If the BKPT instruction is executed with C_DEBUGEN = 0 and MON_EN = 0, it will cause the processor to enter a hard fault exception, with DEBUGEVT in the Hard Fault Status register (HFSR) set to 1, and BKPT in the Debug Fault Status register (DFSR) also set to 1.

The FPB unit can be programmed to generate breakpoint events even if the program memory cannot be altered. However, it is limited to six instruction addresses and two literal addresses. More information about FPB is covered in the next chapter.

15.6 ACCESSING REGISTER CONTENT IN DEBUG

Two more registers are included in the NVIC to provide debug functionality. They are the Debug Core Register Selector register (DCRSR) and the Debug Core Register Data register (DCRDR) (see Tables 15.3 and 15.4). These two registers allow the debugger to access registers of the processors. The register transfer feature can be used only when the processor is halted.

To use these registers to read register contents, the following procedure must be followed:

1. Make sure the processor is halted.
2. Write to the DCRSR with bit 16 set to 0, indicating it is a read operation.
3. Poll until the S_REGRDY bit in DHCSR (0xE000EDF0) is 1.
4. Read the DCRDR to get the register content.

Table 15.3 Debug Core Register Selector Register (0xE000EDF4)

Bits	Name	Type	Reset Value	Description
16	REGWnR	W	—	Direction of data transfer: Write = 1, Read = 0
15:5	Reserved	—	—	—
4:0	REGSEL	W	—	Register to be accessed: 00000 = R0 00001 = R1 . . . 01111 = R15 10000 = xPSR/flags 10001 = Main Stack Pointer (MSP) 10010 = Process Stack Pointer (PSP) 10100 = Special registers: [31:24] Control [23:16] FAULTMASK [15:8] BASEPRI [7:0] PRIMASK Other values are reserved

Table 15.4 Debug Core Register Data Register (0xE000EDF8)

Bits	Name	Type	Reset Value	Description
31:0	Data	R/W	—	Data register to hold register read result or to write data into selected register

Similar operations are needed for writing to a register:

1. Make sure the processor is halted.
2. Write data value to the DCRDR.
3. Write to the DCRSR with bit 16 set to 1, indicating it is a write operation.
4. Poll until the S_REGRDY bit in DHCSR (0xE000EDF0) is 1.

The DCRSR and the DCRDR registers can only transfer register values during halt mode debug. For debugging using a debug monitor handler, the contents of some of the register can be accessed from the stack memory; the others can be accessed directly within the monitor exception handler.

The DCRDR can also be used for semihosting if suitable function libraries and debugger support are available. For example, when an application executes a *printf* statement, the text output could be generated by a number of *putc* (put character) function calls. The *putc* function calls can be implemented as functions that store the output character and status to the DCRDR and then trigger the debug mode. The debugger can then detect the core halt and collect the output character for display. This operation, however, requires the core to halt, whereas the semihosting solution using ITM does not have this limitation.

15.7 OTHER CORE DEBUGGING FEATURES

The NVIC also contains a number of other features for debugging. These include the following:

- *External debug request signal*: The NVIC provides an external debug request signal that allows the Cortex-M3 processor to enter debug mode through an external event such as debug status of other processors in a multiprocessor system. This feature is very useful for debugging a multiprocessor system. In simple microcontrollers, this signal is likely to be tied low.

- *DFSR*: Because of the various debug events available on the Cortex-M3, a DFSR (Debug Fault Status Register) is available for the debugger to determine the debug event that has taken place.

- *Reset control*: During debugging, the processor core can be restarted using the VECTRESET control bit or SYSRESETREQ control bit in the NVIC Application Interrupt and Reset Control register (0xE000ED0C). Using this reset control register, the processor can be reset without affecting the debug components in the system.

- *Interrupt masking*: This feature is very useful during stepping. For example, if you need to debug an application but do not want the code to enter the interrupt service routine during the stepping, the interrupt request can be masked. This is done by setting the C_MASKINTS bit in the DHCSR (0xE000EDF0).

- *Stalled bus transfer termination*: If a bus transfer is stalled for a very long time, it is possible to terminate the stalled transfer by an NVIC control register. This is done by setting the C_SNAPSTALL bit in the DHCSR (0xE000EDF0). This feature can be used only by a debugger during halt.

Debugging Components

16

IN THIS CHAPTER

Introduction..255
Trace Components: DWT 256
Trace Components: ITM..258
Trace Components: ETM...260
Trace Components: TPIU 261
The Flash Patch and Breakpoint Unit 262
The Advanced High-Performance Bus Access Port 264
ROM Table..265

16.1 INTRODUCTION

The Cortex™-M3 processor comes with a number of debugging components used to provide debugging features such as breakpoint, watchpoint, Flash Patch, and trace. If you are an application developer, there might be a chance that you'll never need to know the details about these debugging components because they are normally used only by debugger tools. This chapter will introduce you to the basics of each debug component. If you want to know the details about things such as the actual programmer's model, refer to the *Cortex-M3 Technical Reference Manual* [Ref. 1].

All the debug trace components, as well as the Flash Patch and Breakpoint (FPB), can be programmed through the Cortex-M3 Private Peripheral Bus (PPB). In most cases, the components will only be programmed by the debugging host. It is not recommended for applications to try accessing the debug components (except stimulus port registers in the Instrumentation Trace Macrocell [ITM]) because this could interfere with the debugger's operation.

16.1.1 The Trace System in the Cortex-M3

The Cortex-M3 trace system is based on the CoreSight architecture. Trace results are generated in the form of packets, which can be of various lengths (in terms of number of bytes). The trace components transfer the packets using Advanced Trace Bus (ATB) to the Trace Port Interface Unit (TPIU), which

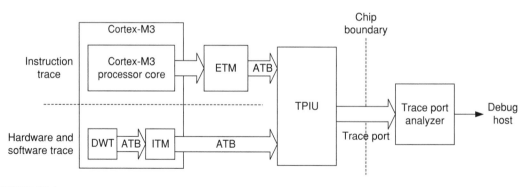

FIGURE 16.1

The Cortex-M3 Trace System.

formats the packets into Trace Interface Protocol. The data is then captured by an external trace capture device such as a Trace Port Analyzer (TPA), as shown in Figure 16.1.

There are up to three trace sources in a standard Cortex-M3 processor: Embedded Trace Macrocell (ETM), ITM, and Data Watchpoint and Trace (DWT). Note that the ETM in the Cortex-M3 is optional, so some Cortex-M3 products do not have instruction trace capability. During operation, each trace source is assigned a 7-bit ATB Trace ID value (ATID), which is transferred along the trace packets during merging in the ATB so that the packets can be separated back into multiple trace streams when they reach the debug host.

Unlike many other standard CoreSight components, the debug components in the Cortex-M3 processor include the functionality of merging ATB streams, whereas in standard CoreSight systems, ATB packet merger, called *ATB funnel*, is a separate block.

Before using the trace system, the Trace Enable (TRCENA) bit in the Debug Exception and Monitor Control register (DEMCR) must be set to 1 (see Table 15.2 or D.38). Otherwise, the trace system will be disabled. In normal operations that do not require tracing, clearing the TRCENA bit can disable some of the trace logic and reduce the power consumption.

16.2 TRACE COMPONENTS: DWT

The DWT has a number of debugging functionalities:

1. It has four comparators, each of which can be configured as follows:
   a. Hardware watch point (generates a watch point event to the processor to invoke debug modes such as halt or debug monitor)
   b. ETM trigger (causes the ETM to emit a trigger packet in the instruction trace stream)
   c. PC sampler event trigger
   d. Data address sampler trigger

   The first comparator can also be used to compare against the clock cycle counter (CYCCNT) instead of comparing with a data address.

2. Counters for counting the following:
   a. Clock cycles (CYCCNT)
   b. Folded instructions
   c. Load store unit operations
   d. Sleep cycles
   e. Cycles per instruction
   f. Interrupt overhead

3. PC sampling at regular intervals

4. Interrupt events trace

When used as a hardware watchpoint or ETM trigger, the comparator can be programmed to compare either data addresses or program counters. When programmed as other functions, it compares the data addresses.

Each of the comparators has three corresponding registers, which are as follows:

- COMP (compare) register
- MASK register
- FUNCTION control register

The COMP register is a 32-bit register that the data address (or program counter value, or CYCCNT) compares to. The MASK register determines whether any bit in the data address will be ignored during the compare (see Table 16.1).

By using the mask register, it is possible to trace data access in an address range of 32 KB maximum size. However, because of the limited first in/first out (FIFO) size in the DWT and the ITM, it is not practical to trace lots of data transfers as this will cause trace overflow and result in loss of trace data.

The comparator's FUNCTION register determines its function. To avoid unexpected behavior, the MASK register and the COMP register should be programmed before this register is set. If the comparator's function is to be changed, you should disable the comparator by setting FUNCTION to 0 (disable), then program the MASK and COMP registers, and then enable the FUNCTION register in the last step.

The rest of the DWT counters are typically used for profiling the application codes. They can be programmed to emit events (in the form of trace packets) when the counter overflows. One typical application is to use the CYCCNT register to count the number of clock cycles required for a specific task, for benchmarking purposes.

Table 16.1 Encoding of the DWT Mask Registers

MASK	Ignore Bit
0	All bits are compared
1	Ignore bit [0]
2	Ignore bit [1:0]
3	Ignore bit [2:0]
…	…
15	Ignore bit [14:0]

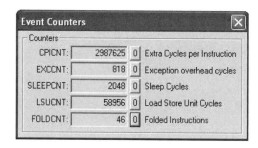

FIGURE 16.2

Program Execution Statistics in Keil μVision Using DWT Counters.

For example, the Keil μVision development tool can use these profiling counters to generate statistical information (see Figure 16.2). These counters trigger event packets to be generated and are collected by the debugger through the Serial Wire Viewer (SWV) output.

The TRCENA bit in the DEMCR must be set to 1 before the DWT is used. If the DWT is being used to generate a trace, the DWT enable (DWTEN) bit in the ITM control register should also be enabled.

16.3 TRACE COMPONENTS: ITM

The ITM has the following functionalities, as shown in Figure 16.3:

- Software can directly write console messages to ITM stimulus ports and output them as trace data.
- The DWT can generate trace packets and output them through the ITM.
- The ITM can generate timestamp packets that are inserted into a trace stream to help the debugger find out the timing of events.

Because the ITM uses a trace port to output data, if the microcontroller or system-on-chip (SoC) does not have TPIU support, the traced information cannot be output. Therefore, it is necessary to check whether the microcontroller or SoC has all the required features before you use the ITM. In the

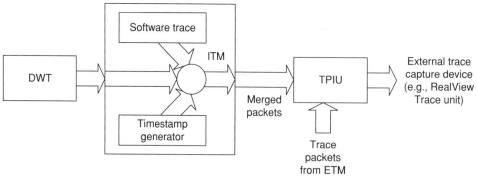

FIGURE 16.3

Merging of Trace Packets on the ITM and TPIU.

worst case, if these features are not available, you can still use semihosting via the Nested Vectored Interrupt Controller (NVIC) debug register (supported by ARM RealView Debugger) or a universal asynchronous receiver/transmitter (UART) to output console messages.

To use the ITM, the TRCENA bit in the DEMCR must be set to 1. Otherwise, the ITM will be disabled and ITM registers cannot be accessed.

In addition, there is also a lock register in the ITM. You need to write the access key 0xC5ACCE55 (CoreSight ACCESS) to this register before programming the ITM. Otherwise, all write operations to the ITM will be ignored.

Finally, the ITM itself is another control register to control the enabling of individual features. The control register also contains the ATID field, which is an ID value for the ITM in the ATB. This ID value must be unique from the IDs for other trace sources, so that the debug host receiving the trace packet can separate the ITM's trace packets from other trace packets.

16.3.1 Software Trace with the ITM

One of the main uses of the ITM is to support debug message output (such as *printf*). The ITM contains 32 stimulus ports, allowing different software processes to output to different ports, and the messages can be separated later at the debug host. Each port can be enabled or disabled by the Trace Enable register and can be programmed (in groups of eight ports) to allow or disallow user processes to write to it.

Unlike UART-based text output, using the ITM to output does not cause much delay for the application. A FIFO buffer is used inside the ITM, so writing output messages can be buffered. However, it is still necessary to check whether the FIFO is full before you write to it.

The output messages can be collected at the trace port interface or the Serial Wire Viewer interface (SWV) on the TPIU. There is no need to remove code that generates the debug messages from the final code because if the TRCENA control bit is low, the ITM will be inactive and debug messages will not be output. You can also switch on the output message in a "live" system and use the Trace Enable register in the ITM to limit which ports are enabled so that only some of the messages can be output.

For example, the Keil µVision development tool can collect and display the text output using the ITM viewer shown in Figure 16.4.

FIGURE 16.4

µVision ITM Viewer Display Shows the Software Generated ITM Text Output.

16.3.2 Hardware Trace with ITM and DWT

The ITM is used in output of hardware trace packets. The packets are generated from the DWT, and the ITM acts as a trace packet merging unit. To use DWT trace, you need to enable the DWTEN bit in the ITM control register; the rest of the DWT trace settings still need to be programmed at the DWT.

16.3.3 ITM Timestamp

ITM has a timestamp feature that allows trace capture tools to find out timing information by inserting delta timestamp packets into the traces when a new trace packet enters the FIFO inside the ITM. The timestamp packet is also generated when the timestamp counter overflows.

The timestamp packets provide the time difference (delta) with previous events. Using the delta timestamp packets, the trace capture tools can then establish the timing of when each packet is generated, and hence reconstruct the timing of various debug events.

Combining the trace functionality of DWT and ITM, we can collect a lot of useful information. For example, the exception trace windows in the Keil µVision development tool can tell you what exceptions have been carried out and how much time was spend on the exceptions, as shown in Figure 16.5.

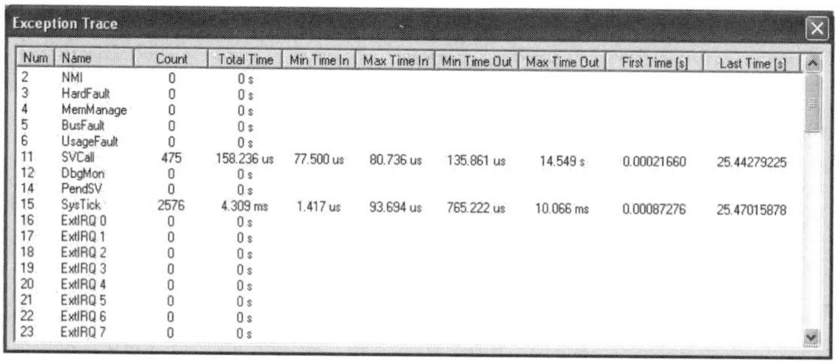

Num	Name	Count	Total Time	Min Time In	Max Time In	Min Time Out	Max Time Out	First Time [s]	Last Time [s]
2	NMI	0	0 s						
3	HardFault	0	0 s						
4	MemManage	0	0 s						
5	BusFault	0	0 s						
6	UsageFault	0	0 s						
11	SVCall	475	158.236 us	77.500 us	80.736 us	135.861 us	14.549 s	0.00021660	25.44279225
12	DbgMon	0	0 s						
14	PendSV	0	0 s						
15	SysTick	2576	4.309 ms	1.417 us	93.694 us	765.222 us	10.066 ms	0.00087276	25.47015878
16	ExtIRQ 0	0	0 s						
17	ExtIRQ 1	0	0 s						
18	ExtIRQ 2	0	0 s						
19	ExtIRQ 3	0	0 s						
20	ExtIRQ 4	0	0 s						
21	ExtIRQ 5	0	0 s						
22	ExtIRQ 6	0	0 s						
23	ExtIRQ 7	0	0 s						

FIGURE 16.5

µVision Exception Trace Output.

16.4 TRACE COMPONENTS: ETM

The ETM block is used for providing instruction traces. It is optional and might not be available on some Cortex-M3 products. When it is enabled and when the trace operation starts, it generates instruction trace packets. A FIFO buffer is provided in the ETM to allow enough time for the trace stream to be captured.

To reduce the amount of data generated by the ETM, it does not always output exactly what address the processor has reached/executed. It usually outputs information about program flow and outputs full addresses only if needed (e.g., if a branch has taken place). Because the debugging host should have a copy of the binary image, it can then reconstruct the instruction sequence the processor has carried out.

The ETM also interacts with other debugging components such as the DWT. The comparators in the DWT can be used to generate trigger events in the ETM or to control the trace start/stop.

Unlike the ETM in traditional ARM processors, the Cortex-M3 ETM does not have its own address comparators, because the DWT can carry out the comparison for ETM. Furthermore, because the data trace functionality is carried out by the DWT, the ETM design in the Cortex-M3 is quite different from traditional ETM for other ARM cores.

To use the ETM in the Cortex-M3, the following setup is required (handled by debug tools):

1. The TRCENA bit in the DEMCR must be set to 1 (refer to Table 15.2 or D.38).
2. The ETM needs to be unlocked so that its control registers can be programmed. This can be done by writing the value 0xC5ACCE55 to the ETM LOCK_ACCESS register.
3. The ATB ID register (ATID) should be programmed to a unique value so that the trace packet output through the TPIU can be separated from packets from other trace sources.
4. The Non-Invasive Debug Enable (NIDEN) input signal of the ETM must be set to high. The implementation of this signal is device specific. Refer to the datasheet from your chip's manufacturer for details.
5. Program the ETM control registers for trace generation.

16.5 TRACE COMPONENTS: TPIU

The TPIU is used to output trace packets from the ITM, DWT, and ETM to the external capture device (for example, a TPA). The Cortex-M3 TPIU supports two output modes:

- Clocked mode, using up to 4-bit parallel data output ports
- SWV mode, using single-bit SWV output[1]

In clocked mode, the actual number of bits being used on the data output port can be programmed to different sizes. This will depend on the chip package as well as the number of signal pins available for trace output in the application. The maximum trace port size supported by the chip can be determined from one of the registers in the TPIU. In addition, the speed of trace data output can also be programmed.

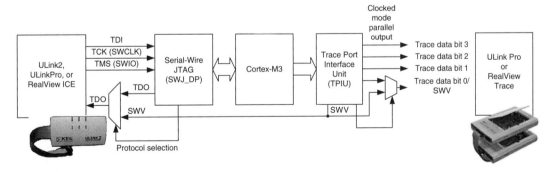

FIGURE 16.6

Pin Sharing of SWV Output.

---

[1]Available in Cortex-M3 products based on Cortex-M3 revision 1 and onwards.

In SWV mode, a one-bit serial protocol is used and this reduces the number of output signal to 1, but the maximum bandwidth for trace output will also be reduced. When combining SWV with Serial-Wire debug protocol, the Text Data Output (TDO) pin normally used for Joint Test Action Group (JTAG) protocol can be shared with SWV (see Figure 16.6). For example, the trace output in SWV mode can be collected using a standard debug connector for JTAG using a Keil U-Link2 module.

Alternatively, the SWV output mode can also share a pin with the trace output pin in clocked mode. The trace data (either in clocked mode or SWV mode) can be collected by external TPA like the ARM RealView Trace.

When instruction trace (using ETM) is required, the clocked mode is more suitable than SWV mode as it provides higher trace bandwidth. For simple data trace and event trace (e.g., tracing of exception events), the SWV mode is usually sufficient and can be used with less connection pins.

To use the TPIU, the TRCENA bit in the DEMCR must be set to 1, and the protocol (mode) selection register and trace port size control registers need to be programmed by the trace capture software.

16.6 THE FLASH PATCH AND BREAKPOINT UNIT

The FPB has the following two functions:

- Hardware breakpoint (generates a breakpoint event to the processor to invoke debug modes such as halt or debug monitor)
- Patch instruction or literal data from Code memory space to Static Random Access Memory (SRAM) memory region

16.6.1 Breakpoint Feature

The breakpoint function is fairly easy to understand—during debugging, you can set one or multiple breakpoints to program addresses or literal constant addresses. If the program code at the breakpoint addresses get executed, or if the literal constant addresses get accessed, then this triggers the breakpoint debug event and causes the program execution to halt (for halt mode debug) or triggers the debug monitor exception (if debug monitor is used). Then, you can examine the register's content, memory contents, debug using single stepping, and so on.

16.6.2 Flash Patch Feature

The Flash Patch function allows using a small programmable memory in the system to apply patches to a program memory which cannot be modified. For products to be produced in high volume, using mask ROM or one-time-programmable ROM can reduce the cost of the product. But, if a software bug is found after the device is programmed, it could be costly to replace the devices. By integrating a small reprogrammable memory, for example, a very small Flash or Electrically Erasable Programmable Read Only Memory (EEPROM), patches can be made to the original software programmed in the device. For microcontrollers that only use Flash to store software, Flash Patch is not required as the whole Flash can be erased and reprogrammed easily.

16.6.3 **Comparators**

The FPB contains eight comparators:

- Six instruction comparators
- Two literal comparators

The comparators can be used either for breakpoint function or Flash Patch function, but both do not function at the same time.

The FPB has a Flash Patch control register that contains an enable bit to enable the FPB. In addition, each comparator comes with a separate enable bit in its comparator control register. Both of the enable bits must be set to 1 for a comparator to operate.

The comparators can be programmed to remap addresses from Code space to the SRAM memory region. When this function is used, the REMAP register needs to be programmed to provide the base address of the remapped contents. The upper three bits of the REMAP register (bit[31:29]) is hardwired to b001, limiting the remap base address location to within 0x20000000 to 0x3FFFFF80, which is always within the SRAM memory region.

When the instruction address or the literal address hits the address defined by the comparator, the read access is remapped to the table pointed to by the REMAP register (see Figure 16.7).

WHAT ARE LITERAL LOADS?

When we program in assembler language, very often we need to set up immediate data values in a register. When the value of the immediate data is large, the operation cannot be fitted into one instruction space. For example,

```
LDR    R0, =0xE000E400  ; External Interrupt Priority Register
                        ; starting address
```

Because no instruction has an immediate value space of 32, we need to put the immediate data in a different memory space, usually after the program code region, and then use a PC relative load instruction to read the immediate data into the register. So what we get in the compiled binary code will be something like the following:

```
LDR    R0, [PC, #<immed_8>*4]
       ; immed_8 = (address of literal value - PC)/4
...
; literal pool
...
DCD    0xE000E400
...
```

or with Thumb®-2 instructions:

```
LDR.W R0, [PC, #+/- <offset_12>]
      ; offset_12 = address of literal value - PC
...
; literal pool
...
DCD    0xE000E400
...
```

Because we are likely to use more than one literal value in our code, the assembler or compiler will usually generate a block of literal data, commonly called the *literal pool*.

In Cortex-M3, the literal loads are data read operations carried out on the data bus (D-Code bus or System bus depending on memory location).

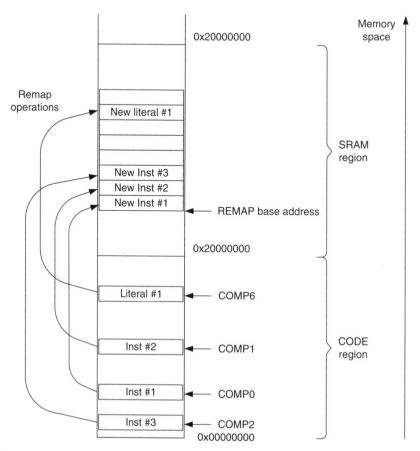

FIGURE 16.7

Flash Patch: Remap of Instructions and Literal Read.

Using the remap function, it is possible to create some "what-if" test cases in which the original instruction or a literal value is replaced by a different one; even the program code is in ROM or Flash memory. An example use is to allow execution of a program or subroutine in the SRAM region by patching program ROM in the Code region so that a branch to the test program or subroutine can take place. This makes it possible to debug a ROM-based device.

Alternatively, the six instruction address comparators can be used to generate breakpoints as well as to invoke halt mode debug or debug monitor exceptions.

16.7 THE ADVANCED HIGH-PERFORMANCE BUS ACCESS PORT

The Advanced High-Performance Bus Access Port (AHB-AP) is a bridge between the debug interface module (Serial-Wire JTAG Debug Port or Serial-Wire Debug Port) and the Cortex-M3 memory system (see Figure 16.8). For the most basic data transfers between the debug host and the Cortex-M3 system, the following three registers in the AHB-AP are used:

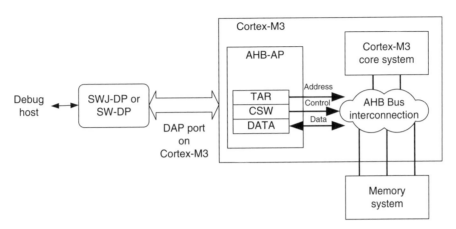

FIGURE 16.8

Connection of the AHB-AP in the Cortex-M3.

- Control and Status Word (CSW)
- Transfer Address register (TAR)
- Data Read/Write (DRW)

The CSW register can control the transfer direction (read/write), transfer size, transfer types, and so on. The TAR register is used to specify the transfer address, and the DRW register is used to carry out the data transfer operation (transfer starts when this register is accessed).

The data register DRW represents exactly what is shown on the bus. For half word and byte transfers, the required data will have to be manually shifted to the correct byte lane by debugger software. For example, if you want to carry out a data transfer of half word size to address 0x1002, you need to have the data on bit [31:16] of the DRW register. The AHB-AP can generate unaligned transfers, but it does not rotate the result data based on the address offset. So, the debugger software will have to either rotate the data manually or split an unaligned data access into several accesses if needed.

Other registers in the AHB-AP provide additional features. For example, the AHB-AP provides four banked registers and an automatic address increment function so that access to memory within close range or sequential transfers can be speeded up. The AHB-AP also contains a register called *base address* to indicate the address of ROM table.

In the CSW register, there is one bit called *MasterType*. This is normally set to 1 so that hardware receiving the transfer from AHB-AP knows that it is from the debugger. However, the debugger can pretend to be the core by clearing this bit. In this case, the transfer received by the device attached to the AHB system should behave as though it is accessed by the processor. This is useful for testing peripherals with FIFO that can behave differently when accessed by the debugger.

16.8 ROM TABLE

The ROM table is used to allow autodetection of debug components inside a Cortex-M3 chip. The Cortex-M3 processor is the first product based on ARM v7-M architecture. It has a defined memory map and includes a number of debug components. However, in newer Cortex-M devices or if the chip designers modified the default debug components, the memory map for the debug devices could be different. To allow debug tools to detect the components in the debug system, a ROM table is included; it provides information on the NVIC and debug block addresses.

The ROM table is located in address 0xE00FF000. Using contents in the ROM table, the memory locations of system and debug components can be calculated. The debug tool can then check the ID registers of the discovered components and determine what is available on the system.

For the Cortex-M3, the first entry in the ROM table (0xE00FF000) should contain the offset to the NVIC memory location. (The default value in the ROM table's first entry is 0xFFF0F003; bit[1:0] means that the device exists and there is another entry in the ROM table following. The NVIC offset can be calculated as 0xE00FF000 + 0xFFF0F000 = 0xE000E000.)

The default ROM table for the Cortex-M3 is shown in Table 16.2. However, because chip manufacturers can add, remove, or replace some of the optional debug components with other CoreSight debug components, the value you find on your Cortex-M3 device could be different.

The lowest two bits of the value indicate whether the device exists. In normal cases, the NVIC, DWT, and FPB should always be there, so the last two bits are always 1. However, the TPIU and the

Table 16.2 Cortex-M3 Default ROM Table Values

Address	Value	Name	Description
0xE00FF000	0xFFF0F003	NVIC	Points to the NVIC base address at 0xE000E000
0xE00FF004	0xFFF02003	DWT	Points to the DWT base address at 0xE0001000
0xE00FF008	0xFFF03003	FPB	Points to the FPB base address at 0xE0002000
0xE00FF00C	0xFFF01003	ITM	Points to the ITM base address at 0xE0000000
0xE00FF010	0xFFF41003 / 0xFFF41002	TPIU	Points to the TPIU base address at 0xE0040000
0xE00FF014	0xFFF42003 / 0xFFF42002	ETM	Points to the ETM base address at 0xE0041000
0xE00FF018	0	End	End-of-table marker
0xE00FFFCC	0x1	MEMTYPE	Indicates that system memory can be accessed on this memory map
0xE00FFFD0	0 / 0x04	PID4	Peripheral ID space; reserved
0xE00FFFD4	0 / 0x00	PID5	Peripheral ID space; reserved
0xE00FFFD8	0 / 0x00	PID6	Peripheral ID space; reserved
0xE00FFFDC	0 / 0x00	PID7	Peripheral ID space; reserved
0xE00FFFE0	0 / 0xC3	PID0	Peripheral ID space; reserved
0xE00FFFE4	0 / 0xB4	PID1	Peripheral ID space; reserved
0xE00FFFE8	0 / 0x0B	PID2	Peripheral ID space; reserved
0xE00FFFEC	0 / 0x00	PID3	Peripheral ID space; reserved
0xE00FFFF0	0 / 0x0D	CID0	Component ID space; reserved
0xE00FFFF4	0 / 0x10	CID1	Component ID space; reserved
0xE00FFFF8	0 / 0x05	CID2	Component ID space; reserved
0xE00FFFFC	0 / 0xB1	CID3	Component ID space; reserved

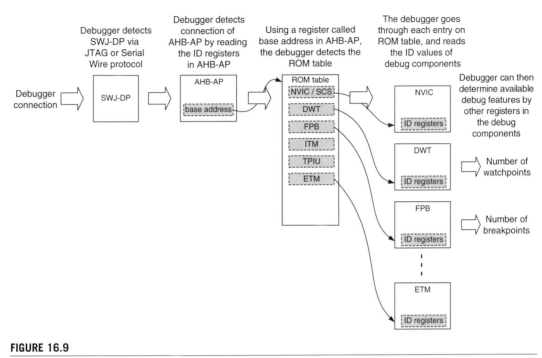

FIGURE 16.9

Automatic Detection of Components via CoreSight Technology.

ETM could be taken out by the chip manufacturer and might be replaced with other debugging components from the CoreSight product family.

The upper part of the value indicates the address offset from the ROM table base address. For example,

```
NVIC address = 0xE00FF000 + 0xFFF0F000 = 0xE000E000 (truncated to 32-bit)
```

For debug tool development using CoreSight technology, it is necessary to determine the address of debug components from the ROM table. Some Cortex-M3 devices might have a different setup of the debug component connection that can result in additional base addresses. By calculating the correct device address from this ROM table, the debugger can determine the base address of the provided debug component, and then from the component ID of those components the debugger can determine the type of debug components that are available (see Figure 16.9).

Getting Started with the Cortex-M3 Processor

IN THIS CHAPTER

Choosing a Cortex-M3 Product 269
Development Tools 270
Differences between the Cortex-M3 Revision 0 and Revision 1 272
Differences between the Cortex-M3 Revision 1 and Revision 2 274
Benefits and Effects of the Revision 2 New Features 277
Differences between the Cortex-M3 and Cortex-M0 278

17.1 CHOOSING A CORTEX-M3 PRODUCT

Aside from memory, peripheral options, and operation speed, a number of other factors make one Cortex™-M3 product different from another. The Cortex-M3 design supplied by ARM contains a number of features that are configurable, such as

- Number of external interrupts
- Number of interrupt priority levels (width of priority-level registers)
- With Memory Protection Unit (MPU) or without MPU
- With Embedded Trace Macrocell (ETM) or without ETM
- Choice of debug interface (Serial-Wire (SW), Joint Test Action Group (JTAG), or both)

In most projects, the features and specification of the microcontroller will certainly affect your choice of Cortex-M3 product. For example,

- *Peripherals*: For many applications, peripheral support is the main criterion. More peripherals might be good, but this also affects the microcontroller's power consumption and price.
- *Memory*: Cortex-M3 microcontrollers can have Flash memory from several kilobytes to several megabytes. In addition, the size of the internal memory might also be important. Usually these factors will have a direct impact on the price.
- *Clock speed*: The Cortex-M3 design from ARM can easily reach more than 100 MHz, even in 0.18 μm processes. However, manufacturers might specify a lower operation speed due to limitations of memory access speed.

- *Footprint*: The Cortex-M3 can be available in many different packages, depending on the chip manufacturer's decision. Many Cortex-M3 devices are available in low pin count packages, making them ideal for low-cost manufacturing environments.

Currently, a number of microcontroller vendors are already shipping Cortex-M3-based microcontrollers, and a number of other vendors will also soon be shipping Cortex-M3 products. Here, we list a number of them:

Texas Instruments (formerly LuminaryMicro): The Stellaris Cortex-M3 microcontroller product range has over 100 devices in the family, including devices with 10/100 Ethernet MAC and PHY, USB, CAN, SPI, I²C, I²S, and so on.

ST Microelectronics: The ST's Cortex-M3 products included several product lines as follows:
- STM32 connectivity line for feature-packed device supporting USB On-The-Go (USB OTG), Ethernet, and memory cards.
- STM32L for ultralow-power applications with analog and LCD interface support.
- STM32 value line for low-cost applications.

Toshiba: The TX03 product series targets various application areas including industrial and automotive applications, consumer applications, and supporting interfaces, including USB, CAN, Ethernet, and analog interface on a number of devices.

Atmel: The SAM3U product family supports high-speed USB interface, dual bank flash, communication interface (SPI, Secure Digital Input Output (SDIO), and Serial Synchronous Controller (SSC)), memory card interface, as well as ADC.

Energy Micro: The EFM®32 is a highly energy efficient product family with innovative peripherals that can react and respond without central processing unit (CPU) intervention. Interface support includes I²C, LCD, ADC, DAC, and special features like Advanced Encryption Standard (AES) support.

NXP: There are two Cortex-M3 product lines in the NXP LPC® product family: the LPC1700 product line and the LPC1300 product line. The LPC1700 is targeted at high-performance application with support for fast communication, motor control, and industrial control (e.g., USB, CAN, I²S, and DAC). LPC1300 is targeted at low-power and mixed signal applications.

17.2 DEVELOPMENT TOOLS

To start using the Cortex-M3, you'll need a number of tools. Typically, they will include the following:

- *A compiler and/or assembler*: Software to compile your C or assembler application codes. Almost all C compiler suites come with an assembler.
- *Instruction set simulator*: Software to simulate the instruction execution for debugging in early stages of software development. This is optional.
- *In-circuit emulator (ICE) or debug probe*: A hardware device to connect your debug host (usually a personal computer) to the target circuit. The interface can be either JTAG or SW.
- *A development board*: A circuit board that contains the microcontroller.
- *Trace capture*: An optional hardware and software package for capturing instruction traces or output from Data Watchpoint and Trace (DWT) and Instrumentation Trace Macrocell (ITM)

modules that outputs them to human-readable format. Sometimes the trace capture feature can be build-in as a part of the ICE.

- *An embedded operating system*: An operating system (OS) running on the microcontroller. This is optional; many applications do not require an OS.

17.2.1 **C Compiler and Debuggers**

A number of C compiler suites and development tools are already available for the Cortex-M3 (see Table 17.1).

Table 17.1 Examples of Development Tools Supporting Cortex-M3	
Company	**Product[1]**
ARM (www.arm.com)	The Cortex-M3 is supported by RealView Development Suite (RVDS). RealView-ICE (RVI) is available for connecting debug target to debug environment. Note that older products, such as ARM Development Suite (ADS) and Software Development Toolkit (SDT), do not support the Cortex-M3.
Keil, an ARM company (www.keil.com)	The Cortex-M3 is supported in Microcontroller Development Kit (MDK-ARM). The ULINK™ USB-JTAG adapters are available for connecting debug target to debug Integrated Development Environment (IDE).
CodeSourcery (www.codesourcery.com)	GNU Tool Chain for ARM processors is now available at www.codesourcery.com/gnu_toolchains/arm/
Rowley Associates (www.rowley.co.uk)	CrossWorks for ARM is a GNU C compiler (GCC)-based development suite supporting the Cortex-M3 (www.rowley.co.uk/arm/index.htm).
IAR Systems (www.iar.com)	IAR Embedded Workbench for ARM and Cortex provides a C/C++ compiler and debug environment (v4.40 or above). A KickStart kit is also available, based on the Luminary Micro LM3S102 microcontroller, including debugger and a J-Link Debug Probe for connecting the target board to debug IDE.
Lauterbach (www.lauterbach.com)	JTAG debugger and trace utilities are available from Lauterbach.
Segger (www.segger.com)	Segger J-Link and J-Trace support Cortex-M3 microcontrollers for debug and trace operations.
Signum (www.signum.com)	JTAGJet and JTAGJet-Trace are in-circuit debuggers supporting Cortex-M3 microcontrollers for debug and trace operations.
Code Red (www.code-red-tech.com)	Red Suite™ 2 provides a GCC-based development environment, and Red Probe is for debug operations supporting JTAG and SW protocols.
National Instrument (www.ni.com)	LabVIEW Embedded Module for ARM Microcontroller
Raisonance (www.raisonance.com)	RKit-ARM Software Toolset is a GCC-based toolchain for ARM Microcontroller. Raisonance also provides debugger (RLink) and starter kit (STM32 Primer)
GNU GCC (www.gcc.gnu.org)	The support for the Cortex-M3 processor has been added to the GCC.

[1]Product names are registered trademarks of the companies listed on the left-hand side of the table.

Table 17.2 Examples of Embedded OSs Supporting Cortex-M3

Company	Product[2]
FreeRTOS (www.freertos.org)	FreeRTOS
Express Logic (www.rtos.com)	ThreadX(TM) RTOS
Micrium (www.micrium.com)	µC/OS-II
Mentor Graphics (www.mentor.com)	Nucleus/Nucleus Plus
Pumpkin Inc. (www.pumpkininc.com)	Salvo RTOS
CMX Systems (www.cmx.com)	CMX-RTX
Keil (www.keil.com)	ARM RTX
Segger (www.segger.com)	emboss
IAR Systems (www.iar.com)	IAR PowerPac for ARM
eCosCentric (www.ecoscentric.com, www.ecos.sourceware.org)	eCos
Interniche Technologies Inc. (www.nichetask.com)	NicheTask
Green Hills Software (www.ghs.com)	µVelOSity
Open source (www.linux-arm.org/LinuxKernel/LinuxM3)	µCLinux
Quadros System (www.quadros.com)	RTXC
ENEA (www.enea.com)	OSE Epsilon RTOS
Raisonance (www.stm32circle.com/projects/circleos.php)	CircleOS
SCIOPTA (www.sciopta.com)	SCIOPTA RTOS
Micro Digital (www.smxrtos.com)	SMX RTOS

Free GCC-based C compilers are available from Gnu's Not Unix (GNU) web site and companies including CodeSourcery and Raisonance. You can also get evaluation versions of some commercial tools, such as Keil MDK-ARM, which are fully functional.

17.2.2 Embedded OS Support

Many applications require an OS to handle multithreading and resource management. Many OSs are developed for the embedded market. Currently, a number of these OSs are supported on the Cortex-M3 (see Table 17.2).

17.3 DIFFERENCES BETWEEN THE CORTEX-M3 REVISION 0 AND REVISION 1

Early versions of Cortex-M3 products were based on revision 0 of the Cortex-M3 processor. Products based on Cortex-M3 revision 1 have been available since the third quarter of 2006. When this book was first published, all new Cortex-M3-based products should have been based on revision 1. Revision 2 of the Cortex-M3 was released in 2008 with products based on this release available in 2009. It could be important to know the revision of the chip you are using because there are a number of changes and improvements in these releases.

---

[2]Product names are registered trademarks of the companies listed on the left-hand side of the table.

For revision 1, the visible changes in the programmer's model and development features include the following:

- From revision 1, the stacking of registers when an exception occurs can be configured such that it is forced to begin from a double word aligned memory address. This is done by setting the STKALIGN bit in the Nested Vectored Interrupt Controller (NVIC) Configuration Control register.
- For that reason, the NVIC Configuration Control register has the STKALIGN bit.
- Release r1p1 includes the new AUXFAULT (Auxiliary Fault) status register (optional).
- Additional features include data value matching added to the DWT.
- ID register value changes due to the revision fields update.

Changes invisible to end users include the following:

- The memory attribute for Code memory space is hardwired to cacheable, allocated, nonbufferable, and nonshareable. This affects the I-Code Advanced High-Performance Bus (AHB) and the D-Code AHB interface but not the system bus interface. The change only affect caching and buffering behavior outside the processor (e.g. level 2 cache or memory controllers with cache). The processor internal write buffer behavior does not change and this modification has no effect on most microcontroller products.
- Supports bus multiplexing operation mode between I-Code AHB and D-Code AHB. Under this operation mode, the I-Code and D-Code buses can be merged using a simple bus multiplexer (the previous solution used an ADK BusMatrix component). This can lower the total gate count.
- Added new output port for connection to the AHB Trace Macrocell (HTM, a CoreSight debug component from ARM) for complex data trace operations.
- Debug components or debug control registers can be accessed even during system reset; only during power-on reset are those registers inaccessible.
- The Trace Port Interface Unit (TPIU) has Serial-Wire Viewer (SWV) operation mode support. This allows trace information to be captured with low-cost hardware.
- In revision 1, the VECTPENDING field in the NVIC Interrupt Control and Status register can be affected by the C_MASKINTS bit in the NVIC Debug Halting Control and Status register. If C_MASKINTS is set, the VECTPENDING value could be zero if the mask is masking a pending interrupt.
- The JTAG-DP debug interface module has been changed to the SW JTAG-Debug Port (SWJ-DP) module (see the next section, "Revision 1 Change: Moving from JTAG-DP to SWJ-DP"). Chip manufacturers can continue to use JTAG-DP, which is still a product in the CoreSight product family.

Since revision 0 of the Cortex-M3 does not have a double word stack alignment feature in its exception sequence, some compiler tools, such as ARM RealView Development Suite (RVDS) and the Keil RealView Microcontroller Development Kit, have special options to allow software adjustment of stacking, which allow the developed application to be embedded-application binary interface (EABI) compliant. This could be important if it has to work with other EABI-compliant development tools.

To determine which revision of the Cortex-M3 processor is used inside the microcontroller or system-on-chip (SoC), you can use the CPU ID Base register in the NVIC. The revision and variant number indicates which version of the Cortex-M3 it is, as shown in Table 17.3.

Individual debug components inside the Cortex-M3 processor also carry their own ID registers, and the revision field might also be different between revision 0 and revision 1.

Table 17.3 CPU ID Base Register (0xE000ED00)

	Implementer [31:24]	Variant [23:20]	Constant [19:16]	Part No [15:4]	Revision [3:0]
Revision 0 (r0p0)	0x41	0x0	0xF	0xC23	0x0
Revision 1 (r1p0)	0x41	0x0	0xF	0xC23	0x1
Revision 1 (r1p1)	0x41	0x1	0xF	0xC23	0x1
Revision 2 (r2p0)	0x41	0x2	0xF	0xC23	0x0

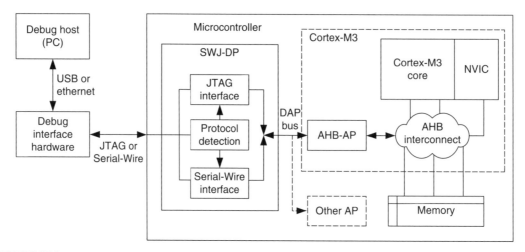

FIGURE 17.1

SWJ-DP: Combining JTAG-DP and SW-DP Functionalities.

17.3.1 Revision 1 Change: Moving from JTAG-DP to SWJ-DP

The JTAG-DP provided in some earlier Cortex-M3 products is replaced with the SWJ-DP. The SWJ-DP combines the function of the SW-DP and the JTAG-DP and with protocol detection (see Figure 17.1). Using this component, a Cortex-M3 device can support debugging with both SW and JTAG interfaces.

17.4 DIFFERENCES BETWEEN THE CORTEX-M3 REVISION 1 AND REVISION 2

In mid-2008, Revision 2 (r2p0) of the Cortex-M3 was released to silicon vendors. Products using revision 2 arrived on the market in 2009. Revision 2 has a number of new features, most of them targeted at reducing power consumption and offering better debug flexibility. Changes that are visible in the programmer's model include the following:

17.4.1 Default Configuration of Double Word Stack Alignment

The double word stack alignment feature for exception stacking is now enabled by default (silicon vendors may select to retain revision 1 behavior). This reduces the start up overhead for most C applications (removing the need to set the STKALIGN bit in the NVIC Configuration Control register).

17.4.2 Auxiliary Control Register

An Auxiliary Control register is added to NVIC to allow fine tuning of the processor behavior. For example, for debugging purposes, it is possible to switch off the write buffers in Cortex-M3 so that bus faults will be synchronous to the memory access instruction (precise). In this way, the faulting instruction can be pinpointed from the stacked return address (stacked PC) easily.

The details of the Auxiliary Control register are shown in Table 17.4.

Table 17.4 Auxiliary Control Register (0xE000E008)

Bits	Name	Type	Reset Value	Description
2	DISFOLD	R/W	0	Disable IT folding (prevent overlap of IT instruction execution phase with next instruction)
1	DISDEFWBUF	R/W	0	Disable write buffer for default memory map (memory accesses in MPU-mapped regions are not affected)
0	DISMCYCINT	R/W	0	Disable interruption of multiple cycle instructions like Load Multiple instruction (LDM), store multiple instruction, (STM), and 64-bit multiply and divide instructions.

17.4.3 ID Register Values Updates

Various ID registers in NVIC and debug components have been updated. For example, the CPU ID register in the NVIC is changed to 0x412FC230 (refer to Table 17.3).

17.4.4 Debug Features

For debug features, revision 2 has a number of improvements as follows:

- Watchpoint triggered data trace in the DWT now supports tracing of read transfers only and write transfers only. This can reduce the trace data bandwidth required as you can specify the data are traced only when it is changed, or only when the data are read.
- Higher flexibility in implementation of debug features. For example, the number of breakpoints and watchpoints can be reduced to reduce the size of the design in very low-power designs.
- Better support in multiprocessor debugging. A new interface is introduced to allow simultaneous restart and single stepping of multiple processors (not visible in the programmer's view).

17.4.5 Sleep Features

On the system design level, the existing sleep features have also improved (see Figure 17.2). In revision 2, it is possible for the wake up of the processor to be delayed. This allows more parts of the chip to be powered down, and the power management system can resume the program execution when the system is ready. This is needed in some microcontroller designs where some parts of the system are powered down during sleep, as it might take some time for the voltage supply to be stabilized after power is restored.

Aside from sleep extension, new technology has been used to allow the design to push the power consumption lower. In previous versions of the Cortex-M3, to allow the processor to wake up from sleep mode via interrupts, the free running clock of the core needed to be active. Although the free running clock only drives a small part of the system, it is still better to have this turned off completely.

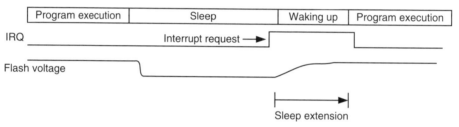

FIGURE 17.2

Sleep Extension Capability Added in Cortex-M3 Revision 2.

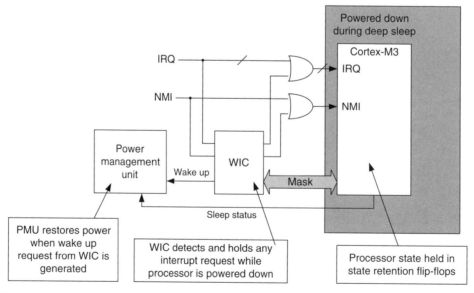

FIGURE 17.3

WIC Added in Cortex-M3 Revision 2.

To solve this problem, a simple interrupt controller is added externally to the processor. This controller, called Wake up Interrupt Controller (WIC), mirrors the interrupt masking function inside the NVIC during deep sleep and can tell the power management system when a wake up is required. By doing this, all the clock signals going to the Cortex-M3 processor could be stopped (see Figure 17.3).

In addition to stopping the clocks, the new design methodology also allows most parts of the processor to be powered down, with the state of the processor saved in special logic cells. When an interrupt arrives, the WIC sends a request to Power Management Unit (PMU) to power up the system. After the processor is powered up, the previous state of the processor is restored from the special logic cells and the processor is then ready to process the interrupt.

This power down feature reduces the power consumption of the design during sleep. However, this feature depends on the silicon manufacturing process being used and might not be available on some of the revision 2 products.

17.5 BENEFITS AND EFFECTS OF THE REVISION 2 NEW FEATURES

So what does that means in terms of developing embedded product development?

First, it means lower power consumption for the embedded products and better battery life. When WIC mode deep sleep is used, only a very small portion of the design needs to be active. Also, in designs targeted at extremely low power, the silicon vendor can reduce the size of the design by reducing the number of breakpoint and watchpoints.

Second, it provides better flexibility in debugging and troubleshooting. Beside from the improved data trace feature that can be used by debugger, we can also use the new Auxiliary Control register to force write transfers to be nonbufferable to pinpoint faulting instruction or disable interrupts during multiple cycle instructions so that each multiple load/store instruction will be completed before the exception is taken, this can make the analysis of memory contents easier. For systems of multiple Cortex-M3, revision 2 also brings the capability of simultaneous restarting and stepping of multiple cores.

In addition, revision 2 has a number of internal optimizations to allow higher performance and better interface features. This allows silicon vendors to develop faster Cortex-M3 products with more features. However, there are a few things that embedded programmers need to be aware of. They are as follows:

1. *Double word stack alignment for exception stack frame*: The exception stack frames will be aligned to double word memory location by default. Some assembly applications written for revision 0 or revision 1 that use stack to transfer data to exception handlers could be affected. The exception handler should determine if stack alignment has been carried out by reading the bit 9 of the stacked Program Status register (PSR) in the stack frame, then it can determine the address of the stacked data before the exception. Alternatively, the application can program the STKALIGN bit to 0 to get the same stacking behavior as revision 0 and revision 1. Applications compliant to EABI standard (e.g., C-Code compiled using an EABI compliant compiler) are not affected.

2. *SYSTICK Timer might stop in deep sleep*: If the Cortex-M3 microcontroller included power down feature or the core clocks are completely stopped for deep sleep mode, then the SYSTICK Timer might not be running during deep sleep. Embedded applications that use OS will need to use a timer externally to the processor core to wake up the processor for event scheduling.

3. *Debug and power-down feature*: The new power down feature is disabled when the processor is connected to a debugger. This is because the debugger needs to access to the processor's debug registers during a debug session. During a debug session, the core will still need to be able to be halted or enter sleep modes, but it should not trigger the power down sequence even if the power down feature is enabled. For testing of power down operations, the device under test should be disconnected from the debugger.

17.6 DIFFERENCES BETWEEN THE CORTEX-M3 AND CORTEX-M0

Some of you might have heard of the ARM Cortex-M0 processor. The programmer's model of the Cortex-M0 processor is quite similar to the Cortex-M3. However, it is smaller, supports fewer instructions and is based on the Von Neumann architecture. The Cortex-M0 processor is developed for ultralow-power designs and mixed signal applications, where logic gate count is critical. In minimum configuration, the Cortex-M0 processor is only 12 k logic gates in size, smaller than most 16-bit processors and some of the high-end 8-bit processors. However, the performance of the Cortex-M0 is 0.9 DMIPS per MHz, more than double of most 16-bit processors, and nearly 10 times of modern 8-bit microcontrollers. This makes the Cortex-M0 the most energy efficient 32-bit processor for microcontrollers.

17.6.1 Programmer's Model

There are a number of differences between the programmer's model of Cortex-M3 and Cortex-M0 processors (see Figure 17.4). The unprivileged level is only available in the Cortex-M3 processor. In addition, the FAULTMASK and BASEPRI special registers are not present in the Cortex-M0 either.

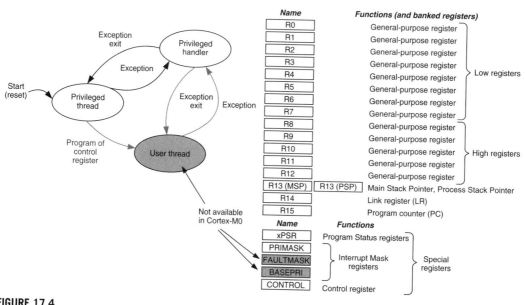

FIGURE 17.4

Programmer's Model Differences between the Cortex-M3 and Cortex-M0.

The xPSR also has some minor differences. The Q bit in the Application Program Status register (APSR) and the Interrupt Continuable Instruction/IF-THEN (ICI/IT) bit fields in the Execution Program Status register (EPSR) are not available in the Cortex-M0. The Cortex-M0 processor does not support the IT instruction block, and interruption of a multiple load or store instruction will result in the instruction being cancelled and restarted when the interrupt handler completes.

17.6.2 Exceptions and NVIC

The exception operation of the Cortex-M0 processor is the same as in the Cortex-M3. The interrupts and exceptions are vectored, and the NVIC handles nested exceptions automatically. Some of the system exceptions on the Cortex-M3 are not available on the Cortex-M0 processor. These include the bus fault, usage fault, memory management fault, and the debug monitor exceptions. If a fault occurs in a Cortex-M0 application, the hard fault handler is executed. Some of the fault status registers available on the Cortex-M3 are also not available on the Cortex-M0.

The priority registers in the Cortex-M0 are only 2 bits. As a result, only four priority levels are available for interrupts and system exceptions with configurable priority. There is no dynamic priority switching support in the Cortex-M0 processor, so the priority of interrupt and exceptions are normally programmed at the beginning of the application and remain unchanged afterwards. The NVIC in the Cortex-M0 has a similar programmer's model compared with the one in the Cortex-M3. However, the registers are word accessible only. So if the priority level of an interrupt needs to be changed, it might be necessary to read the whole word, modify the priority level for the interrupt, and then write it back. A number of registers in the Cortex-M3 NVIC are also not available in the Cortex-M0 NVIC as follows:

- *Vector Table Offset register*: The vector table is fixed at address 0x0. However, microcontrollers might feature memory map switching features to allow changing of exception vectors at run time.
- *Software Trigger Interrupt register*: To generate an exception by software, the Interrupt Set Pending register is used instead.
- *Interrupt Active Status register*
- *Interrupt Controller Type register*

17.6.3 Instruction Set

The Cortex-M0 processor is based on ARMv6-M architecture. It supports 16-bit Thumb® instructions and a few 32-bit Thumb instructions (branch with link [BL], instruction synchronization barrier [ISB], data synchronization barrier [DSB], data memory barrier [DMB], MRS, and MSR). A number of instructions in Cortex-M3 are not supported on the Cortex-M0. For example,

- IT instruction block
- Compare and branch (compare and branch if zero [CBZ] and compare and branch if nonzero [CBNZ])
- Multiple accumulate instructions (multiply accumulate [MLA], multiply and subtract [MLS], signed multiply accumulate long [SMLAL], and unsigned multiply accumulate long [UMLAL]) and multiply instructions with 64-bit results (unsigned multiply long [UMULL] and signed multiply long [SMULL])

- Hardware divide instructions (unsigned divide [UDIV] and signed divide [SDIV]) and saturation (signed saturate [SSAT] and unsigned saturate [USAT])
- Table branch instruction (Table Branch Half word [TBH] and Table Branch Byte [TBB])
- Exclusive access instructions
- Bit field processing instructions (unsigned bit field extract [UBFX], signed bit field extract [SBFX], Bit Field Insert [BFI], and Bit Field Clear [BFC])
- Some data processing instructions (count leading zero [CLZ], rotate right extended [RRX], and reverse bit [RBIT])
- Load/store instructions with address modes or register combinations that are only supported with 32-bit instruction format
- Load/store instructions with translate (load word data from memory to register with unprivileged access [LDRT] and store word to memory with unprivileged access [STRT])

17.6.4 Memory System Features

Both the memory maps from the Cortex-M3 and Cortex-M0 are divided into a number of regions including CODE, SRAM, peripherals, and so on. However, a number of memory system features on the Cortex-M3 are not available in the Cortex-M0. These include the following:

- Bit band regions
- Unaligned transfer support
- MPU (optional in Cortex-M3 processor)
- Exclusive accesses

17.6.5 Debug Features

The Cortex-M0 processor does not include any trace feature (no ETM, ITM). When compared with the Cortex-M3, it supports a smaller number of breakpoints and watchpoints. (See Table 17.5.)

In most cases, the debug connection of Cortex-M0 microcontrollers only supports one type of debug communication protocol (SW debug or JTAG). While for Cortex-M3 microcontroller products, the debug interface normally supports both SW and JTAG protocols and allows dynamic switching between the two.

17.6.6 Compatibility

The Cortex-M0 processor is upwards compatible with the Cortex-M3 processor. Programs compiled for the Cortex-M0 can be used on Cortex-M3 directly. However, programs compiled for Cortex-M3

Table 17.5 Debug Feature Comparison

	Cortex-M0	Cortex-M3
Breakpoints	Up to 4	Up to 8
Watchpoints	Up to 2	Up to 4

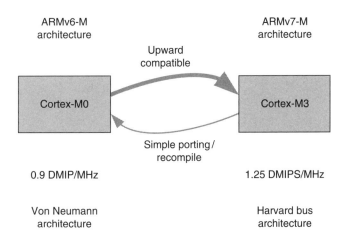

FIGURE 17.5

Compatibility between the Cortex-M3 and Cortex-M0 Processors.

cannot be used on Cortex-M0 (see Figure 17.5). Due to similarities between the two processors, most microcontroller applications can be written so that they can be used on either processor (provided that the memory map and the peripherals are compatible).

If an embedded application is to be used on both Cortex-M3 and Cortex-M0 products, there are various areas we need to pay attention to so that the application can be compatible with both processors as follows:

1. Access to the NVIC registers should use word access only or use core access functions in the CMSIS compliant device drivers.

2. To not use any unaligned data: In the Cortex-M3 processor, you can set the UNALIGN_TRP bit in the Configuration Control register to detect accidental generation of unaligned transfers. On the Cortex-M0 processor, attempts to generate unaligned transfers always result in a hard fault exception.

3. Since bit-band regions are not supported on the Cortex-M0, if the application needs to be used on both processors, the bit-band feature cannot be used. Alternatively, you can add conditionally compiled code to provide a software-based solution on the Cortex-M0.

The compatibility between Cortex-M0 and Cortex-M3 brings many advantages to embedded product developers. Besides from making software porting easy, it also allows embedded developers to debug their Cortex-M0 applications on a Cortex-M3 platform, which has more debug features like instruction trace and event trace. To make the behavior of the Cortex-M3 more like Cortex-M0, we can use the Auxiliary Control register in Cortex-M3 to disable buffered writes. However, since the Cortex-M3 processor has different instruction timing in a number of instructions compared with the Cortex-M0, there may still be execution cycle differences between the two.

Porting Applications from the ARM7 to the Cortex-M3

18

IN THIS CHAPTER

Overview 283
System Characteristics 283
Assembly Language Files 286
C Program Files 288
Precompiled Object Files 288
Optimization 289

18.1 OVERVIEW

For many engineers, porting existing program code to new architecture is a typical task. With the Cortex™-M3 products starting to emerge on the market, many of us have to face the challenge of porting ARM7TDMI (referred to as ARM7 in the following text) code to the Cortex-M3. This chapter evaluates a number of aspects involved in porting applications from the ARM7 to the Cortex-M3.

There are several areas to consider when you're porting from the ARM7 to the Cortex-M3. They are as follows:

- System characteristics
- Assembly language files
- C language files
- Optimization

Overall, low-level code, such as hardware control, task management, and exception handlers, requires the most changes whereas application codes normally can be ported with minor modification and recompiling.

18.2 SYSTEM CHARACTERISTICS

There are a number of system characteristic differences between ARM7-based systems and Cortex-M3-based systems (e.g., memory map, interrupts, Memory Protection Unit [MPU], system control, and operation modes).

18.2.1 Memory Map

The most obvious target of modification in porting programs between different microcontrollers is their memory map differences. In the ARM7, memory and peripherals can be located in almost any address whereas the Cortex-M3 processor has a predefined memory map. Memory address differences are usually resolved in compile and linking stages. Peripheral code porting could be more time consuming because the programmer model for the peripheral could be completely different. In that case, device driver codes might need to be completely rewritten.

Many ARM7 products provide a memory remap feature so that the vector table can be remapped to the Static Random Access Memory (SRAM) after boot-up. In the Cortex-M3, the vector table can be relocated using the NVIC register so that memory remapping is no longer needed. Therefore, the memory remap feature might be unavailable in many Cortex-M3 products.

Big endian support in the ARM7 is different from such support in the Cortex-M3. Program files can be recompiled to the new big endian system, but hardcoded lookup tables might need to be converted during the porting process.

In ARM720T, and some later ARM processors like ARM9, a feature called high vector is available, which allows the vector table to be located to 0xFFFF0000. This feature is for supporting Windows CE and is not available in the Cortex-M3.

18.2.2 Interrupts

The second target is the difference in the interrupt controller being used. Program code to control the interrupt controller, such as enabling or disabling interrupts, will need to be changed. In addition, new code is required for setting up interrupt priority levels and vector addresses for various interrupts.

The interrupt return method is also changed. This requires modification of interrupt return in assembler code or, if C language is used, it might be necessary to make adjustments on compile directives.

Enable and disable of interrupts, previously done by modifying Current Program Status register (CPSR), must be replaced by setting up the Interrupt Mask register. In addition, in the ARM7T-DMI, it is possible to reenable interrupt at the same time as interrupt return due to restore of CPSR from Saved Program Status register (SPSR). In the Cortex-M3, if interrupts are disabled during an interrupt handler by setting PRIMASK, the PRIMASK should be cleared manually before interrupt return. Otherwise, the interrupts will remain disabled.

In the Cortex-M3, some registers are automatically saved by the stacking and unstacking mechanisms. Therefore, some of the software stacking operations could be reduced or removed. However, in the case of the Fast Interrupt request (FIQ) handler, traditional ARM cores have separate registers for FIQ (R8–R11). Those registers can be used by the FIQ without the need to push them into the stack. However, in the Cortex-M3, these registers are not stacked automatically, so when an FIQ handler is ported to the Cortex-M3, either the registers being used by the handler must be changed or a stacking step will be needed.

Code for nested interrupt handling can be removed. In the Cortex-M3, the NVIC has built-in nested interrupt handling.

There are also differences in error handling. The Cortex-M3 provides various fault status registers so that the cause of faults can be located. In addition, new fault types are defined in the Cortex-M3 (e.g., stacking and unstacking faults, memory management faults, and hard faults). Therefore, fault handlers will need to be rewritten.

18.2.3 MPU

The MPU programming model is another system block that needs new program code set up. Microcontroller products based on the ARM7TDMI/ARM7TDMI-S do not have MPUs, so moving the application code to the Cortex-M3 should not be a problem. However, products based on the ARM720T have a Memory Management Unit (MMU), which has different functionalities to the MPU in Cortex-M3. If the application needs to use the MMU (as in a virtual memory system), it cannot be ported to the Cortex-M3.

18.2.4 System Control

System control is another key area to look into when you're porting applications. The Cortex-M3 has built-in instructions for entering sleep mode. In addition, the system controller inside Cortex-M3 products is likely to be completely different from that of the ARM7 products, so function code that involves system management features will need to be rewritten.

18.2.5 Operation Modes

In the ARM7, there are seven operation modes; in the Cortex-M3, these have been changed to difference exceptions (see Table 18.1).

The FIQ in the ARM7 can be ported as a normal interrupt request (IRQ) in the Cortex-M3 because in the Cortex-M3, we can set up the priority for a particular interrupt to be highest; thus it will be able to preempt other exceptions, just like the FIQ in the ARM7. However, due to the difference between banked FIQ registers in the ARM7 and the stacked registers in the Cortex-M3, the registers being used in the FIQ handler must be changed, or the registers used by the handler must be saved to the stack manually.

Table 18.1 Mapping of ARM7TDMI Exceptions and Modes to the Cortex-M3

Modes and Exceptions in the ARM7	Corresponding Modes and Exceptions in the Cortex-M3
Supervisor (default)	Privileged, thread
Supervisor (software interrupt)	Privileged, Supervisor Call (SVC)
FIQ	Privileged, interrupt
Interrupt request (IRQ)	Privileged, interrupt
Abort (prefetch)	Privileged, bus fault exception
Abort (data)	Privileged, bus fault exception
Undefined	Privileged, usage fault exception
System	Privileged, thread
User	User access (nonprivileged), thread

FIQ AND NONMASKABLE INTERRUPT

Many engineers might expect the FIQ in the ARM7 to be directly mapped to the nonmaskable interrupt (NMI) in the Cortex-M3. In some applications it is possible, but a number of differences between the FIQ and the NMI need special attention when you're porting applications using the NMI as an FIQ.

First, the NMI cannot be disabled whereas on the ARM7, the FIQ can be disabled by setting the F-bit in the CPSR. So it is possible in the Cortex-M3 for an NMI handler to start right at boot-up time whereas in the ARM7, the FIQ is disabled at reset.

Second, you cannot use SVC in an NMI handler on the Cortex-M3 whereas you can use software interrupt (SWI) in an FIQ handler on the ARM7. During execution of an FIQ handler on the ARM7, it is possible for other exceptions to take place (except IRQ, because the I-bit is set automatically when the FIQ is served). However, on the Cortex-M3, a fault exception inside the NMI handler can cause the processor to lock up.

18.3 ASSEMBLY LANGUAGE FILES

Porting assembly files depends on whether the file is for ARM state or Thumb® state.

18.3.1 Thumb State

If the file is for Thumb state, the situation is much easier. In most cases, the file can be reused without a problem. However, a few Thumb instructions in the ARM7 are not supported in the Cortex-M3 as follows:

- Any code that tries to switch to ARM state
- SWI is replaced by SVC (note that the program code for parameters passing and return need to be updated.)

Finally, make sure that the program accesses the stack only in full descending stack operations. It is possible, though uncommon, to implement a different stacking model differently (e.g., full ascending) in ARM7TDMI.

18.3.2 ARM State

The situation for ARM code is more complicated. There are several scenarios as follows:

- *Vector table*: In the ARM7, the vector table starts from address 0x0 and consists of branch instructions. In the Cortex-M3, the vector table contains the initial value for the stack pointer and reset vector address, followed by addresses of exception handlers. Due to these differences, the vector table will need to be completely rewritten.
- *Register initialization*: In the ARM7, it is often necessary to initialize different registers for different modes. For example, there are banked stack pointers (R13), a link register (R14), and an SPSR in the ARM7. Since the Cortex-M3 has a different programmer's model, the register initialization code will have to be changed. In fact, the register initialization code on the Cortex-M3 will be much simpler because there is no need to switch the processor into a different mode.
- *Mode switching and state switching codes*: Since the operation mode definition in the Cortex-M3 is different from that of the ARM7, the code for mode switching needs to be removed. The same applies to ARM/Thumb state switching code.

- *Interrupt enabling and disabling*: In the ARM7, interrupts can be enabled or disabled by clearing or setting the I-bit in the CPSR. In the Cortex-M3, this is done by clearing or setting an Interrupt Mask register, such as PRIMASK or FAULTMASK. Furthermore, there is no F-bit in the Cortex-M3 because there is no FIQ input.

- *Coprocessor accesses*: There is no coprocessor support on the Cortex-M3, so this kind of operation cannot be ported.

- *Interrupt handler and interrupt return*: In the ARM7, the first instruction of the interrupt handler is in the vector table, which normally contains a branch instruction to the actual interrupt handler. In the Cortex-M3, this step is no longer needed. For interrupt returns, the ARM7 relies on manual adjustment of the return program counter. In the Cortex-M3, the correctly adjusted program counter is saved into the stack and the interrupt return is triggered by loading EXC_RETURN into the program counter. Instructions, such as MOVS and SUBS, should not be used as interrupt returns on the Cortex-M3. Because of these differences, interrupt handlers and interrupt return codes need modification during porting.

- *Nested interrupt support code*: In the ARM7, when a nested interrupt is needed, usually the IRQ handler will need to switch the processor to system mode and re-enable the interrupt. This is not required in the Cortex-M3.

- *FIQ handler*: If an FIQ handler is to be ported, you might need to add an extra step to save the contents of R8–R11 to stack memory. In the ARM7, R8–R12 are banked, so the FIQ handler can skip the stack push for these registers. However, on the Cortex-M3, R0–R3 and R12 are saved onto the stack automatically, but R8–R11 are not.

- *SWI handler*: The SWI is replaced with an SVC. However, when porting an SWI handler to an SVC, the code to extract the passing parameter for the SWI instruction needs to be updated. The calling SVC instruction address can be found in the stacked PC, which is different from the SWI in the ARM7, where the program counter address has to be determined from the link register.

- *SWP instruction (swap)*: There is no swap instruction (SWP) in the Cortex-M3. If the SWP was used for semaphores, the exclusive access instructions should be used as replacement. This requires rewriting the semaphores code. If the instruction was used purely for data transfers, this can be replaced by multiple memory access instructions.

- *Access to CPSR and SPSR*: The CPSR in the ARM7 is replaced with Combined Program Status registers (xPSR) in the Cortex-M3, and the SPSR has been removed. If the application would like to access the current values of processor flags, the program code can be replaced with read access to the APSR. If an exception handler would like to access the Program Status register (PSR) before the exception takes place, it can find the value in the stack memory because the value of xPSR is automatically saved to the stack when an interrupt is accepted. So there is no need for an SPSR in the Cortex-M3.

- *Conditional execution*: In the ARM7, conditional execution is supported for many ARM instructions whereas most Thumb-2 instructions do not have the condition field inside the instruction coding. When porting these codes to the Cortex-M3, the assembly tool might automatically convert these conditional codes to use the IF-THEN (IT) instruction block; alternatively, we can manually insert the IT instructions or insert branches to produce conditionally executed codes. One potential issue with replacing conditional execution code with IT instruction blocks is that it could increase the code size and, as a result, could cause minor problems, such as load/store operations in some part of the program could exceed the access range of the instruction.

- *Use of the program counter value in code that involves calculation using the current program counter*: In running ARM code on the ARM7, the read value of the PC during an instruction is the address of the instruction plus 8. This is because the ARM7 has three pipeline stages and, when reading the PC during the execution stage, the program counter has already incremented twice, 4 bytes at a time. When porting code that processes the PC value to the Cortex-M3, since the code will be in Thumb, the offset of the program counter will only be 4.

- *Use of the R13 value*: In the ARM7, the stack pointer R13 has 32 bits; in the Cortex-M3 processor, the lowest 2 bits of the stack pointer are always forced to zero. Therefore, in the unlikely case that R13 is used as a data register, the code has to be modified because the lowest 2 bits would be lost.

For the rest of the ARM program code, we can try to compile it as Thumb/Thumb-2 and see if further modifications are needed. For example, some of the preindex and postindex memory access instructions in the ARM7 are not supported in the Cortex-M3 and have to be recoded into multiple instructions. Some of the code might have long branch range or large immediate data values that cannot be compiled as Thumb code and so must be modified to Thumb-2 code manually.

18.4 C PROGRAM FILES

Porting C program files is much easier than porting assembly files. In most cases, application code in C can be recompiled for the Cortex-M3 without a problem. However, there are still a few areas that potentially need modification, which are as follows:

- *Inline assemblers*: Some C program code might have inline assembly code that needs modification. This code can be easily located via the __asm keyword. If RealView Development Suite (RVDS)/RealView Compilation Tools (RVCT) 3.0 or later is used, it should be changed to Embedded Assembler.

- *Interrupt handler*: In the C program you can use __irq to create interrupt handlers that work with the ARM7. Due to the difference between the ARM7 and the Cortex-M3 interrupt behaviors, such as saved registers and interrupt returns, depending on development tools being used, the __irq keyword might need to be removed. (However, in ARM development tools including RVDS and RVCT, support for the Cortex-M3 is added to the __irq, and use of the __irq directive is recommended for reasons of clarity.)

ARM C compiler pragma directives like "#pragma arm" and "#pragma thumb" should be removed.

18.5 PRECOMPILED OBJECT FILES

Most C compilers will provide precompiled object files for various function libraries and startup code. Some of those (such as startup code for traditional ARM cores) cannot be used on the Cortex-M3 due to the difference in operation modes and states. A number of them will have source code available and can be recompiled using Thumb-2 code. Refer to your tool vendor documentation for details.

18.6 OPTIMIZATION

After getting the program to work with the Cortex-M3, you might be able to further improve it to obtain better performance and lower memory use. A number of areas should be explored:

- *Use of the Thumb-2 instruction*: For example, if a 16-bit Thumb instruction transfers data from one register to another and then carries a data processing operation on it, it might be possible to replace the sequence with a single Thumb-2 instruction. This can reduce the number of clock cycles required for the operation.

- *Bit band*: If peripherals are located in bit-band regions, access to control register bits can be greatly simplified by accessing the bit via a bit-band alias.

- *Multiply and divide*: Routines that require divide operations, such as converting values into decimals for display, can be modified to use the divide instructions in the Cortex-M3. For multiplication of larger data, the multiple instructions in the Cortex-M3, such as unsigned multiply long (UMULL), signed multiply long (SMULL), multiply accumulate (MLA), multiply and subtract (MLS), unsigned multiply accumulate long (UMLAL), and signed multiply accumulate long (SMLAL) can be used to reduce complexity of the code.

- *Immediate data*: Some of the immediate data that cannot be coded in Thumb instructions can be produced using Thumb-2 instructions.

- *Branches*: Some of the longer distance branches that cannot be coded in Thumb code (usually ending up with multiple branch steps) can be coded with Thumb-2 instructions.

- *Boolean data*: Multiple Boolean data (either 0 or 1) can be packed into a single byte/half word/ word in bit-band regions to save memory space. They can then be accessed via the bit-band alias.

- *Bit-field processing*: The Cortex-M3 provides a number of instructions for bit-field processing, including unsigned bit field extract (UBFX), signed bit field extract (SBFX), Bit Field Insert (BFI), Bit Field Clear (BFC), and reverse bits (RBIT). They can simplify many program codes for peripheral programming, data packet formation, or extraction and serial data communications.

- *IT instruction block*: Some of the short branches might be replaceable by the IT instruction block. By doing that, we could avoid wasting clock cycles when the pipeline is flushed during branching.

- *ARM/Thumb state switching*: In some situations, ARM developers divide code into various files so that some of them can be compiled to ARM code and others compiled to Thumb code. This is usually needed to improve code density where execution speed is not critical. With Thumb-2 features in the Cortex-M3, this step is no longer needed, so some of the state switching overhead can be removed, producing short code, less overhead, and possibly fewer program files.

Starting Cortex-M3 Development Using the GNU Tool Chain

IN THIS CHAPTER

Background 291
Getting the GNU Tool Chain 292
Development Flow 292
Examples 294
Accessing Special Registers 304
Using Unsupported Instructions 305
Inline Assembler in the GNU C Compiler 305

19.1 BACKGROUND

Many people use Gnu's Not Unix (GNU) tool chain for ARM product development, and a number of development tools for ARM are based on the GNU tool chain. The GNU tool chains supporting the Cortex™-M3 are available from the GNU gcc source, as well as from a number of vendors providing precompiled ready-to-use tool chains.

One of the vendors providing the GNU that supports the Cortex-M3 processor is CodeSourcery. The CodeSourcery ARM compiler is available in various packages:

CodeSourcery G++ Lite: This is freely available from the CodeSourcery web site (*www.codesourcery. com*). This free version provides command-line tools only and limited debug support.

CodeSourcery G++ Personal Edition: A popular choice as it is low-cost and has support features, including the following:
- Integration of Eclipse Integrated Development Environment (IDE) environment
- Support for a wide range of ARM microcontrollers, including CS3 support for these microcontrollers (e.g., linker scripts and debug configurations)
- Evaluation board support including Luminary Micro (Texas Instrument) Stellaris and STMicroelectronics STM32 evaluation boards. This allows full browsing of peripheral registers. The list of supported boards grows with every release.

DOI: 10.1016/B978-1-85617-963-8.00022-3

- Large collection of design examples
- Integrated support for multiple debug interfaces including
  - ARMUSB (built into Stellaris parts)
  - Segger J-Link
  - Keil ULINK2
- Board Builder Wizard to set up support for custom boards
  - Clone board definitions
  - Modify memory layout
  - Modify reset and start-up sequence
  - Debug configurations
- Support for importing StellarisWare examples

CodeSourcery G++ Professional Edition: All features in Personal Edition plus addition libraries and unlimited support.

The examples in this chapter are based on the command-line tools in CodeSourcery G++ Lite, as these areas of information are common to most GNU-based tool chains. This chapter introduces only the most basic steps in using the GNU tool chain. Detailed uses of the tool chain are available from documentation from tool vendors and on the Internet and are outside the scope of this book.

Assembler syntax for GNU assembler (AS in the GNU tool chain) is a bit different from ARM assembler. These differences include declarations, compile directives, comments, and the like. Therefore, assembly codes for ARM RealView Development Tools need modification before being used with the GNU tool chain.

19.2 GETTING THE GNU TOOL CHAIN

The compiled version of the GNU tool chain can be downloaded from *www.codesourcery.com/sgpp/ lite/arm*. A number of binary builds are available. For the simplest uses, let's select one with embedded-application binary interface (EABI) and without a specific embedded OS as the target platform. The tool chain is available for various development platforms such as Windows and Linux. The examples shown in this chapter should work with either version.

19.3 DEVELOPMENT FLOW

As with ARM tools, the GNU tool chain contains a compiler, an assembler, a linker, and additional utilities. The tools allow projects that contain source code in both C and assembly language (see Figure 19.1).

With GNU C compiler, the linking stage is normally invoked by the C compiler during the compile stage. This ensures that correct libraries and settings are passed on to the linker. If the linker is used directly without correct parameters from the C compiler, it might not be able to link the object files. In addition, without the compiler help routine, the linker might generate an output image that is not EABI compliant.

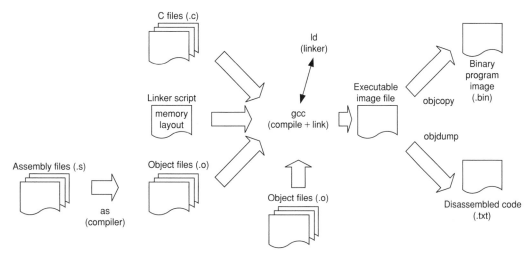

FIGURE 19.1

Example Development Flow Based on the CodeSourcery G++ Tool Chain.

Table 19.1 Command Name of the CodeSourcery Tool Chain

Function	Command (EABI Version)
Assembler	*arm-none-eabi-as*
C Compiler	*arm-none-eabi-gcc*
Linker	*arm-none-eabi-ld*
Binary image generator	*arm-none-eabi-objcopy*
Disassembler	*arm-none-eabi-objdump*

Notice the command names of tool chains differ from other vendors.

There are versions of the tool chain for different application environments (Symbian, Linux, EABI, and so on). The filenames of the programs usually have a prefix, depending on your tool chain target options. For example, if the EABI[1] environment is used, the Gnu C Compiler (GCC) command could be arm-xxxx-eabi-gcc. The following examples use the commands from the CodeSourcery GNU ARM Tool Chain, as shown in Table 19.1.

If your project is developed completely in assembler, then you could link the objects by using the linker directly (see Figure 19.2).

---

[1]EABI for the ARM architecture—executables must conform to this specification in order for them to be used with various development tool sets.

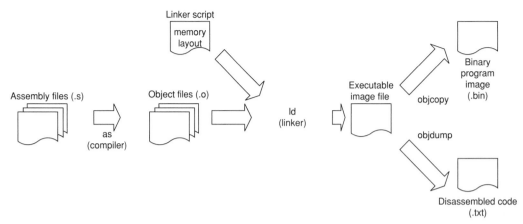

FIGURE 19.2

Example Development Flow for Assembly Projects.

19.4 **EXAMPLES**

Let's look at a few examples using the GNU tool chain.

19.4.1 **Example 1: The First Program**

For a start, let's try a simple assembly program that we covered in Chapter 10 that calculates $10 + 9 + 8 \ldots + 1$:

```
========== example1.s ==========
/* define constants */
        .equ    STACK_TOP, 0x20000800
        .text
        .syntax unified
        .thumb
        .global _start
        .type start, %function
_start:
        .word STACK_TOP, start
        /* Start of main program */
start:
        movs r0, #10
        movs r1, #0
        /* Calculate 10+9+8...+1 */
loop:
        adds r1, r0
        subs r0, #1
        bne  loop
        /* Result is now in R1 */
deadloop:
        b       deadloop
        .end
========== end of file ==========
```

- The *.word* directive here helps us define the starting stack pointer value as 0x20000800 and the reset vector as start.
- *.text* is a predefined directive indicating that it is a program region that needs to be assembled.
- *.syntax unified* indicates that the unified assembly language syntax is used.
- *.thumb* indicates that the program code is in Thumb® instruction set. Alternatively, you can use *.code16* for legacy Thumb instruction syntax.
- *.global* allows the label *_start* to be shared with other object files if needed.
- *_start* is a label indicating the starting point of the program region.
- *start* is a separate label indicating the reset handler.
- *.type start, %function* declares that the symbol *start* is a function. This is necessary for all the exception vectors in the vector table. Otherwise, the assembler sets the least significant bit (LSB) of the vector to zero.
- *.end* indicates the end of this program file.

Unlike ARM assembler, labels in GNU assemblers are followed by a colon (:). Comments are quoted with /* and */, and directives are prefixed by a period (.).

Notice that the reset vector (start) is defined as a function (.type start, %function) within thumb code (.thumb). The reason for this is to force the LSB of the reset vector to 1 to indicate that it starts in Thumb state. Otherwise, the processor will try starting in ARM mode, resulting in a hard fault. To assemble this file, we can use as in the following command:

```
$> arm-none-eabi-as -mcpu=cortex-m3 -mthumb example1.s -o example1.o
```

This creates the object file example1.o. The options *-mcpu* and *-mthumb* define the instruction set to be used. The linking stage can be done by *ld* as follows:

```
$> arm-none-eabi-ld -Ttext 0x0 -o example1.out example1.o
```

Then, the binary file can be created using Object Copy (*objcopy*) as follows:

```
$> arm-none-eabi-objcopy -Obinary example1.out example1.bin
```

We can examine the output by creating a disassembled code listing file using Object Dump (*objdump*):

```
$> arm-none-eabi-objdump -S example1.out > example1.list
```

which looks like this:

```
example1.out:  file format elf32-littlearm
Disassembly of section .text:
00000000 <_start>:
   0:   20000800   .word 0x20000800
   4:   00000009   .word 0x00000009

00000008 <start>:
   8:   200a   movs   r0, #10
   a:   2100   movs   r1, #0

0000000c <loop>:
   c:   1809   adds   r1, r1, r0
   e:   3801   subs   r0, #1
  10:   d1fc   bne.n  c <loop>

00000012 <deadloop>:
  12:   e7fe   b.n    12 <deadloop>
```

19.4.2 Example 2: Linking Multiple Files

As mentioned before, we can create multiple object files and link them together. Here, we have an example of two assembly files: example2a.s and example2b.s; example2a.s contains the vector table only, and example2b.s contains the program code. The *.global* is used to pass the address from one file to another:

```
========== example2a.s ==========
/* define constants */

        .equ      STACK_TOP, 0x20000800
        .syntax unified
        .global vectors_table
        .global start
        .global nmi_handler
        .thumb

vectors_table:
        .word STACK_TOP, start, nmi_handler, 0x00000000
        .end
========== end of file ==========

========== example2b.s ==========
/* Main program */
        .text
        .syntax unified
        .thumb
        .type start, %function
        .type nmi_handler, %function
        .global _start
        .global start
        .global nmi_handler
_start:
        /* Start of main program */
start:
        movs    r0, #10
        movs    r1, #0
        /* Calculate 10+9+8...+1 */
loop:
        adds    r1, r0
        subs    r0, #1
        bne     loop
        /* Result is now in R1 */
deadloop:
        b       deadloop
        /* Dummy NMI handler for illustration */
nmi_handler:
        bx      lr
        .end
========== end of file ==========
```

To create the executable image, the following steps are used:

1. Assemble example2a.s:

```
$> arm-none-eabi-as -mcpu=cortex-m3 -mthumb example2a.s -o example2a.o
```

2. Assemble example2b.s:

```
$> arm-none-eabi-as -mcpu=cortex-m3 -mthumb example2b.s -o example2b.o
```

3. Link the object files to a single image. Note that the order of the object files in the command line affects the order of the objects in the final executable image:

```
$> arm-none-eabi-ld -Ttext 0x0 -o example2.out example2a.o example2b.o
```

4. The binary file can then be generated as follows:

```
$> arm-none-eabi-objcopy -Obinary example2.out example2.bin
```

5. As in the previous example, we generate a list file to check that we have a correctly assembled image:

```
$> arm-none-eabi-objdump -S example2.out > example2.list
```

As the number of files increases, the compile process can be simplified using a UNIX *makefile*. Individual development suites may also have built-in facilities to make the compile process easier.

19.4.3 Example 3: A Simple "Hello World" Program

To be a bit more ambitious, let's now try the "Hello World" program. (*Note*: We skipped the universal asynchronous receiver/transmitter [UART] initialization here; you need to add your own UART initialization code to try this example. An example of UART initialization in C language is provided in Chapter 20.)

```
========== example3a.s ==========
/* define constants */
        .equ     STACK_TOP, 0x20000800
        .syntax unified
        .thumb
        .global vectors_table
        .global _start
vectors_table:
        .word STACK_TOP, _start
        .end
========== end of file ==========

========== example3b.s ==========
        .text
        .syntax unified
        .thumb
        .global _start
        .type _start, %function
_start:
        /* Start of main program */
        movs r0, #0
        movs r1, #0
        movs r2, #0
        movs r3, #0
        movs r4, #0
        movs r5, #0
```

```
        ldr        r0,=hello
        bl         puts
        movs       r0, #0x4
        bl         putc
deadloop:
        b          deadloop
hello:
        .asciz     "Hello\n"
        .align

puts:   /* Subroutine to send string to UART */
        /* Input r0 = starting address of string */
        /* The string should be null terminated */
        push {r0, r1, lr}   /* Save registers */
        mov r1, r0          /* Copy address to R1, because */
                            /* R0 will be used as input for */
                            /* putc */
putsloop:
        ldrb.w r0,[r1],#1
        /* Read one character and increment address */
        cbz r0, putsloopexit /* if character is null, goto end */
        bl   putc
        b    putsloop
putsloopexit:
        pop {r0, r1, pc} /* return */

.equ UART0_DATA, 0x4000C000
.equ UART0_FLAG, 0x4000C018

putc:   /* Subroutine to send a character via UART */
        /* Input R0 = character to send */
        push {r1, r2, r3, lr}   /* Save registers */
        LDR    r1,=UART0_FLAG
putcwaitloop:
        ldr        r2,[r1]              /* Get status flag */
        tst.w      r2, #0x20            /* Check transmit buffer full flag bit */
        bne        putcwaitloop         /* If busy then loop */
        ldr        r1,=UART0_DATA       /* otherwise output data to transmit buffer */
        str        r0, [r1]
        pop        {r1, r2, r3, pc}     /* Return */
        .end
========== end of file ==========
```

In this example, we used *.asciz* to create a null terminated string. This is equivalent to using *.ascii* to define the string and following *.byte* to create a byte with a value of null. After defining the string, we used *.align* to ensure that the next instruction starts in the right place. Otherwise, the assembler might put the next instruction in an unaligned location.

To compile the program, create the binary image and disassemble outputs, the following steps can be used:

```
$> arm-none-eabi-as -mcpu=cortex-m3 -mthumb example3a.s -o example3a.o
$> arm-none-eabi-as -mcpu=cortex-m3 -mthumb example3b.s -o example3b.o
```

```
$> arm-none-eabi-ld -Ttext 0x0 -o example3.out example3a.o example3b.o
$> arm-none-eabi-objcopy -Obinary example3.out example3.bin
$> arm-none-eabi-objdump -S example3.out > example3.list
```

19.4.4 Example 4: Data in RAM

Very often we will store the data in Static Random Access Memory (SRAM). The following simple example shows the required setup:

```
========== example4.s ==========
      .equ        STACK_TOP, 0x20000800
      .text
      .syntax unified
      .thumb
      .global _start
      .type start, %function
_start:
      .word       STACK_TOP, start
       /* Start of main program */
start:
      movs r0, #10
      movs r1, #0
      /* Calculate 10+9+8...+1 */
loop:
      adds r1, r0
      subs r0, #1
      bne loop
      /* Result is now in R1 */
      ldr  r0,=Result
      str  r1,[r0]
deadloop:
      b     deadloop
      /* Data in LC - Local Common section */
      .lcomm Result 4 /* A 4 byte data called Result */
      .end
========== end of file ==========
```

In the program, the *.lcomm* pseudo-op is used to create an uninitialized block of storage inside the "bss" region. Inside this region, a *.word* directive is used to reserve a space labelled *Result*. The program code can then access this space using the defined label *Result*.

To link this program, we need to tell the linker where the RAM is. This can be done using the *-Tbss* option, which sets the data segment to the required location:

```
$> arm-none-eabi-as -mcpu=cortex-m3 -mthumb example4.s -o example4.o
$> arm-none-eabi-ld -Ttext 0x0 -Tbss 0x20000000 -o example4.out example4.o
$> arm-none-eabi-objcopy -Obinary example4.out example4.bin
$> arm-none-eabi-objdump -S example4.out > example4.list
```

19.4.5 **Example 5: C Program**

One of the main components in the GNU tool chain is the C compiler. In this example, the whole executable is coded using C. In addition, a linker script is needed to put the segments in place. First, let's look at the C program file:

```
========== example5.c ==========
// Declare functions
void myputs(char *string1);
void myputc(char mychar);
int  main(void);
void Reset_Handler(void);
void NMI_Handler(void);
void HardFault_Handler(void);
void UartInit(void);
// Declare _start - C startup code
extern void _start(void);
//--------------------------------
void Reset_Handler(void)
{
  // Call the CS3 reset handler
  _start();
}
//--------------------------------
//Dummy handler
void NMI_Handler(void)
{
  return;
}
//--------------------------------
//Dummy handler
void HardFault_Handler(void)
{
  return;
}
//--------------------------------
void UartInit(void)
{
  /* Add your UART initialization code here */
  return;
}
//--------------------------------
// Start of main program
int main(void)
{
#define NVIC_CCR (*((volatile unsigned long *)(0xE000ED14)))
const char *helloworld="Hello world\n";
  NVIC_CCR = NVIC_CCR | 0x200; /* Set STKALIGN in NVIC */
  UartInit();
  myputs(helloworld);
  while(1);
  return(0);
}
```

```
//----------------------------------
// Function to print a string
void myputs(char *string1)
{
char mychar;
int j;
j=0;
do {
  mychar = string1[j];
  if (mychar!=0) {
    myputc(mychar);
    j++;
    }
    } while (mychar != 0);
return;
}
//----------------------------------
void myputc(char mychar)
{
#define UART0_DATA   (*((volatile unsigned long *)(0x4000C000)))
#define UART0_FLAG   (*((volatile unsigned long *)(0x4000C018)))
// Wait until busy flag is clear
while ((UART0_FLAG & 0x20) != 0);
// Output character to UART
UART0_DATA = mychar;
return;
}
========== end of file ==========
```

This program prints the "Hello world" message via a UART interface. Depending on the UART you use, you need to provide your own UART setup code or use the device driver library from a microcontroller vendor to initialize the UART.

After reset, the reset handler calls the _start function, which is the C start-up routine. When the C runtime initialization is done, it executes the *main()* code. The CodeSourcery G++ packages use the CS3 (CodeSourcery Common Start-up Code Sequence) for start-up and vector table handling in microcontrollers. CS3 has a predefined vector table for the Cortex-M3 processor called "__cs3_interrupt_ vector_micro." The vector table is shown in Table 19.2.

The exception handlers we used in the program are mapped into these vector symbols using a linker script. In addition, the memory layout including the vector table positioning is also defined in this file. Users of CodeSourcery G++ Personal and Professional Editions, can find linker scripts for most available Cortex-M3 microcontrollers already included in the installation. For the CodeSourcery G++ Lite edition, a number of generic linker scripts in the arm-none-eabi\lib path can be found. In this example, we use a linker script modified from the generic linker script for Cortex-M processors (*generic-m.ld*). This modified linker script (*cortexm3.ld*) is provided in Appendix F.

The command for the compiler and link process is as follows:

```
$> arm-none-eabi-gcc -mcpu=cortex-m3 -mthumb example5.c
  -T cortexm3.ld -o example5.o
```

Table 19.2 Cortex-M3 Vector Table Definition in CS3

Number	Vector Name	Description
0	__cs3_stack	Initial Main Stack Pointer
1	__cs3_reset	Reset vector
2	__cs3_isr_nmi	Nonmaskable interrupt
3	__cs3_isr_hard_fault	Hard fault
4	__cs3_isr_mpu_fault	Memory management fault
5	__cs3_isr_bus_fault	Bus fault
6	__cs3_isr_usage_fault	Usage fault
7 … 10	__cs3_isr_reserved_7 … 10	Reserved exception types
11	__cs3_isr_svcall	Supervisor Call
12	__cs3_isr_debug	Debug monitor exception
13	__cs3_isr_reserved_13	Reserved exception types
14	__cs3_isr_pendsv	PendSV
15	__cs3_isr_systick	System Tick Timer
16 … 47	__cs3_isr_external_0 … __cs3_isr_external_31	External interrupt

The memory map information is passed on to the linker during the compile stage.

The gcc automatically carried out the linking, so there is no need to carry out a linking stage. Finally, the binary and disassembled list file can be generated:

```
$> arm-none-eabi-objcopy -Obinary example5.out example5.bin
$> arm-none-eabi-objdump -S example5.out > example5.list
```

The use of *Reset_Handler* in this C example is optional. You can point "*__cs_reset*" to the "*_start*" start-up routine in the linker script instead.

19.4.6 Example 6: C with Retargeting

In the last example, we created our own text output function, but in many cases, we would use the text output function provided by the C library. For example, we might need to use "printf" for text outputting. In this case, we need to implement a function to redirect printf output to the UART output routine.

The following example illustrates how to implement retargeting function to support "printf":

```
========== example6.c ==========
#include<stdio.h>
// Declare functions
void myputc(char mychar);
int  main(void);
void Reset_Handler(void);
void NMI_Handler(void);
void HardFault_Handler(void);
void UartInit(void);
// Declare _start - C startup code
extern void _start(void);
//-------------------------------
```

```c
void Reset_Handler(void)
{
  // Call the CS3 reset handler
  _start();
}
//---------------------------------
//Dummy handler
void NMI_Handler(void)
{
  return;
}
//---------------------------------
//Dummy handler
void HardFault_Handler(void)
{
  return;
}
//---------------------------------
void UartInit(void)
{
  /* Add your UART initialization code here */
  return;
}
//---------------------------------
// Retarget function
int _write_r(void *reent, int fd, char *ptr, size_t len)
{
  size_t i;
  for (i=0; i<len; i++)
  {
  myputc(ptr[i]); // call our character output function
  }
  return len;
}
//---------------------------------
// Start of main program
int main(void)
{
#define NVIC_CCR (*((volatile unsigned long *)(0xE000ED14)))
  NVIC_CCR = NVIC_CCR | 0x200; /* Set STKALIGN in NVIC */
  UartInit();
  printf("Hello world\n");
  while(1);
  return(0);
}
//---------------------------------
// Function to output a character
void myputc(char mychar)
{
#define UART0_DATA (*((volatile unsigned long *)(0x4000C000)))
#define UART0_FLAG (*((volatile unsigned long *)(0x4000C018)))
// Wait until busy flag is clear
  while ((UART0_FLAG & 0x20) != 0);
```

```
// Output character to UART
  UART0_DATA = mychar;
return;
}
========== end of file ==========
```

The retargeting is carried out by implementing the "_write_r" function. This function calls our own character output routine to display the "Hello world" message.

19.4.7 Example 7: Implement Your Own Vector Table

If you are not using CodeSourcery G++ tool chain, you might need to implement your own vector table. This can be done by using the following C code:

```
// Define the vector table
__attribute__ ((section("vectors")))
void (* const VectorArray[])(void) = {
  (void (*)(void))((unsigned long) MainStack + sizeof(MainStack)),
  Reset_Handler,
  NMI_Handler,
  HardFault_Handler
  };
```

And the stack memory can be defined using an array:

```
// Reserve 64 words memory space for the main stack
static unsigned long MainStack[64];
```

The vector table can be allocated to the start of the memory using linker script. For example:

```
.text :
{
  CREATE_OBJECT_SYMBOLS
  __cs3_region_start_rom = .;
  *(.cs3.region-head.rom)
  __cs3_interrupt_vector = __cs3_interrupt_vector_micro;
  *(vectors) /* vector table */
```

The "vectors" section needs to match the section name used when we declare the vector table. Otherwise, the vector table will not be allocating to the beginning of the memory correctly.

This method can be useful even when you are using the CodeSourcery tool chain; if you need more than 32 interrupt vectors, you can create extra vectors and place it right after the CS3 vector table.

19.5 ACCESSING SPECIAL REGISTERS

The CodeSourcery GNU ARM tool chain supports access to special registers. The names of the special registers must be in lowercase. For example:

```
msr     control, r1
mrs     r1, control
msr     apsr, R1
mrs     r0, psr
```

19.6 USING UNSUPPORTED INSTRUCTIONS

If you are using another GNU ARM tool chain, there might be cases in which the GNU assembler you are using does not support the assembly instruction that you wanted. In this situation, you can still insert the instruction in form of binary data using .*word*. For example:

```
.equ DW_MSR_CONTROL_R0, 0x8814F380
        ...
        MOV   R0, #0x1
.word   DW_MSR_CONTROL_R0   /* This set the processor in user mode */
        ...
```

19.7 INLINE ASSEMBLER IN THE GNU C COMPILER

As in the ARM C Compiler, the GNU C Compiler supports an inline assembler. The syntax is a little bit different:

```
__asm       ("    inst1 op1, op2... \n"
            "     inst2 op1, op2... \n"
            ...
            "     inst op1, op2...\n"
            : output_operands          /* optional */
            : input_operands           /* optional */
            : clobbered_register_list  /* optional */
            );
```

For example, a simple code to enter sleep mode looks like this:

```
void Sleep(void)
{  // Enter sleep mode using Wait-For-Interrupt
  __asm (
    "WFI\n"
    );
}
```

If the assembler code needs to have an input variable and an output variable—for example, divide a variable by 5 in the following code—it can be written as follows:

```
unsigned int DataIn, DataOut; /* variables for input and output */
...
__asm   ("mov   r0, %0\n"
         "mov   r3, #5\n"
         "udiv  r0, r0, r3\n"
         "mov   %1, r0\n"
         :"=r" (DataOut) : "r" (DataIn) : "cc", "r3" );
```

With this code, the input parameter is a C variable called *DataIn* (*%0* first parameter), and the code returns the result to another C variable called *DataOut* (*%1* second parameter). The inline assembler code manually modifies register *r3* and changes the condition flags *cc* so that they are listed in the clobbered register list.

For more examples of an inline assembler, refer to the GNU tool chain documentation *GCC-Inline-Assembly-HOWTO* on the Internet.

Getting Started with the Keil RealView Microcontroller Development Kit

20

IN THIS CHAPTER

Overview 307
Getting Started with μVision 308
Outputting the "Hello World" Message Via Universal Asynchronous Receiver/Transmitter 314
Testing the Software 317
Using the Debugger 318
The Instruction Set Simulator 325
Modifying the Vector Table...326
Stopwatch Example with Interrupts with CMSIS 327
Porting Existing Applications to Use CMSIS 334

20.1 OVERVIEW

Various commercial development platforms are available for the Cortex™-M3. One of the popular choices is the Keil RealView Microcontroller Development Kit (RealView MDK-ARM). The Real-View MDK-ARM contains various components as follows:

- μVision Integrated Development Environment (IDE)
- Debugger
- Simulator
- RealView Compilation Tools from ARM
  - C/C++ Compiler
  - Assembler
  - Linker and utilities
- RTX Real-Time Kernel
- Detailed start-up code for microcontrollers
- Flash programming algorithms
- Program examples

For learning about the Cortex-M3 with RealView MDK-ARM, it is not necessary to have Cortex-M3 hardware. The µVision environment contains an instruction set simulator that allows testing of simple programs that do not require a development board.

A free evaluation compact disc read-only memory (CD-ROM) for the Keil tool can be requested from the Keil web site (*www.keil.com*). This version is also included in a number of Cortex-M3 evaluation kits from various microcontroller vendors.

20.2 GETTING STARTED WITH µVISION

A number of examples are provided with the RealView MDK-ARM, including some examples for various Cortex-M3 microcontroller products and evaluation boards available on the market. In addition, you can also download device driver libraries from microcontroller vendor web sites, which also contain a number of examples. These examples provide a powerful set of device driver libraries that are ready to use. It's easy to modify the provided examples to start developing your application or you can develop your project from scratch. The following examples illustrate how this is done. The examples shown in this chapter are based on the version 3.80 of Keil MDK-ARM and on Luminary Micro LM3S811 devices.

After installing the RealView MDK, you can start the µVision from the program menu. After installation, the µVision might start with a default project for a traditional ARM processor. We can close the current project and start a new one by selecting New Project in the pull-down menu (see Figure 20.1).

Figure 20.2 shows a new project directory called HelloWorld is created. Now, we need to select the targeted device for this project. In this example, the LM3S811 is selected (see Figure 20.3).

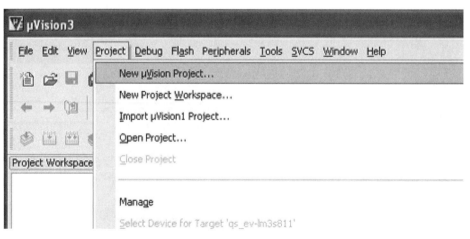

FIGURE 20.1

Selecting a New Project from the Program Menu.

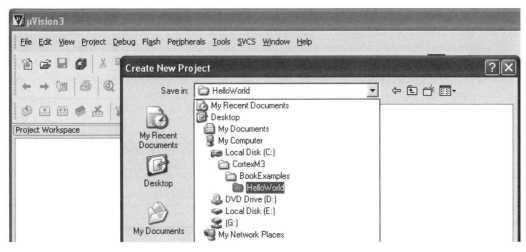

FIGURE 20.2

Choosing the HelloWorld Project Directory.

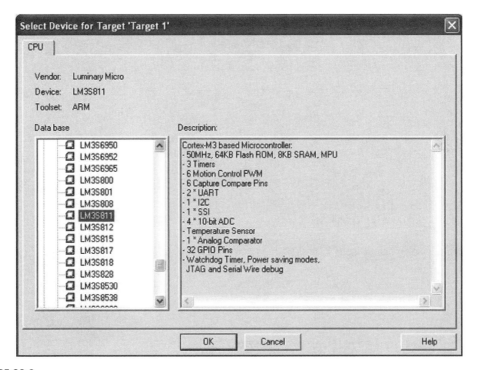

FIGURE 20.3

Selecting the LM3S811 Device.

The software will then ask if you would like to use the default startup code. In this case, we select Yes (see Figure 20.4).

As Figure 20.5 shows, we now have a project called Hello with only one file, called Startup.s. We can create a new C program file containing the main program (see Figure 20.6). A text file is created and saved as hello.c (see Figure 20.7). As Figure 20.8 shows, we now need to add this file to our project by right-clicking Source Group 1.

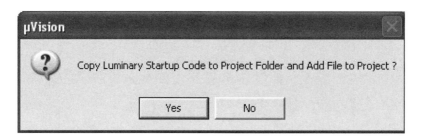

FIGURE 20.4

Choosing to Use the Default Startup Code.

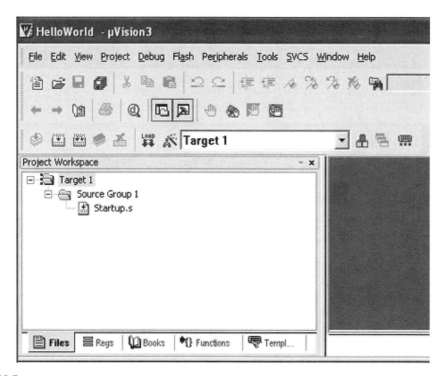

FIGURE 20.5

Project Created with the Default Startup Code.

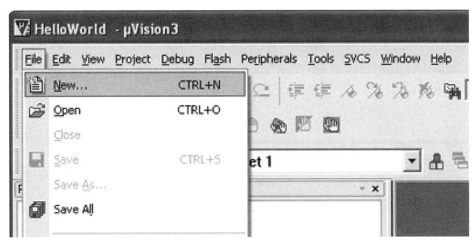

FIGURE 20.6

Creating a New C Program File.

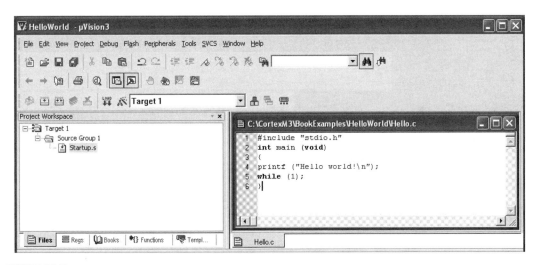

FIGURE 20.7

A HelloWorld C Example.

Select the hello.c that we created and then close the Add File window. Now, the project contains two files (see Figure 20.9).

We can define the project setting by right clicking on the "Target 1" in the Project Workspace and selecting option for target "Target 1." In the Target tab, we find that the memory layout details are

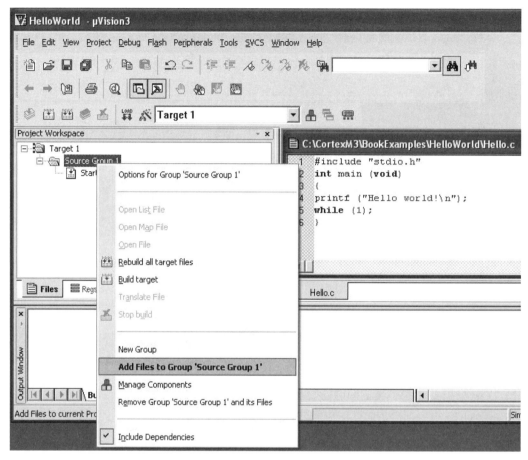

FIGURE 20.8

Adding the Hello.c Example to the Project.

RENAMING THE TARGET AND FILE GROUPS

The target name "Target 1" and file group name "Source Group 1" can be renamed to give a clearer meaning. This is done by clicking Target 1 and Source Group 1 in the Project Workspace and editing the names from there.

---

already set up by the tool automatically (see Figure 20.10). From this dialog, you can also access various other project options by the tabs on the top.

We can now compile the program. This can be done by clicking on the build target icons on the tool bar (see Figure 20.11) or by right-clicking Target 1 and selecting Build target.

You should see the compilation success message in the output window (see Figure 20.12).

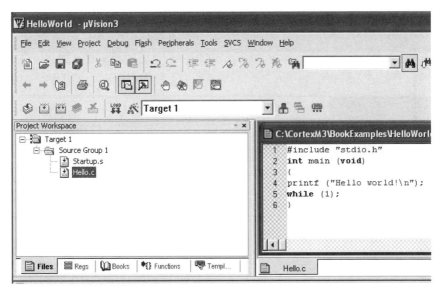

FIGURE 20.9

Project Window After the Hello.c Example is Added.

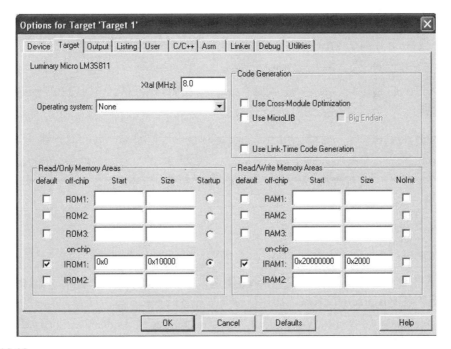

FIGURE 20.10

Project Option Dialog.

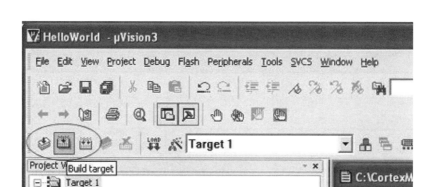

FIGURE 20.11

Starting the Compilation.

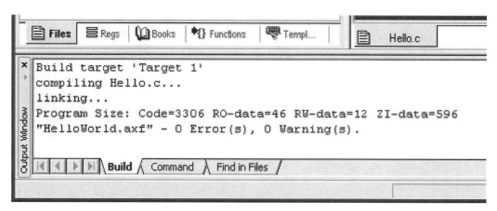

FIGURE 20.12

Compilation Result in the Output Window.

20.3 OUTPUTTING THE "HELLO WORLD" MESSAGE VIA UNIVERSAL ASYNCHRONOUS RECEIVER/TRANSMITTER

In the program code we created, we used the *printf* function in the standard C library. Since the C library does not know about the actual hardware we are using, if we want to output the text message using real hardware, such as the universal asynchronous receiver/transmitter (UART) on a chip, we need additional code.

As mentioned earlier in this book, the implementation of output to actual hardware is often referred to as *retargeting*. Besides creating text output, the retargeting code might also include functions for error handling and program termination. In this example, only the text output retargeting is covered.

For the following code, the "Hello world" message is output to UART 0 of the LM3S811 device. The target system used is the Luminary Micro LM3S811 evaluation board. The board has a 6 MHz crystal as a clock source and an internal Phase-Locked Loop (PLL) module that can step up the clock frequency to 50 MHz after a simple setup process. The baud rate setting is 115 200 and is output to HyperTerminal running on a Windows PC.

To retarget the *printf* message, we need to implement the *fputc* function. In the following code, we have created an *fputc* function that calls the *sendchar* function, which carries out the UART control.

```
hello.c
#include "stdio.h"
#pragma import(__use_no_semihosting_swi)
struct __FILE { int handle; };
FILE __stdout;

#define CR      0x0D // Carriage return
#define LF      0x0A // Linefeed
void Uart0Init(void);
void SetClockFreq(void);
int sendchar(int ch);
// Comment out the following line to use 6MHz clock
#define CLOCK50MHZ
// Register addresses
#define SYSCTRL_RCC     *((volatile unsigned long *)(0x400FE060))
#define SYSCTRL_RIS     *((volatile unsigned long *)(0x400FE050))
#define SYSCTRL_RCGC1   *((volatile unsigned long *)(0x400FE104))
#define SYSCTRL_RCGC2   *((volatile unsigned long *)(0x400FE108))
#define GPIOPA_AFSEL    *((volatile unsigned long *)(0x40004420))
#define UART0_DATA      *((volatile unsigned long *)(0x4000C000))
#define UART0_FLAG      *((volatile unsigned long *)(0x4000C018))
#define UART0_IBRD      *((volatile unsigned long *)(0x4000C024))
#define UART0_FBRD      *((volatile unsigned long *)(0x4000C028))
#define UART0_LCRH      *((volatile unsigned long *)(0x4000C02C))
#define UART0_CTRL      *((volatile unsigned long *)(0x4000C030))
#define UART0_RIS       *((volatile unsigned long *)(0x4000C03C))
#define NVIC_CCR        *((volatile unsigned long *)(0xE000ED14))
int main (void)
{ // Simple code to output hello world message
NVIC_CCR = NVIC_CCR | 0x200; // Set STKALIGN
SetClockFreq(); // Setup clock setting (50MHz/6MHz)
Uart0Init();     // Initialize Uart0
printf ("Hello world!\n");
while (1);
}
void SetClockFreq(void)
{
#ifdef    CLOCK50MHZ
// Set BYPASS, clear USRSYSDIV and SYSDIV
SYSCTRL_RCC = (SYSCTRL_RCC & 0xF83FFFFF) | 0x800 ;
```

Continued

```
// Clr OSCSRC, PWRDN and OEN
SYSCTRL_RCC = (SYSCTRL_RCC & 0xFFFFCFCF);
// Change SYSDIV, set USRSYSDIV and Crystal value
SYSCTRL_RCC = (SYSCTRL_RCC & 0xF87FFC3F) | 0x01C002C0;
// Wait until PLLLRIS is set
while ((SYSCTRL_RIS & 0x40)==0); // wait until PLLLRIS is set
// Clear bypass
SYSCTRL_RCC = (SYSCTRL_RCC & 0xFFFFF7FF) ;
#else
// Set BYPASS, clear USRSYSDIV and SYSDIV
SYSCTRL_RCC = (SYSCTRL_RCC & 0xF83FFFFF) | 0x800 ;
#endif
return;
}
void Uart0Init(void)
{
SYSCTRL_RCGC1 = SYSCTRL_RCGC1 | 0x0003; // Enable UART0 & UART1 clock
SYSCTRL_RCGC2 = SYSCTRL_RCGC2 | 0x0001; // Enable PORTA clock
UART0_CTRL = 0; // Disable UART
#ifdef  CLOCK50MHZ
UART0_IBRD = 27; // Program baud rate for 50MHz clock
UART0_FBRD = 9;
#else
UART0_IBRD = 3; // Program baud rate for 6MHz clock
UART0_FBRD = 17;
#endif
UART0_LCRH = 0x60; // 8 bit, no parity
UART0_CTRL = 0x301; // Enable TX and RX, and UART enable
GPIOPA_AFSEL = GPIOPA_AFSEL | 0x3; // Use GPIO pins as UART0
return;
}
/* Output a character to UART0 (used by printf function to output data) */
int sendchar (int ch) {
if (ch == '\n') {
   while ((UART0_FLAG & 0x8));    // Wait if it is busy
   UART0_DATA = CR;          // output extra CR to get correct
}                    // display on HyperTerminal
while ((UART0_FLAG & 0x8));    // Wait if it is busy
return (UART0_DATA = ch);      // output data
}
/* Retargetting code for text output */
int fputc(int ch, FILE *f) {
return (sendchar(ch));
}
void _sys_exit(int return_code) {
/* dummy exit */
label: goto label; /* endless loop */
}
```

The *SetupClockFreq* routine sets the system clock to 50 MHz. The setup sequence is device dependent. The subroutine can also be used to set the clock frequency to 6 MHz if the "CLOCK50MHZ" compile directive is not set.

The UART initialization is carried out inside the Uart0Init subroutine. The setup process includes setting up the baud rate generator to provide a baud rate of 115 200; configuring the UART to 8 bits, no parity, and 1 stop bit; and switching the General Purpose Input/Output (GPIO) port to alternate function because the UART pins are shared with the GPIO port A. Before accessing the UART and the GPIO, the clocks for these blocks must be turned on. This is done by writing to `SYSCTRL_RCGC1` and `SYSCTRL_RCGC2`.

The retargeting code is carried out by *fputc*, a predefined function name for character outputs. This function calls the *sendchar* function to output the character to the UART. The *sendchar* function outputs an extra carriage return character as a new line is detected. This is needed to get the text output correct on HyperTerminal; otherwise, the new text in the next line will overwrite the previous line of text.

After the hello.c program is modified to include the retargeting code, the program is compiled again.

20.4 TESTING THE SOFTWARE

If you've got the Luminary Micro LM3S811 evaluation board, you can try out the example by downloading the compile program into Flash and getting the "Hello world" message display output from the HyperTerminal. Assuming that you have set up the software drivers that come with the evaluation board, you can download and test the program by following these steps.

First, set up the Flash download option. This can be accessed from the *Flash* pull-down menu, as shown in Figure 20.13.

Inside this menu, we select the Luminary Evaluation Board as the download target for this example (see Figure 20.14). In this menu, we also can see that μVision supports a number of different debug hardware.

After selecting the flash download target, we might need to click on the Setting button to ensure that a suitable setting is used. (The settings are shown in Figure 20.15.)

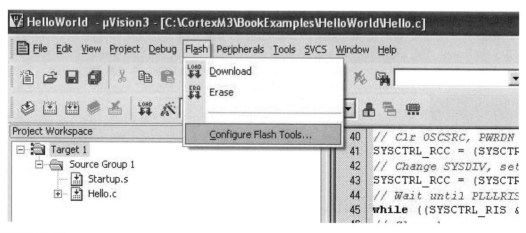

FIGURE 20.13

Setting Up Flash Programming Configuration.

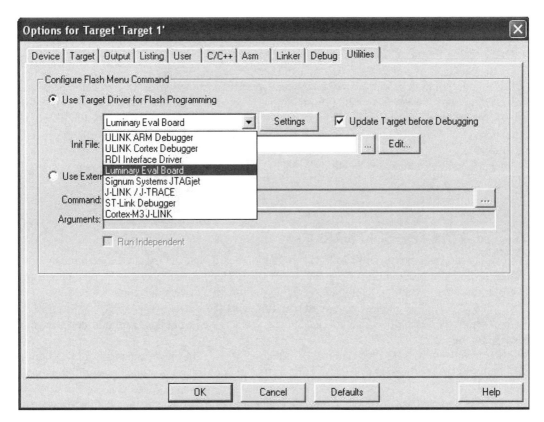

FIGURE 20.14

Selecting Flash Programming Driver.

Then, we can download the program to the Flash on a chip by selecting Download in the *Flash* pull-down menu. The message shown in Figure 20.16 will appear, indicating that the download is complete. *Note*: If you have the board already running with HyperTerminal, you might need to close HyperTerminal, disconnect the Universal Serial Bus (USB) cable from the PC, and reconnect before programming the Flash.

After the programming is completed, you can start HyperTerminal and connect to the board using the Virtual COM Port driver (via USB connection) and get the text display from the program running on the microcontroller (see Figure 20.17).

20.5 USING THE DEBUGGER

The debugger in μVision supports a number of in-circuit debuggers including the ULINK products (ULINK-2, ULINK-Pro, and ULINK-ME) from Keil (see Figure 20.18) and a number of debugger products from third parties.

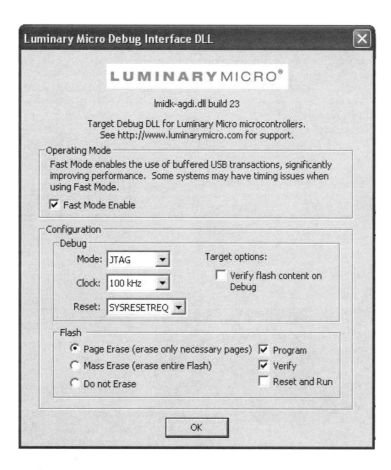

FIGURE 20.15

Flash Download Option.

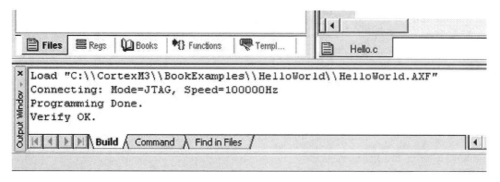

FIGURE 20.16

Report of the Download Process in the Output Window.

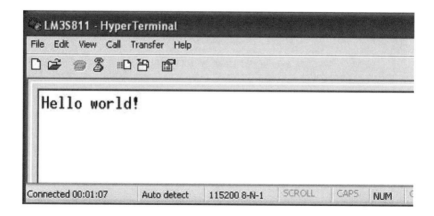

FIGURE 20.17

Output of the "Hello World" Example from HyperTerminal Console.

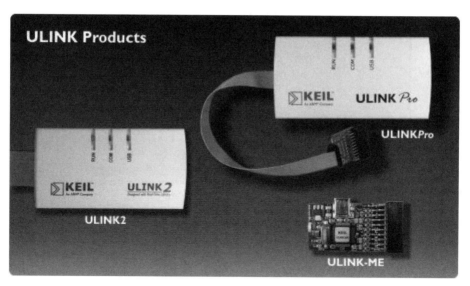

FIGURE 20.18

Keil ULINK Products.

Third party in-circuit debugger support includes the following:

- Signum JTAGJet and JTAGJet-Trace
- Segger J-Link and J-Trace
- Luminary Micro Evaluation board
- ST ST-Link

For this example, we will use the debugger in µVision to connect to the Luminary Evaluation Board to debug the application. By right-clicking the project Target 1 and selecting Options, we can access the debug option. In this case, we select to use the Luminary Eval Board for debugging (see Figure 20.19).

If you click on the "Settings" button next to the debug target selection, you can see the setting dialog as a Flash download option shown in Figure 20.15. You can select the debug protocol being used, debug clock speed and a few other settings. If you are using Keil ULINK-2/ULINK-Pro debugger for development, you can find more options (see Figure 20.20) via the setting button.

From here, you can select debug protocol (JTAG/Serial-Wire), debug communication clock speed, and trace and flash download options.

We can then start the debug session from the pull-down menu (see Figure 20.21). *Note*: If you are using virtual COM port with FTDI device driver and have the board already running with HyperTerminal, you might need to close HyperTerminal, disconnect the USB cable from the PC, and reconnect before starting the debug session.

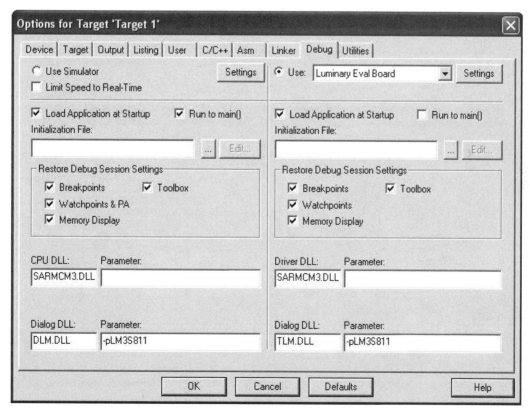

FIGURE 20.19

Configuring to Use Luminary Micro Evaluation Board with µVision Debugger.

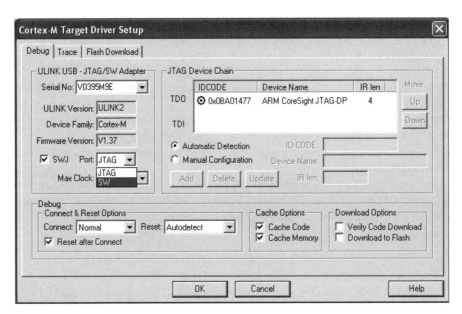

FIGURE 20.20

ULINK-2 Debug Options.

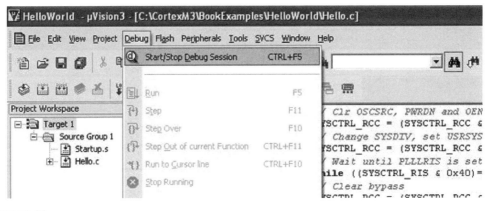

FIGURE 20.21

Starting a Debugger Session in μVision.

When the debugger starts, the IDE provides a register view to display register contents. You can also get the disassemble code window and see the current instruction address. In Figure 20.22, we can see that the core is halted at the *Reset_Handler*, the first instruction of the program execution.

For demonstration, a breakpoint is set to stop the program execution at the beginning of main. This can be done by right-clicking on the first line of code in the main program in the program code

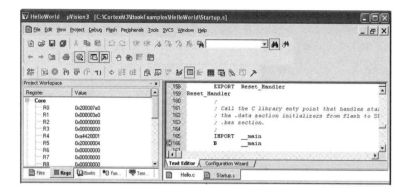

FIGURE 20.22

µVision Debug Environment.

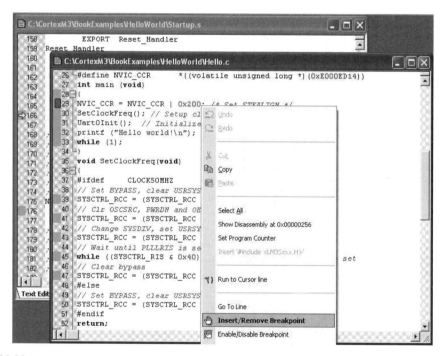

FIGURE 20.23

Insert or Remove Breakpoint.

window and selecting Insert/Remove Breakpoint (see Figure 20.23). *Note*: We could also use the *Run to main( )* feature in the debug option to get the program execution to stop at the beginning of main.

After the breakpoint is added, a red marker will be shown in the left side of the code line as shown in line 29 of the code in Figure 20.23. The program execution can then be started using the Run button on the tool bar (see Figure 20.24).

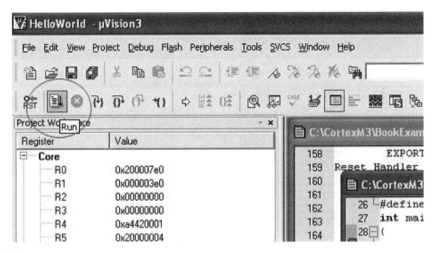

FIGURE 20.24

Starting Program Execution Using the Run Button.

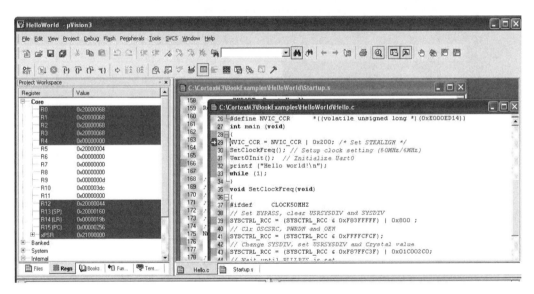

FIGURE 20.25

Program Execution Halted at the Beginning of Main When a Breakpoint Is Hit.

The program execution is then started, and it stops when it gets to the start of the main program (see Figure 20.25).

We can then use the stepping control of the tool bar to test our application and examine the results using the register window.

20.6 **THE INSTRUCTION SET SIMULATOR**

The μVision IDE also comes with an instruction set simulator that can be used for debugging applications. The operation is similar to using the debugger with hardware and is a useful tool for learning the Cortex-M3. To use the instruction set simulator, change the debug option of the project to Use Simulator (see Figure 20.26). Note that the simulator might not be able to simulate all hardware peripheral behaviors for some microcontroller products, so the UART interface code might not simulate correctly.

The simulator in μVision has full device level support for a wide range of Cortex-M3 devices. In some cases where full device simulation is not available, you might need to adjust the memory setting when using the simulator for debugging. This is done by accessing the Memory Map option after starting the debugging session (see Figure 20.27).

For example, you might need to add the UART memory address range to the memory map (see Figure 20.28). Otherwise, you will get an abort exception in the simulation when you try to access the

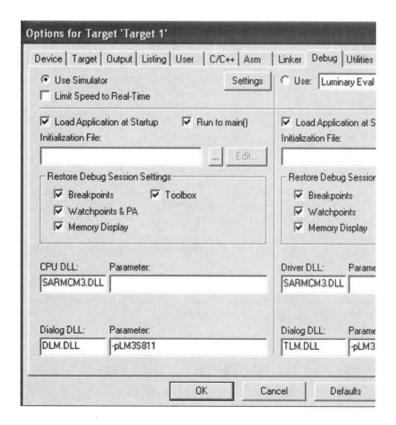

FIGURE 20.26

Selecting Simulator as Debugging Target.

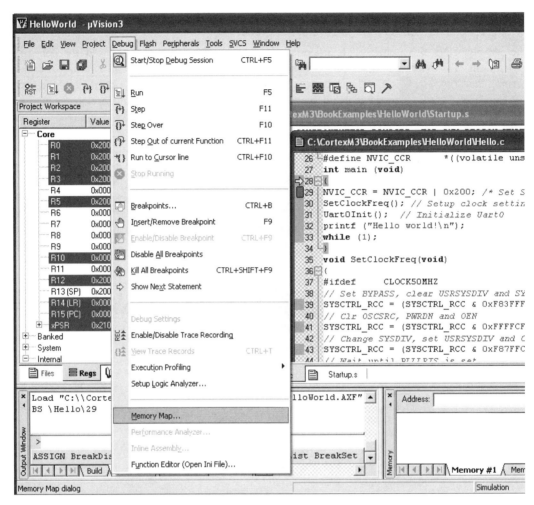

FIGURE 20.27

Accessing the Memory Map Option.

UART. But in most cases, all of the required memory map should have been set up for you when you specified a microcontroller, so usually this is not necessary.

20.7 MODIFYING THE VECTOR TABLE

In the previous example, the vector table is defined inside the file Startup.s, which is a standard startup code the tool prepares automatically. This file contains the vector table, a default reset handler, a default nonmaskable interrupt (NMI) handler, a default hard fault handler, and a default interrupt handler. These exception handlers might need to be customized or modified, depending on your application. For

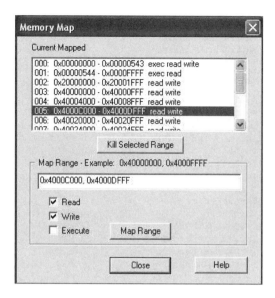

FIGURE 20.28

Adding New Memory Range to Simulator Memory Setup.

example, if a peripheral interrupt is required for your application, you need to change the vector table so that the Interrupt Service Routine (ISR) you created will be executed when the interrupt is triggered.

The default exception handlers are in the form of assembly code inside *Startup.s*. However, the exception handlers can be implemented in C or in a different assembly program file. In these cases, the IMPORT command in the assembler will be required to indicate that the interrupt handler address label is defined in another file.

For example, if we want to add a SYSTICK exception handler and a UART exception handler, we can edit the file startup.s as shown in Figure 20.29 as follows:

- Comment out existing default interrupt handler for the exception.
- Add the IMPORT commands for the two exception vectors that are defined in the C-code. This is required if the handlers are in a separated C or assembly program file.
- Add the exception vectors in the vector table with Define Constant Data (DCD) command.

These modifications of startup code (startup.s) for adding exception handlers are not required with Cortex Microcontroller Software Interface Standard (CMSIS) compliant device driver libraries if the name of the exception handlers match the handler names specific in the CMSIS startup code. The next example illustrates how this is done.

20.8 STOPWATCH EXAMPLE WITH INTERRUPTS WITH CMSIS

This example includes the use of exceptions, such as SYSTICK and the interrupt (UART0). The stopwatch to be developed has three states, as illustrated in Figure 20.30.

```
097     DCD     0                         ; Reserved
098     DCD     IntDefaultHandler         ; SVCall Handler
099     DCD     IntDefaultHandler         ; Debug Monitor Handler
100     DCD     0                         ; Reserved
101     DCD     IntDefaultHandler         ; PendSV Handler
102     ;DCD    IntDefaultHandler         ; SysTick Handler
103     IMPORT  SysTickHandler
104     DCD     SysTickHandler
105     DCD     IntDefaultHandler         ; GPIO Port A
106     DCD     IntDefaultHandler         ; GPIO Port B
107     DCD     IntDefaultHandler         ; GPIO Port C
108     DCD     IntDefaultHandler         ; GPIO Port D
109     DCD     IntDefaultHandler         ; GPIO Port E
110     ;DCD    IntDefaultHandler         ; UART0
111     IMPORT  Uart0Handler
112     DCD     Uart0Handler
113     DCD     IntDefaultHandler         ; UART1
114     DCD     IntDefaultHandler         ; SSI
115     DCD     IntDefaultHandler         ; I2C
116     DCD     IntDefaultHandler         ; PWM Fault
```

FIGURE 20.29

Adding Exception Vectors to the Vector Table by IMPORT and DCD Commands.

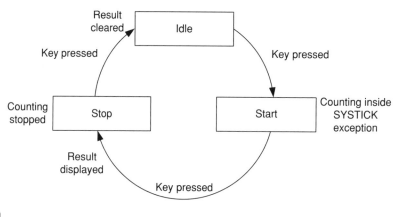

FIGURE 20.30

State Machine Design for Stopwatch.

Based on the previous example, the stopwatch is controlled by the PC using the UART interface. To simplify the example code, we fix the operating speed at 50 MHz.

The timing measurement is carried out by the SYSTICK, which interrupts the processor at 100 Hz. The SYSTICK is running from the core clock frequency at 50 MHz. Every time the SYSTICK exception handler is executed and if the stopwatch is running, the counter variable *TickCounter* increments.

Since display of text via UART is relatively slow, the control of the stopwatch is handled inside the exception handler and the display of the text and stopwatch values is carried out in the main (thread level). A simple software state machine is used to control the start, stop, and clear of the stopwatch.

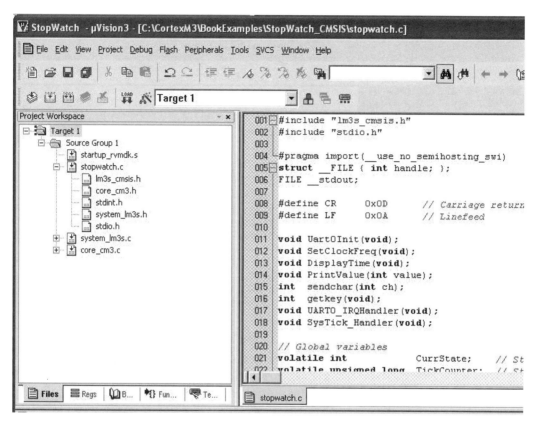

FIGURE 20.31

Stopwatch Project with CMSIS.

The state machine is controlled via the UART handler, which is triggered every time a character is received.

Using the same procedure we used for the "Hello world" example, let's start a new project called *stopwatch*. Instead of having hello.c, a C program file called stopwatch.c is added. In addition, we added a number of other files from the CMSIS in the project.

In Figure 20.31, we can see that the *Startup.s* is replaced by *startup_rvmdk.s*, and *lm3s_cmsis.h* is included in the *stopwatch.c*. In addition, *system_lm3s.c* and *core_cm3.c* are also included in the project. The file *system_lm3s.c* contains *SystemInit()* function that we use in the program. The file *core_cm3.c* contains intrinsic functions. Despite that this example does not use any intrinsic function, this file is included for completeness.

Some of these filenames are microcontroller vendor specific (e.g., startup_rvmdk.s, and lm3s_cmsis.h). These CMSIS files used in this project can be found in CMSIS compliant device driver libraries or from the CMSIS files available from *www.onarm.com*.

Use of CMSIS reduces the complexity of the *stopwatch.c* because peripheral register definitions and a number of system functions are provided by CMSIS.

```
stopwatch.c
#include "lm3s_cmsis.h" // Vendor specific CMSIS header
#include "stdio.h"       // For printf function
#pragma import(__use_no_semihosting_swi)
struct __FILE { int handle; };
FILE __stdout;
// Special characters for display function
#define CR 0x0D       // Carriage return
#define LF 0x0A       // Linefeed

// Function definitions
void Uart0Init(void);            // Initialization of UART0
void SetClockFreq(void);         // Set clock frequency to 50MHz
void DisplayTime(void);          // Display time value
int  sendchar(int ch);           // Output character to UART
int  getkey(void);               // Read charcter from UART
void UART0_IRQHandler(void);     // UART0 interrupt handler
void SysTick_Handler(void);      // SysTick exception handler

// Global variables
volatile int           CurrState;   // State machine
volatile unsigned long TickCounter; // Stop watch value
volatile int           KeyReceived; // Indicate user pressed a key
volatile int           userinput;   // Key pressed by user

#define IDLE_STATE 0                     // Definition of state machine
#define RUN_STATE  1
#define STOP_STATE 2

int main (void)
{
int    CurrStateLocal; // A local copy of current state

SystemInit();     // System initialization - part of CMSIS standard
                  // Not required with CMSIS v1.30 or after
                  // because this is done in startup code.
SetClockFreq();   // Setup clock setting (50MHz)

// Initialize global variable
CurrState = 0;
KeyReceived = 0;

// Initialization of hardware
SCB->CCR = SCB->CCR | 0x200; // Set STKALIGN
Uart0Init();                 // Initialize Uart0
SysTick_Config(499999);      // Initialize Systick (CMSIS function)
printf ("Stop Watch\n");

while (1) {
  CurrStateLocal = CurrState; // Make a local copy because the
  // value could change by UART handler at any time.
  switch (CurrStateLocal) {
    case (IDLE_STATE):
      printf ("\nPress any key to start\n");
      break;
    case (RUN_STATE):
      printf ("\nPress any key to stop\n");
      break;
    case (STOP_STATE):
      printf ("\nPress any key to clear\n");
      break;
```

```
    default:
      CurrState = IDLE_STATE;
      break;
    } // end of switch
    while (KeyReceived == 0) {
      if (CurrStateLocal==RUN_STATE){
        DisplayTime();
        }
      }; // Wait for user input
      if (CurrStateLocal==STOP_STATE) {
        TickCounter=0;
        DisplayTime(); // Display to indicate result is cleared
        }
      else if (CurrStateLocal==RUN_STATE) {
        DisplayTime(); // Display result
        }
    if (KeyReceived!=0) KeyReceived=0;
  }; // end of while loop
} // end of main
void SetClockFreq(void) // Set processor and UART clock
{
// Set BYPASS, clear USRSYSDIV and SYSDIV
SYSCTL->RCC = (SYSCTL->RCC & 0xF83FFFFF) | 0x800 ;
// Clr OSCSRC, PWRDN and OEN
SYSCTL->RCC = (SYSCTL->RCC & 0xFFFFCFCF);
// Change SYSDIV, set USRSYSDIV and Crystal value
SYSCTL->RCC = (SYSCTL->RCC & 0xF87FFC3F) | 0x01C002C0;
// Wait until PLLLRIS is set
while ((SYSCTL->RIS & 0x40)==0); // wait until PLLLRIS is set
// Clear bypass
SYSCTL->RCC = (SYSCTL->RCC & 0xFFFFF7FF);
return;
}
// UART0 initialization
void Uart0Init(void)
{ // Clock for UART functions
SYSCTL->RCGC1 = SYSCTL->RCGC1 | 0x0003; // Enable UART0 & UART1 clock
SYSCTL->RCGC2 = SYSCTL->RCGC2 | 0x0001; // Enable PORTA clock
  // UART setup
UART0->CTL = 0;        // Disable UART
UART0->IBRD = 27;    // Program baud rate for 50MHz clock
UART0->FBRD = 9;
UART0->LCRH = 0x60;  // 8 bit, no parity
UART0->CTL = 0x301;   // Enable TX and RX, and UART enable
UART0->IM = 0x10;     // Enable UART interrupt for receive data
GPIOA->AFSEL = GPIOA->AFSEL | 0x3; // Use GPIO pins as UART0
NVIC_EnableIRQ(UART0_IRQn); // Enable UART interrupt at NVIC
                           //(CMSIS function)
return;
}
// SYSTICK exception handler
void SysTick_Handler(void) // Function name is conform to CMSIS
{
if (CurrState==RUN_STATE) {
  TickCounter++;
  }
```

Continued

```c
return;
}
// UART0 RX interrupt handler
void UART0_IRQHandler(void) // Function name defined in CMSIS startup code
{
userinput = getkey();
// Indicate a key has been received
KeyReceived++;
// De-assert UART interrupt
UART0->ICR = 0x10;
// Switch state
switch (CurrState) {
  case (IDLE_STATE):
    CurrState = RUN_STATE;
    break;
  case (RUN_STATE):
    CurrState = STOP_STATE;
    break;
  case (STOP_STATE):
    CurrState = IDLE_STATE;
    break;
  default:
    CurrState = IDLE_STATE;
    break;
  } // end of switch
return;
}

// Display the time value
void DisplayTime(void)
{
unsigned long TickCounterCopy;
unsigned long TmpValue;

sendchar(CR);
TickCounterCopy = TickCounter; // Make a local copy because the
// value could change by SYSTICK handler at any time.
TmpValue        = TickCounterCopy / 6000; // Minutes
printf ("%d", TmpValue);
TickCounterCopy = TickCounterCopy - (TmpValue * 6000);
TmpValue        = TickCounterCopy / 100; // Seconds
sendchar(':');
printf ("%d", TmpValue);
TmpValue        = TickCounterCopy - (TmpValue * 100);
sendchar(':');
printf ("%d", TmpValue); // mini-seconds
sendchar(' ');
sendchar(' ');
return;
}

// Output a character to UART0 (used by printf function to output data)
int sendchar (int ch){
  if (ch == '\n'){
    while ((UART0->FR & 0x8)); // Wait if it is busy
    UART0->DR = CR; // output extra CR to get correct
                                // display on hyperterminal
  }
  while ((UART0->FR & 0x8)); // Wait if it is busy
```

```
   return (UART0->DR = ch); // output data
}
// Get user input
int getkey (void) { // Read character from Serial Port
   while (UART0->FR & 0x10); // Wait if RX FIFO empty
   return (UART0->DR);
}
// Retarget text output
int fputc(int ch, FILE *f) {
   return (sendchar(ch));
}

void _sys_exit(int return_code) {
   /* dummy exit */
label: goto label; /* endless loop */
}
```

When compared with the previous "Hello world" example, the UART initialization has changed slightly to enable interrupts when a character is received via the UART interface. To enable the UART interrupt request, the interrupt has to be enabled at the UART Interrupt Mask register, as well as at the NVIC. For the SYSTICK, only the exception control at the SYSTICK Control and Status register needs to be programmed.

In addition, a number of extra functions are added, including the UART and SYSTICK handlers, display functions, and SYSTICK initialization. Depending on the design of the peripheral, an exception/interrupt handler might need to clear the exception/interrupt request. In this case, the UART handler clears the UART interrupt request using the Interrupt Clear register (UART0->ICR).

After the program is compiled and downloaded to the evaluation board, it can then be tested by connecting to a PC running HyperTerminal. Figure 20.32 shows the result.

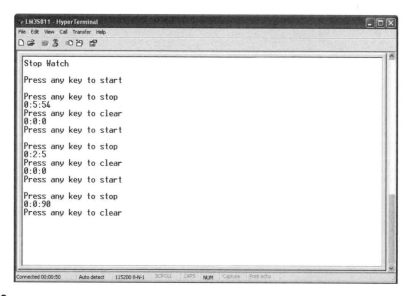

FIGURE 20.32

Output of the Stopwatch Example on the HyperTerminal Console.

20.9 PORTING EXISTING APPLICATIONS TO USE CMSIS

It is easy to port existing Cortex-M3 applications to use CMSIS. The modifications include the following:

- Replace the default startup code with CMSIS startup code for the targeted microcontroller.
- Modify the project setup to include CMSIS file.
- Modify the program to include CMSIS header file
- Modify the register definitions with CMSIS register definitions
- Replace existing processor peripheral access functions with CMSIS processor access functions.
- Name of exception handlers might need to be modified to ensure that they match the exception handler names used by the CMSIS startup code.
- Peripheral setup code could be replaced by device driver library functions if available.

By changing the application code to use CMSIS, the application becomes more portable, as outlined in Chapter 10.

Programming the Cortex-M3 Microcontrollers in NI LabVIEW

21

IN THIS CHAPTER

Overview 335
What Is LabVIEW 335
Development Flow 337
Example of a LabVIEW Project 339
How It Works 343
Additional Features in LabVIEW 344
Porting to Another ARM Processor 345

21.1 OVERVIEW

Besides C language and assembly language, there are other ways to create applications for Cortex™-M3 microcontrollers. One of the possible methods is by using the National Instruments LabVIEW graphical development environment. You can use LabVIEW on PCs as well as ARM microcontrollers, including Cortex-M3 microcontroller products.

21.2 WHAT IS LABVIEW

There are several versions of the LabVIEW development environment, including versions for PC (available for Windows and Linux) and embedded platforms, which support a number of different embedded processors, including the Cortex-M3 processor and ARM7TDMI processor.

The LabVIEW graphical programming language supports all the features that you expect in any programming language such as looping, conditional execution, and the handling of different data types. The main difference in working with LabVIEW is that you design programs in diagrams. For example, a simple loop to compute the sum of 1–10 can be represented by the For Loop shown in Figure 21.1.

The flow of data is represented by connections; different data types are represented in different line styles; and different color and variables are represented by icons. For example, a U32 icon refers to an unsigned 32-bit integer and an I32 icon refers to a signed 32-bit integer.

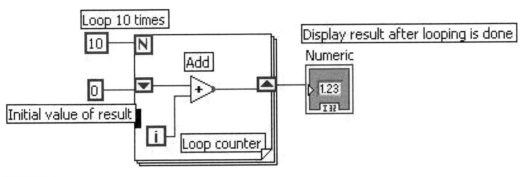

FIGURE 21.1

A Simple For Loop to Add 1–10.

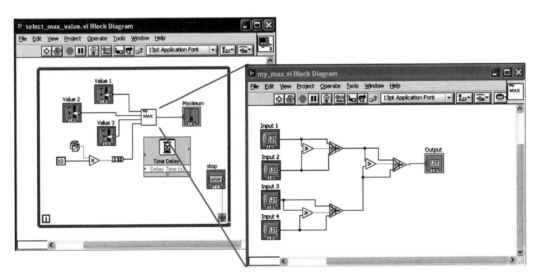

FIGURE 21.2

The VI and Sub-VI Allow Programs to Divide into Hierarchy Levels.

Similar to function calls and subroutines in traditional programming, in LabVIEW, you can design the software into hierarchy of modules called *virtual instruments* (VIs). Data passing into and outputs from sub-VIs indicate connection points so that a higher-hierarchy VI can connect input variables and output variables to it. For example, the right side of Figure 21.2 shows a sub-VI block diagram that takes four 32-bit integer inputs and selects the largest value as output. On the left side, the sub-VI is used to find the largest values from four 32-bit integer value sources: three from the slide bar control and one that is a random number.

The LabVIEW programming environment contains a large number of ready-to-use components that make your software development easier. These include common data-processing functions (e.g., abs—absolute value), signal processing functions, such as filter and spectral analysis, and many user interface components.

21.2.1 **Typical Application Areas**

So how does LabVIEW programming compare with traditional programming?

Ease of use—The LabVIEW programming environment makes it much easier to develop complex applications without the need to learn low-level hardware and software details. This allows different types of users, from students to scientists, to develop their applications without spending time learning processor architecture—they can focus on developing the algorithms and features. Domain-experts can now take advantage of the benefits of microcontroller designs.

Component library—The large number of software components available in LabVIEW also makes it easy to develop complex applications in a short time. The component library includes hundreds of mathematical and signal processing functions to quickly develop algorithms. Applications running on ARM microcontrollers can even connect to a graphical user interface (GUI) running on a PC so users can control the embedded system and observe the output easily.

Multithreading support—The graphical programming environment is concurrent in nature. It allows multiple threads to run at the same time. In contrast, in a traditional programming environment, it takes time for an inexperienced engineer to learn an embedded real time operating system (RTOS) well enough to develop a multithreading application.

Connectivity to test and data acquisition equipment—The LabVIEW programming environment provides easy-to-use interface components to data acquisition equipment communications systems as well as a large number of interface boards for industrial control and educational purposes.

The LabVIEW programming environment is popular in universities, laboratories, and scientific research institutions. It is commonly used in data acquisition, testing system automation, algorithm development and modeling, industrial control, and embedded system prototyping.

21.2.2 **What You Need to Use LabVIEW and ARM**

To start using LabVIEW for the Cortex-M3 microcontroller, you need the NI LabVIEW Embedded Module for ARM Microcontrollers, which includes the Keil ARM Microcontroller Development Kit (MDK-ARM). For evaluation, you can purchase a low-cost kit that includes an ARM microcontroller board (Cortex-M3/ARM7), Keil ULINK2 debugger, and the required development software, as shown in Figure 21.3 (a trial version of MDK-ARM is included on the CD). Details of the products are available on the National Instruments web site (*www.ni.com/arm*).

You can use LabVIEW with other Cortex-M3 microcontrollers or ARM7 microcontrollers. However, LabVIEW does not have device drivers for the peripherals of some of the microcontrollers. For these devices, you may need to develop the interface code in C and access it in LabVIEW.

21.3 **DEVELOPMENT FLOW**

The development of a LabVIEW application typically involves the steps below and shown in Figure 21.4.

- *Create a project and a VI*: You can use the project wizard to create an ARM project easily. This includes setting up the targeted platform and device drivers. After the project creation, a VI (usually a blank one) is created.

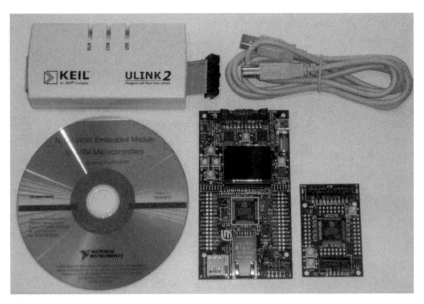

FIGURE 21.3

LabVIEW Evaluation Kit for the Cortex-M3 Processor.

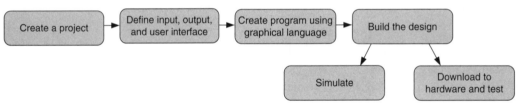

FIGURE 21.4

Example Design Flow.

- *Define inputs and outputs*: These can be hardware interfaces on the microcontroller or user interfaces running on the PC connected to the system. For hardware interfaces, you need to define the inputs and outputs as elemental input/output (I/O) in the project before you can use them in your VI design.
- *Create* the application using graphical programming.
- *Build* the design.
- *Simulate*: It is possible to download the created execution image to the device simulation in the Keil MDK-ARM to test your application.
- *Download to the microcontroller and test*: By default, the program is downloaded to the microcontroller as a test when the compilation is done. You can use the LabVIEW interface to pause, stop, or single-step the execution. You can also probe the variable value by clicking the connection during execution.

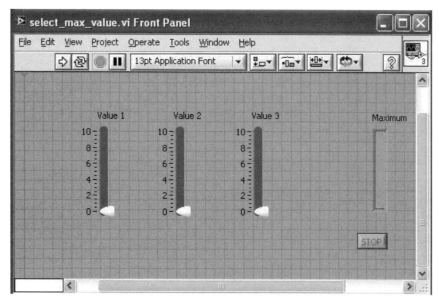

FIGURE 21.5

Simple Front Panel Design for the VI in Figure 2.

Each VI has two views:

- *Front panel view*: Containing a GUI of the VI.
- *Block diagram view*: Serving as a work-space for graphical programming.

When you create a VI, it has a blank front panel by default. You can then add control elements and indicator elements to define the inputs and outputs of the system. For instance, in the previous example of getting the largest integer from three inputs and a random value, the front panel may be similar to the one shown in Figure 21.5.

You can choose from a variety of control and indicator components in the VI libraries. After you add these controls and indicators, they become visible in the block diagram view. You can then customize properties like data types and create your graphical program by connecting them with various LabVIEW functions.

21.4 EXAMPLE OF A LABVIEW PROJECT

21.4.1 Create the Project

This example creates a simple application that samples analog signals and displays on the Organic LED (OLED) screen as a waveform. The first step is to use the project wizard to create a new project. From the LabVIEW startup screen, select "more" and then "ARM project," as shown in Figure 21.6. Then you can create a blank VI, or use an existing one. For this example, create a new VI. Now you have the choice of target type. For this example, select the LM3S8962 evaluation kit (Figure 21.7).

In the final step, the project wizard builds the specification that allows you to run the project on a simulator. For this example, run it in hardware instead of selecting the simulator option. The project creation wizard also saves the project and a default blank VI.

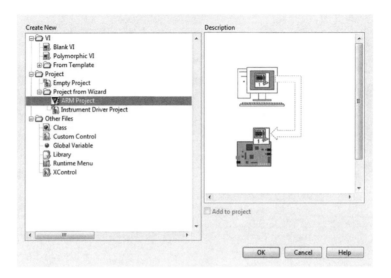

FIGURE 21.6

Create New ARM Project via the Project Wizard.

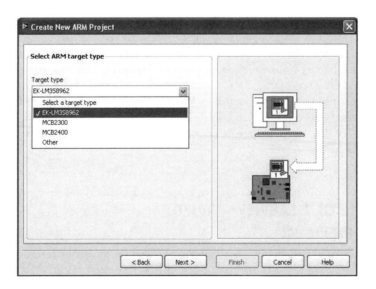

FIGURE 21.7

Select Target Type.

21.4.2 Define Inputs and Outputs

The next step of the project is to define the inputs and outputs for the applications. In this example, you need to define only the ADC0 input (the OLED display is controlled by library components, so you don't have to define it here). To complete that, right click on the target in the project window and select "New-> Elemental I/O" (Figure 21.8). Then define the input used in the project in the New Elemental I/O window, as shown in Figure 21.9.

In this window, a number of other I/O options are available. For example, the push buttons, light emitting diode (LED), General Purpose Input/Output (GPIO), and Pulse Width Modulator (PWM) interfaces can help you simplify your project development.

21.4.3 Create the Program

Then you can create the application code in the block diagram view. LabVIEW features many functions and it is impossible to cover all of these in this chapter. LabVIEW has detailed documentation on how to create graphical programming elements, and each element has context-sensitive help information. In general, you can right click on the block diagram area to browse the available components (see Figure 21.10). For example, a number of the OLED screen controls are available in the ARM category.

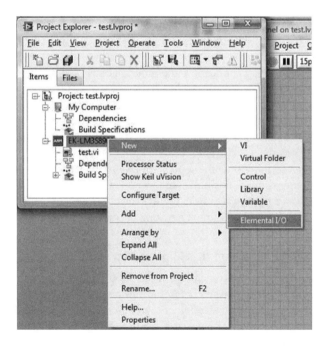

FIGURE 21.8

Define Elemental I/O.

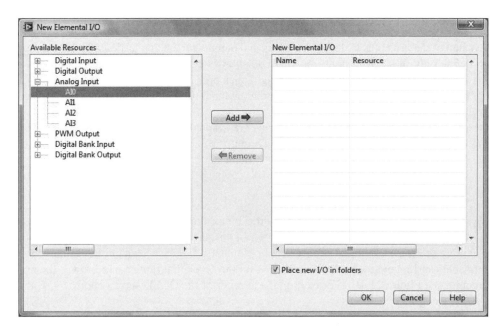

FIGURE 21.9

Adding Elemental I/O to Your Project.

In addition to the OLED control, you can find the elemental I/O, Controller Area Network (CAN), Inter-Integrated Circuit (I²C), Serial Peripherial Interface (SPI), and interrupt control functions in the ARM category.

In the analog waveform display application, the program code is divided into two parts: on the left side of the block diagram, the code initializes the OLED screen, displays a startup screen, delays for a short period and then clears the screen and displays "Analog Input 0" at the top of the screen. Use this sub-VI, which is specific to the LM3S8962 evaluation board you are using, for the OLED controls provided in the LabVIEW Embedded Module for ARM Microcontrollers (Figure 21.11).

The right side of the block diagram features a While Loop that samples from ADC0 of the LM3S8962 and then displays the waveform by drawing a pixel on the screen. If the X-position is reached on the right side of the OLED screen, the X-position counter resets to 0.

The calculated Y position value is also stored into an array. The array has a size of 128 integers and is used to remove the old waveform before the new pixel is drawn on the screen.

21.4.4 Build the Design and Test the Application

Once you have completed the design, you can build the design and test the application. First click the arrow button on the tool bar. If the program contains an error, the arrow icon displays as broken to indicate that the program is not ready. You can click on it to report the errors detected in the program.

When you have completed the compile process, the program is downloaded to the board automatically and executed. In this example, the program has executed successfully and generated the waveform for an analog input (see Figure 21.12 on page 345).

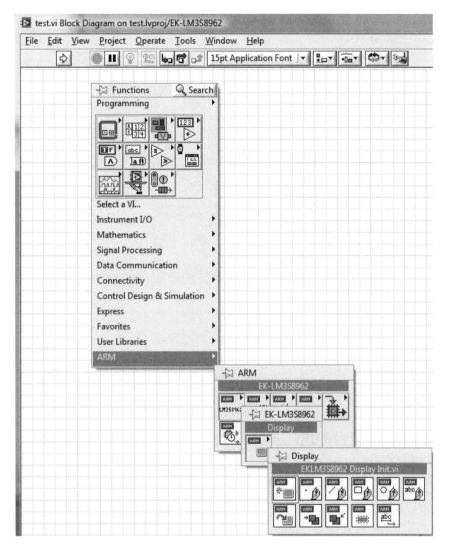

FIGURE 21.10

Access LabVIEW Functions and ARM-Specific Features.

21.5 HOW IT WORKS

When you build the program, LabVIEW generates C code from the VI created. You can then compile C code using the MDK-ARM (Figure 21.13 on page 346). To run VI-s in parallel, the generated C code uses the RealView Real-Time Library (RL-ARM, *www.keil.com/arm/rl-arm/*), which works with the Keil RTX Real Time Kernel (*www.keil.com/arm/rl-arm/kernel.asp*).

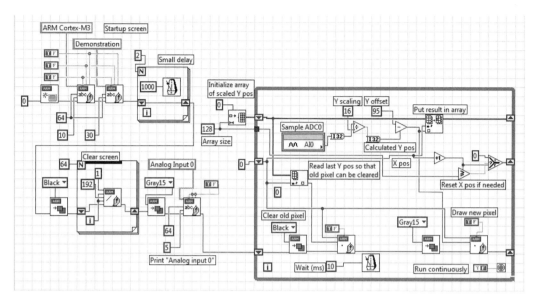

FIGURE 21.11

Block Diagram of the Analog Waveform Display Application.

In addition to the capability of multithreading, using the RL-ARM provides access to a variety of drivers including timing control, Transmission Control Protocol/Internet Protocol (TCP/IP) stack, and CAN protocol stack.

21.6 ADDITIONAL FEATURES IN LABVIEW

Semihosting: LabVIEW supports the creation of GUIs running on the debug host (PC). This is very useful for system prototyping because developing GUIs running on a microcontroller can be a time-consuming task. The LabVIEW library supports user-interface elements including switches, slide bars, and graphs, making it easy to create professional GUIs in just a few mouse clicks. During operation, the GUIs are running on the PC and the data are communicated to the running microcontroller target via the debug connection Joint Test Action Group, TCP/IP, or Serial. This allows you to control and access the results on the hardware in real time.

C code integration: LabVIEW supports embedding C code inside a VI. This is useful for creating new device drivers and handling data-processing tasks that you cannot complete using elements in LabVIEW.

Debugging: Debugging is easy in the LabVIEW environment. During design simulation or test, you can use the VI design environment as a debugger by showing the values on data variables when it stops. The VI block diagram view provides pause, run, and single-stepping controls as well as value probing (Figure 21.14 on page 346).

In addition, because the Keil μVision comes with a full-featured debugger, you can debug the application code you generated on LabVIEW using the μVision debugger (Figure 21.15 on page 347).

FIGURE 21.12

LabVIEW Application Running on the Evaluation Kit.

21.7 PORTING TO ANOTHER ARM PROCESSOR

In addition to running your LabVIEW application using the evaluation kit, you can port the application to other Cortex-M3 microcontrollers or other ARM microcontrollers that are supported by the RTX Real-Time Kernel. This typically involves the following steps:

- *Port the RTX Real-Time Kernel*: This step is not required for Cortex-M3 devices because the processor contains the required support for the RTX Real-Time Kernel. However, if you choose to use ARM7 microcontrollers, you may need this. A port for the RTX Real-Time Kernel may already be available in the Keil MDK-ARM installation options. You can determine if an RTX port is available for your ARM microcontroller by browsing the \Keil\ARM\Startup directory. If an RTX_Conf*.c is already available for the microcontroller target, then the RTX Real-Time Kernel has already been ported.

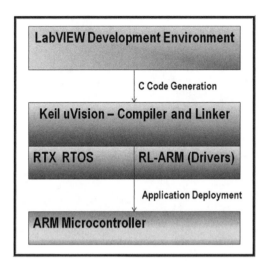

FIGURE 21.13

Design Flow Using LabVIEW and Keil μVision.

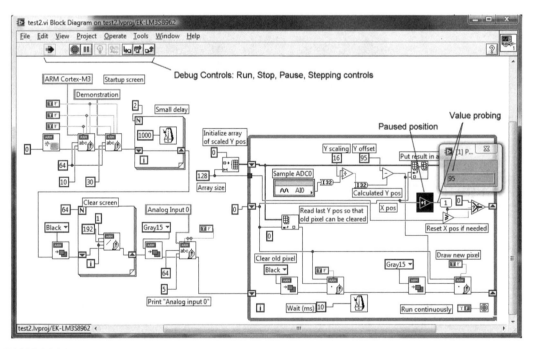

FIGURE 21.14

Debugging in a VI Block Diagram.

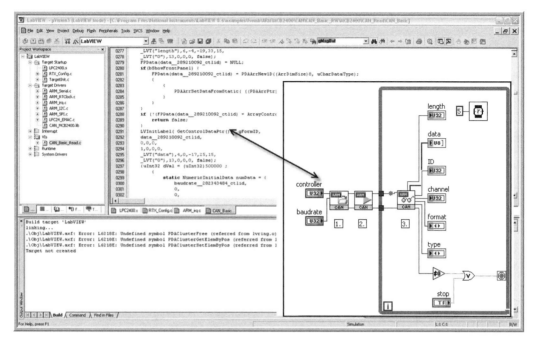

FIGURE 21.15

Using µVision to Debug Code Generated in LabVIEW.

- *Create the target in LabVIEW and incorporate the Keil toolchain*: You can do this by manually creating the target folder.

- *Integrate the Real-Time Agent Module for debugging*: You may need to customize the RTX_config.c to include the Real-Time Agent Configuration option.

- *Develop peripheral and I/O drivers*: You can implement this step using the Element I/O Device Editor.

For more detailed information on the porting process, view the National Instruments online tutorial titled *LabVIEW Embedded for ARM Porting Guide* (*http://zone.ni.com/devzone/cda/tut/p/id/6994*).

The Cortex-M3 Instruction Set, Reference Material

This appendix is the Cortex™-M3 instruction set description from the ARM Cortex-M3 user guide reference material; it is reproduced with permission from ARM. The following sections give general information:

- *Instruction set summary* on page 349
- *About the instruction descriptions* on page 353

Each of the following sections describes a functional group of Cortex-M3 instructions. Together they describe all the instructions supported by the Cortex-M3 processor:

- *Memory access instructions* on page 361
- *General data-processing instructions* on page 373
- *Multiply and divide instructions* on page 383
- *Saturating instructions* on page 386
- *Bitfield instructions* on page 388
- *Branch and control instructions* on page 391
- *Miscellaneous instructions* on page 397

A.1 INSTRUCTION SET SUMMARY

The processor implements a version of the Thumb® instruction set. Table A.1 lists the supported instructions.

Table A.1 Cortex-M3 Instructions

Mnemonic	Operands	Brief Description	Flags	Page
ADC, ADCS	{Rd,} Rn, Op2	Add with Carry	N,Z,C,V	Page 374
ADD, ADDS	{Rd,} Rn, Op2	Add	N,Z,C,V	Page 374
ADD, ADDW	{Rd,} Rn, #imm12	Add	N,Z,C,V	Page 374
ADR	Rd, label	Load PC-relative address	—	Page 362
AND, ANDS	{Rd,} Rn, Op2	Logical AND	N,Z,C	Page 376
ASR, ASRS	Rd, Rm, <Rs|#n>	Arithmetic Shift Right	N,Z,C	Page 377

Continued

DOI: 10.1016/B978-1-85617-963-8.00025-9

Table A.1 Cortex-M3 Instructions *Continued*

Mnemonic	Operands	Brief Description	Flags	Page
B	Label	Branch	—	Page 391
BFC	Rd, #lsb, #width	Bit Field Clear	—	Page 388
BFI	Rd, Rn, #lsb, #width	Bit Field Insert	—	Page 388
BIC, BICS	{Rd,} Rn, Op2	Bit Clear	N,Z,C	Page 376
BKPT	#imm	Breakpoint	—	Page 397
BL	Label	Branch with Link	—	Page 391
BLX	Rm	Branch indirect with Link	—	Page 391
BX	Rm	Branch indirect	—	Page 391
CBNZ	Rn, label	Compare and Branch if Nonzero	—	Page 393
CBZ	Rn, label	Compare and Branch if Zero	—	Page 393
CLREX	—	Clear Exclusive	—	Page 372
CLZ	Rd, Rm	Count leading zeros	—	Page 378
CMN	Rn, Op2	Compare Negative	N,Z,C,V	Page 378
CMP	Rn, Op2	Compare	N,Z,C,V	Page 378
CPSID	iflags	Change Processor State, Disable Interrupts	—	Page 398
CPSIE	iflags	Change Processor State, Enable Interrupts	—	Page 398
DMB	—	Data Memory Barrier	—	Page 398
DSB	—	Data Synchronization Barrier	—	Page 399
EOR, EORS	{Rd,} Rn, Op2	Exclusive OR	N,Z,C	Page 376
ISB	—	Instruction Synchronization Barrier	—	Page 399
IT	—	If-Then condition block	—	Page 393
LDM	Rn{!}, reglist	Load Multiple registers, increment after	—	Page 368
LDMDB, LDMEA	Rn{!}, reglist	Load Multiple registers, decrement before	—	Page 368
LDMFD, LDMIA	Rn{!}, reglist	Load Multiple registers, increment after	—	Page 368
LDR	Rt, [Rn, #offset]	Load Register with word	—	Page 362
LDRB, LDRBT	Rt, [Rn, #offset]	Load Register with byte	—	Page 362
LDRD	Rt, Rt2, [Rn, #offset]	Load Register with 2 bytes	—	Page 362
LDREX	Rt, [Rn, #offset]	Load Register Exclusive	—	Page 371
LDREXB	Rt, [Rn]	Load Register Exclusive with byte	—	Page 371
LDREXH	Rt, [Rn]	Load Register Exclusive with halfword	—	Page 371
LDRH, LDRHT	Rt, [Rn, #offset]	Load Register with halfword	—	Page 362
LDRSB, LDRSBT	Rt, [Rn, #offset]	Load Register with signed byte	—	Page 362

Table A.1 Cortex-M3 Instructions *Continued*

Mnemonic	Operands	Brief Description	Flags	Page
LDRSH, LDRSHT	Rt, [Rn, #offset]	Load Register with signed halfword	—	Page 362
LDRT	Rt, [Rn, #offset]	Load Register with word	—	Page 362
LSL, LSLS	Rd, Rm, <Rs|#n>	Logical Shift Left	N,Z,C	Page 377
LSR, LSRS	Rd, Rm, <Rs|#n>	Logical Shift Right	N,Z,C	Page 377
MLA	Rd, Rn, Rm, Ra	Multiply with Accumulate, 32-bit result	—	Page 383
MLS	Rd, Rn, Rm, Ra	Multiply and Subtract, 32-bit result	—	Page 383
MOV, MOVS	Rd, Op2	Move	N,Z,C	Page 379
MOVT	Rd, #imm16	Move Top	—	Page 381
MOVW, MOV	Rd, #imm16	Move 16-bit constant	N,Z,C	Page 379
MRS	Rd, spec_reg	Move from special register to general register	—	Page 400
MSR	spec_reg, Rm	Move from general register to special register	N,Z,C,V	Page 400
MUL, MULS	Rd, Rn, Rm	Multiply, 32-bit result	N,Z	Page 383
MVN, MVNS	Rd, Op2	Move NOT	N,Z,C	Page 379
NOP	—	No Operation	—	Page 401
ORN, ORNS	{Rd,} Rn, Op2	Logical OR NOT	N,Z,C	Page 376
ORR, ORRS	{Rd,} Rn, Op2	Logical OR	N,Z,C	Page 376
POP	reglist	Pop registers from stack	—	Page 370
PUSH	reglist	Push registers onto stack	—	Page 370
RBIT	Rd, Rn	Reverse bits	—	Page 381
REV	Rd, Rn	Reverse byte order in a word	—	Page 381
REV16	Rd, Rn	Reverse byte order in each halfword	—	Page 381
REVSH	Rd, Rn	Reverse byte order in bottom halfword and sign extend	—	Page 381
ROR, RORS	Rd, Rm, <Rs|#n>	Rotate Right	N,Z,C	Page 377
RRX, RRXS	Rd, Rm	Rotate Right with Extend	N,Z,C	Page 377
RSB, RSBS	{Rd,} Rn, Op2	Reverse Subtract	N,Z,C,V	Page 374
SBC, SBCS	{Rd,} Rn, Op2	Subtract with Carry	N,Z,C,V	Page 374
SBFX	Rd, Rn, #lsb, #width	Signed Bit Field Extract	—	Page 389
SDIV	{Rd,} Rn, Rm	Signed Divide	—	Page 386
SEV	—	Send Event	—	Page 402
SMLAL	RdLo, RdHi, Rn, Rm	Signed Multiply with Accumulate ($32 \times 32 + 64$), 64-bit result	—	Page 385
SMULL	RdLo, RdHi, Rn, Rm	Signed Multiply (32×32), 64-bit result	—	Page 385

Continued

Table A.1 Cortex-M3 Instructions *Continued*

Mnemonic	Operands	Brief Description	Flags	Page
SSAT	Rd, #n, Rm {,shift #s}	Signed Saturate	Q	Page 386
STM	Rn{!}, reglist	Store Multiple registers, increment after	—	Page 368
STMDB, STMEA	Rn{!}, reglist	Store Multiple registers, decrement before	—	Page 368
STMFD, STMIA	Rn{!}, reglist	Store Multiple registers, increment after	—	Page 368
STR	Rt, [Rn, #offset]	Store Register word	—	Page 362
STRB, STRBT	Rt, [Rn, #offset]	Store Register byte	—	Page 362
STRD	Rt, Rt2, [Rn, #offset]	Store Register two words	—	Page 362
STREX	Rd, Rt, [Rn, #offset]	Store Register Exclusive	—	Page 371
STREXB	Rd, Rt, [Rn]	Store Register Exclusive byte	—	Page 371
STREXH	Rd, Rt, [Rn]	Store Register Exclusive halfword	—	Page 371
STRH, STRHT	Rt, [Rn, #offset]	Store Register halfword	—	Page 362
STRT	Rt, [Rn, #offset]	Store Register word	—	Page 362
SUB, SUBS	{Rd,} Rn, Op2	Subtract	N,Z,C,V	Page 374
SUB, SUBW	{Rd,} Rn, #imm12	Subtract	N,Z,C,V	Page 374
SVC	#imm	Supervisor Call	—	Page 402
SXTB	Rd, Rm {,ROR #n}	Sign extend a byte	—	Page 390
SXTH	Rd, Rm {,ROR #n}	Sign extend a halfword	—	Page 390
TBB	[Rn, Rm]	Table Branch Byte	—	Page 395
TBH	[Rn, Rm, LSL #1]	Table Branch Halfword	—	Page 395
TEQ	Rn, Op2	Test Equivalence	N,Z,C	Page 382
TST	Rn, Op2	Test	N,Z,C	Page 382
UBFX	Rd, Rn, #lsb, #width	Unsigned Bit Field Extract	—	Page 389
UDIV	{Rd,} Rn, Rm	Unsigned Divide	—	Page 386
UMLAL	RdLo, RdHi, Rn, Rm	Unsigned Multiply with Accumulate ($32 \times 32 + 64$), 64-bit result	—	Page 385
UMULL	RdLo, RdHi, Rn, Rm	Unsigned Multiply (32×32), 64-bit result	—	Page 385
USAT	Rd, #n, Rm {,shift #s}	Unsigned Saturate	Q	Page 386
UXTB	Rd, Rm {,ROR #n}	Zero extend a byte	—	Page 390
UXTH	Rd, Rm {,ROR #n}	Zero extend a halfword	—	Page 390
WFE	—	Wait For Event	—	Page 403
WFI	—	Wait For Interrupt	—	Page 403

Note: Angle brackets, <>, enclose alternative forms of the operand; braces, {}, enclose optional operands; the Operands column is not exhaustive; Op2 is a flexible second operand that can be either a register or a constant; most instructions can use an optional condition code suffix.

For more information on the instructions and operands, see the instruction descriptions.

A.2 ABOUT THE INSTRUCTION DESCRIPTIONS

The following sections give more information about using the instructions:

- *Operands* on page 353
- *Restrictions when using PC or SP* on page 353
- *Flexible second operand* on page 353
- *Shift Operations* on page 354
- *Address alignment* on page 357
- *PC-relative expressions* on page 358
- *Conditional execution* on page 358
- *Instruction width selection* on page 360

A.2.1 Operands

An instruction operand can be an ARM register, a constant, or another instruction-specific parameter. Instructions act on the operands and often store the result in a destination register. When there is a destination register in the instruction, it is usually specified before the operands.

Operands in some instructions are flexible in that they can either be a register or a constant; see "Flexible Second Operand" section.

A.2.2 Restrictions When Using PC or SP

Many instructions have restrictions on whether you can use the *program counter* (PC) or *stack pointer* (SP) for the operands or destination register. See *instruction descriptions* for more information.

Note
Bit[0] of any address you write to the PC with a BX, BLX, LDM, LDR, or POP instruction must be 1 for correct execution, because this bit indicates the required instruction set, and the Cortex-M3 processor only supports Thumb instructions.

A.2.3 Flexible Second Operand

Many general data-processing instructions have a flexible second operand. This is shown as *Operand2* in the descriptions of the syntax of each instruction.

Operand2 can be a

- *Constant*
- *Register with optional shift* on page 354.

Constant
You specify an Operand2 constant in the form:

```
#constant
```

where `constant` can be

- Any constant that can be produced by shifting an 8-bit value left by any number of bits within a 32-bit word
- Any constant of the form `0x00XY00XY`
- Any constant of the form `0xXY00XY00`
- Any constant of the form `0xXYXYXYXY`.

Note

In the constants shown above, `X` and `Y` are hexadecimal digits.

In addition, in a small number of instructions, `constant` can take a wider range of values. These are described in the individual instruction descriptions.

When an Operand2 constant is used with the instructions `MOVS`, `MVNS`, `ANDS`, `ORRS`, `ORNS`, `EORS`, `BICS`, `TEQ`, or `TST`, the carry flag is updated to bit[31] of the constant, if the constant is greater than 255 and can be produced by shifting an 8-bit value. These instructions do not affect the carry flag if Operand2 is any other constant.

Instruction Substitution

Your assembler might be able to produce an equivalent instruction in cases where you specify a constant that is not permitted. For example, an assembler might assemble the instruction `CMP Rd, #0xFFFFFFFE` as the equivalent instruction `CMN Rd, #0x2`.

Register with Optional Shift

You specify an Operand2 register in the form:

`Rm {, shift}`

where

`Rm`	is the register holding the data for the second operand.
`shift`	is an optional shift to be applied to `Rm`. It can be one of the following:

`ASR #n`	Arithmetic Shift Right n bits, $1 \leq n \leq 32$
`LSL #n`	Logical Shift Left n bits, $1 \leq n \leq 31$
`LSR #n`	Logical Shift Right n bits, $1 \leq n \leq 32$
`ROR #n`	Rotate Right n bits, $1 \leq n \leq 31$
`RRX`	Rotate Right 1 bit, with Extend
-	If omitted, no shift occurs, equivalent to `LSL #0`

If you omit the shift, or specify `LSL #0`, the instruction uses the value in `Rm`.

If you specify a shift, the shift is applied to the value in `Rm`, and the resulting 32-bit value is used by the instruction. However, the contents in the register `Rm` remain unchanged. Specifying a register with shift also updates the carry flag when used with certain instructions. For information on the shift operations and how they affect the carry flag, see "Shift Operations" section.

A.2.4 Shift Operations

Register shift operations move the bits in a register left or right by a specified number of bits, the *shift length*. Register shift can be performed

- Directly by the instructions ASR, LSR, LSL, ROR, and RRX, and the result is written to a destination register.
- During the calculation of *Operand2* by the instructions that specify the second operand as a register with shift; see "Flexible Second Operand" section on page 353. The result is used by the instruction.

The permitted shift lengths depend on the shift type and the instruction; see the individual instruction description or "Flexible Second Operand" section on page 353. If the shift length is 0, no shift occurs. Register shift operations update the carry flag except when the specified shift length is 0. The following subsections describe the various shift operations and how they affect the carry flag. In these descriptions, *Rm* is the register containing the value to be shifted, and *n* is the shift length.

ASR

Arithmetic Shift Right by *n* bits moves the left-hand $32-n$ bits of the register *Rm* to the right by *n* places, into the right-hand $32-n$ bits of the result. And it copies the original bit[31] of the register into the left-hand *n* bits of the result; see Figure A.1.

You can use the ASR #*n* operation to divide the value in the register *Rm* by 2^n, with the result being rounded toward negative-infinity.

When the instruction is ASRS or when ASR #*n* is used in *Operand2* with the instructions MOVS, MVNS, ANDS, ORRS, ORNS, EORS, BICS, TEQ, or TST, the carry flag is updated to the last bit shifted out, bit[$n-1$], of the register *Rm*.

Note

- If *n* is 32 or more, then all the bits in the result are set to the value of bit[31] of *Rm*.
- If *n* is 32 or more and the carry flag is updated, it is updated to the value of bit[31] of *Rm*.

LSR

Logical Shift Right by *n* bits moves the left-hand $32-n$ bits of the register *Rm*, to the right by *n* places, into the right-hand $32-n$ bits of the result. And it sets the left-hand *n* bits of the result to 0. See Figure A.2.

You can use the LSR #*n* operation to divide the value in the register *Rm* by 2^n, if the value is regarded as an unsigned integer.

When the instruction is LSRS or when LSR #*n* is used in *Operand2* with the instructions MOVS, MVNS, ANDS, ORRS, ORNS, EORS, BICS, TEQ or TST, the carry flag is updated to the last bit shifted out, bit[$n-1$], of the register *Rm*.

FIGURE A.1

ASR #3.

FIGURE A.2

LSR #3.

FIGURE A.3

LSL #3.

Note
- If n is 32 or more, then all the bits in the result are cleared to 0.
- If n is 33 or more and the carry flag is updated, it is updated to 0.

LSL

Logical Shift Left by n bits moves the right-hand $32-n$ bits of the register Rm, to the left by n places, into the left-hand $32-n$ bits of the result. And it sets the right-hand n bits of the result to 0. See Figure A.3.

You can use the LSL #n operation to multiply the value in the register Rm by 2^n, if the value is regarded as an unsigned integer or a two's complement signed integer. Overflow can occur without warning.

When the instruction is LSLS or when LSL #n, with nonzero n, is used in *Operand2* with the instructions MOVS, MVNS, ANDS, ORRS, ORNS, EORS, BICS, TEQ, or TST, the carry flag is updated to the last bit shifted out, bit[$32-n$], of the register Rm. These instructions do not affect the carry flag when used with LSL #0.

Note
- If n is 32 or more, then all the bits in the result are cleared to 0.
- If n is 33 or more and the carry flag is updated, it is updated to 0.

ROR

Rotate Right by n bits moves the left-hand $32-n$ bits of the register Rm, to the right by n places, into the right-hand $32-n$ bits of the result. And it moves the right-hand n bits of the register into the left-hand n bits of the result. See Figure A.4.

FIGURE A.4

ROR #3.

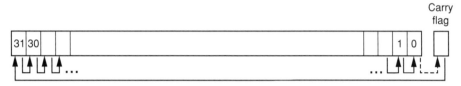

FIGURE A.5

RRX.

When the instruction is RORS or when ROR #n is used in *Operand2* with the instructions MOVS, MVNS, ANDS, ORRS, ORNS, EORS, BICS, TEQ, or TST, the carry flag is updated to the last bit rotation, bit[n-1], of the register Rm.

Note
- If n is 32, then the value of the result is the same as the value in Rm, and if the carry flag is updated, it is updated to bit[31] of Rm.
- ROR with shift length, n, more than 32 is the same as ROR with shift length n-32.

RRX
Rotate Right with Extend moves the bits of the register Rm to the right by 1 bit. And it copies the carry flag into bit[31] of the result; see Figure A.5.

When the instruction is RRXS or when RRX is used in *Operand2* with the instructions MOVS, MVNS, ANDS, ORRS, ORNS, EORS, BICS, TEQ, or TST, the carry flag is updated to bit[0] of the register Rm.

A.2.5 Address Alignment

An aligned access is an operation where a word-aligned address is used for a word, dual word, or multiple word access, or where a halfword-aligned address is used for a halfword access. Byte accesses are always aligned.

The Cortex-M3 processor supports unaligned access only for the following instructions:

- LDR, LDRT
- LDRH, LDRHT
- LDRSH, LDRSHT

- STR, STRT
- STRH, STRHT

All other load and store instructions generate a usage fault exception if they perform an unaligned access, and therefore, their accesses must be address aligned.

Unaligned accesses are usually slower than aligned accesses. In addition, some memory regions might not support unaligned accesses. Therefore, ARM recommends that programmers ensure that accesses are aligned. To trap accidental generation of unaligned accesses, use the UNALIGN_TRP bit in the Configuration and Control register to trigger usage fault exception for all unaligned accesses.

A.2.6 PC-Relative Expressions

A PC-relative expression or *label* is a symbol that represents the address of an instruction or literal data. It is represented in the instruction as the PC value plus or minus a numeric offset. The assembler calculates the required offset from the label and the address of the current instruction. If the offset is too big, the assembler produces an error.

Note
- For B, BL, CBNZ, and CBZ instructions, the value of the PC is the address of the current instruction plus 4 bytes.
- For most other instructions that use labels, the value of the PC is the address of the current instruction plus 4 bytes, with bit[1] of the result cleared to 0 to make it word-aligned.
- Your assembler might permit other syntaxes for PC-relative expressions, such as a label plus or minus a number, or an expression of the form [PC, #number].

A.2.7 Conditional Execution

Most data-processing instructions can optionally update the condition flags in the *Application Program Status Register* (APSR) according to the result of the operation. Some instructions update all flags, and some only update a subset. If a flag is not updated, the original value is preserved. See the *instruction descriptions* for the flags they affect.

You can execute an instruction conditionally, based on the condition flags set in another instruction, either immediately after the instruction that updated the flags or after any number of intervening instructions that have not updated the flags.

Conditional execution is available by using conditional branches or by adding condition code suffixes to instructions. See Table A.2 for a list of the suffixes to add to instructions to make them conditional instructions. The condition code suffix enables the processor to test a condition based on the flags. If the condition test of a conditional instruction fails, the instruction

- Does not execute
- Does not write any value to its destination register
- Does not affect any of the flags
- Does not generate any exception

Conditional instructions, except for conditional branches, must be inside an If-Then instruction block. See "IT" section on page 393 for more information and restrictions when using the IT instruction.

Table A.2 Condition Code Suffixes

Suffix	Flags	Meaning
EQ	Z = 1	Equal
NE	Z = 0	Not equal
CS or HS	C = 1	Higher or same, unsigned ≥
CC or LO	C = 0	Lower, unsigned <
MI	N = 1	Negative
PL	N = 0	Positive or zero
VS	V = 1	Overflow
VC	V = 0	No overflow
HI	C = 1 and Z = 0	Higher, unsigned >
LS	C = 0 or Z = 1	Lower or same, unsigned ≤
GE	N = V	Greater than or equal, signed ≥
LT	N != V	Less than, signed <
GT	Z = 0 and N = V	Greater than, signed >
LE	Z = 1 and N != V	Less than or equal, signed ≤
AL	Can have any value	Always; default when no suffix is specified

Depending on the vendor, the assembler might automatically insert an IT instruction if you have conditional instructions outside the IT block.

Use the CBZ and CBNZ instructions to compare the value of a register against zero and branch on the result.

This section describes

- *The condition flags* on page 359
- *Condition code suffixes* on page 360.

The Condition Flags

The APSR contains the following condition flags:

N   Set to 1 when the result of the operation was negative, cleared to 0 otherwise.
Z   Set to 1 when the result of the operation was zero, cleared to 0 otherwise.
C   Set to 1 when the operation resulted in a carry, cleared to 0 otherwise.
V   Set to 1 when the operation caused overflow, cleared to 0 otherwise.

A carry occurs

- If the result of an addition is greater than or equal to 2^{32}
- If the result of a subtraction is positive or zero
- As the result of an inline barrel shifter operation in a move or logical instruction.

Overflow occurs when the sign of the result, in bit[31], does not match the sign of the result had the operation been performed at infinite precision, for example:

- if adding two negative values results in a positive value
- if adding two positive values results in a negative value

- if subtracting a positive value from a negative value generates a positive value
- if subtracting a negative value from a positive value generates a negative value

The Compare operations are identical to subtracting, for CMP, or adding, for CMN, except that the result is discarded. See the instruction descriptions for more information.

Note
Most instructions update the status flags only if the S suffix is specified; see the *instruction descriptions* for more information.

Condition Code Suffixes
The instructions that can be conditional have an optional condition code, shown in syntax descriptions as {cond}. Conditional execution requires a preceding IT instruction. An instruction with a condition code is only executed if the condition code flags in the APSR meet the specified condition. Table A.2 shows the condition codes to use.

You can use conditional execution with the IT instruction to reduce the number of branch instructions in code. Table A.2 also shows the relationship between condition code suffixes and the N, Z, C, and V flags.

Example A.1 shows the use of a conditional instruction to find the absolute value of a number. $R0 = ABS(R1)$.

Example A.2 shows the use of conditional instructions to update the value of R4 if the signed values R0 is greater than R1 and R2 is greater than R3.

Example A.1 Absolute Value

```
MOVS    R0, R1          ; R0 = R1, setting flags
IT      MI              ; IT - Skip next instruction if value 0 or positive
RSBMI   R0, R1, #0      ; If negative, R0 = -R1
```

Example A.2 Compare and Update Value

```
CMP     R0, R1          ; Compare R0 and R1, setting flags
ITT     GT              ; IT - Skip next two instructions unless GT condition holds
CMPGT   R2, R3          ; If 'greater than', compare R2 and R3, setting flags
MOVGT   R4, R5          ; If still 'greater than', do R4 = R5
```

A.2.8 Instruction Width Selection
There are many instructions that can generate either a 16-bit encoding or a 32-bit encoding depending on the operands and destination register specified. For some of these instructions, you can force a specific instruction size by using an instruction width suffix. The .W suffix forces a 32-bit instruction encoding. The .N suffix forces a 16-bit instruction encoding.

If you specify an instruction width suffix and the assembler cannot generate an instruction encoding of the requested width, it generates an error.

Note

In some cases, it might be necessary to specify the .W suffix, for example, if the operand is the label of an instruction or literal data, as in the case of branch instructions. This is because the assembler might not automatically generate the right size encoding.

To use an instruction width suffix, place it immediately after the instruction mnemonic and condition code, if any. Example A.3 shows instructions with the instruction width suffix.

Example A.3 Instruction Width Selection

```
BCS.W label            ; creates a 32-bit instruction even for a short branch
ADDS.W R0, R0, R1      ; creates a 32-bit instruction even though the same
                       ; operation can be done by a 16-bit instruction
```

A.3 MEMORY ACCESS INSTRUCTIONS

Table A.3 shows the memory access instructions.

Table A.3 Memory Access Instructions

Mnemonic	Brief Description	See
ADR	Generate PC-relative address	*ADR* on page 362
CLREX	Clear Exclusive	*CLREX* on page 372
LDM{mode}	Load Multiple registers	*LDM and STM* on page 368
LDR{type}	Load Register using immediate offset	*LDR and STR, Immediate Offset* on page 362
LDR{type}	Load Register using register offset	*LDR and STR, Register Offset* on page 365
LDR{type}T	Load Register with unprivileged access	*LDR and STR, Unprivileged* on page 366
LDR	Load Register using PC-relative address	*LDR, PC-Relative* on page 367
LDREX{type}	Load Register Exclusive	*LDREX and STREX* on page 371
POP	Pop registers from stack	*PUSH and POP* on page 370
PUSH	Push registers onto stack	*PUSH and POP* on page 370
STM{mode}	Store Multiple registers	*LDM and STM* on page 368
STR{type}	Store Register using immediate offset	*LDR and STR, Immediate Offset* on page 362
STR{type}	Store Register using register offset	*LDR and STR, Register Offset* on page 365
STR{type}T	Store Register with unprivileged access	*LDR and STR, Unprivileged* on page 366
STREX{type}	Store Register Exclusive	*LDREX and STREX* on page 371

A.3.1 **ADR**

Generate PC-relative address.

Syntax
```
ADR{cond} Rd, label
```

where

cond	is an optional condition code; see "Conditional Execution" section on page 358.
Rd	is the destination register.
label	is a PC-relative expression. See "PC-Relative Expressions" section on page 358.

Operation
ADR generates an address by adding an immediate value to the PC and writes the result to the destination register.

ADR facilitates the generation of position-independent code because the address is PC-relative. If you use ADR to generate a target address for a BX or BLX instruction, you must ensure that bit[0] of the address you generate is set to 1 for correct execution.

Values of *label* must be within the range of −4095 to +4095 from the address in the PC.

Note
You might have to use the .W suffix to get the maximum offset range or to generate addresses that are not word-aligned; see "Instruction Width Selection" section on page 360.

Restrictions
Rd must not be SP and must not be PC.

Condition Flags
This instruction does not change the flags.

Examples
```
ADR    R1,   TextMessage ; Write address value of a location labelled as
                         ; TextMessage to R1
```

A.3.2 **LDR and STR, Immediate Offset**

Load and Store with immediate offset, preindexed immediate offset or postindexed immediate offset.

Syntax
```
op{type}{cond} Rt, [Rn {, #offset}]     ; immediate offset
op{type}{cond} Rt, [Rn, #offset]!       ; pre-indexed
op{type}{cond} Rt, [Rn], #offset        ; post-indexed
opD{cond} Rt, Rt2, [Rn {, #offset}]     ; immediate offset, two words
opD{cond} Rt, Rt2, [Rn, #offset]!       ; pre-indexed, two words
opD{cond} Rt, Rt2, [Rn], #offset        ; post-indexed, two words
```

where

Op	is one of the following:
	LDR Load Register
	STR Store Register
Type	is one of the following:
	B Unsigned byte, zero extends to 32 bits on loads
	SB Signed byte, sign extends to 32 bits (LDR only)
	H Unsigned halfword, zero extends to 32 bits on loads
	SH Signed halfword, sign extends to 32 bits (LDR only)
	- Omit, for word
cond	is an optional condition code; see "Conditional Execution" section on page 358.
Rt	is the register to load or store.
Rn	is the register on which the memory address is based.
Offset	is an offset from Rn. If offset is omitted, the address is the contents of Rn.
Rt2	is the additional register to load or store for two-word operations.

Operation

LDR instructions load one or two registers with a value from memory. STR instructions store one or two register values to memory.

Load and store instructions with immediate offset can use the following addressing modes.

Offset Addressing

The offset value is added to or subtracted from the address obtained from the register Rn. The result is used as the address for the memory access. The register Rn is unaltered. The assembly language syntax for this mode is

```
[Rn, #offset]
```

Preindexed Addressing

The offset value is added to or subtracted from the address obtained from the register Rn. The result is used as the address for the memory access and written back into the register Rn. The assembly language syntax for this mode is

```
[Rn, #offset]!
```

Postindexed Addressing

The address obtained from the register Rn is used as the address for the memory access. The offset value is added to or subtracted from the address and written back into the register Rn. The assembly language syntax for this mode is

```
[Rn], #offset
```

The value to load or store can be a byte, halfword, word, or two words. Bytes and halfwords can either be signed or be unsigned; see "Address Alignment" section on page 357.

Table A.4 shows the ranges of offset for immediate, preindexed, and postindexed forms.

Table A.4 Offset Ranges

Instruction Type	Immediate Offset	Preindexed	Postindexed
Word, halfword, signed halfword, byte, or signed byte	−255 to 4095	−255 to 255	−255 to 255
Two words	multiple of 4 in the range −1020 to 1020	multiple of 4 in the range −1020 to 1020	multiple of 4 in the range −1020 to 1020

Restrictions

For load instructions:

- *Rt* can be SP or PC for word loads only.
- *Rt* must be different from *Rt2* for two-word loads.
- *Rn* must be different from *Rt* and *Rt2* in the preindexed or postindexed forms.

When *Rt* is PC in a word load instruction:

- bit[0] of the loaded value must be 1 for correct execution.
- A branch occurs to the address created by changing bit[0] of the loaded value to 0.
- If the instruction is conditional, it must be the last instruction in the IT block.

For store instructions:

- *Rt* can be SP for word stores only.
- *Rt* must not be PC.
- *Rn* must not be PC.
- *Rn* must be different from *Rt* and *Rt2* in the preindexed or postindexed forms.

Condition Flags

These instructions do not change the flags.

Examples

```
LDR     R8, [R10]               ; Loads R8 from the address in R10.
LDRNE   R2, [R5, #960]!         ; Loads (conditionally) R2 from a word
                                ; 960 bytes above the address in R5, and
                                ; increments R5 by 960.
STR     R2, [R9,#const-struc]   ; const-struc is an expression evaluating
                                ; to a constant in the range 0-4095.
STRH    R3, [R4], #4            ; Store R3 as halfword data into address in
                                ; R4, then increment R4 by 4
LDRD    R8, R9, [R3, #0x20]     ; Load R8 from a word 32 bytes above the
                                ; address in R3, and load R9 from a word 36
                                ; bytes above the address in R3
STRD    R0, R1, [R8], #-16      ; Store R0 to address in R8, and store R1 to
                                ; a word 4 bytes above the address in R8,
                                ; and then decrement R8 by 16.
```

A.3.3 **LDR and STR, Register Offset**

Load and Store with register offset.

Syntax

```
op{type}{cond} Rt, [Rn, Rm {, LSL #n}]
```

where

op	is one of the following:
	LDR Load Register
	STR Store Register
Type	is one of the following:
	B Unsigned byte, zero extends to 32 bits on loads
	SB Signed byte, sign extends to 32 bits (LDR only)
	H Unsigned halfword, zero extends to 32 bits on loads
	SH Signed halfword, sign extends to 32 bits (LDR only)
	- Omit, for word
Cond	is an optional condition code; see "Conditional Execution" section on page 358.
Rt	is the register to load or store.
Rn	is the register on which the memory address is based.
Rm	is a register containing a value to be used as the offset.
LSL #*n*	is an optional shift, with *n* in the range 0 to 3.

Operation

LDR instructions load a register with a value from memory.

STR instructions store a register value into memory.

The memory address to load from or store to is at an offset from the register *Rn*. The offset is specified by the register *Rm* and can be shifted left by up to 3 bits using LSL.

The value to load or store can be a byte, halfword, or word. For load instructions, bytes and halfwords can either be signed or be unsigned; see "Address Alignment" section on page 357.

Restrictions

In these instructions:

- *Rn* must not be PC.
- *Rm* must not be SP and must not be PC.
- *Rt* can be SP only for word loads and word stores.
- *Rt* can be PC only for word loads.

When *Rt* is PC in a word load instruction:

- bit[0] of the loaded value must be 1 for correct execution, and a branch occurs to this halfword-aligned address.
- If the instruction is conditional, it must be the last instruction in the IT block.

Condition Flags
These instructions do not change the flags.

Examples
```
STR    R0,   [R5, R1]           ; Store value of R0 into an address equal to
                                ; sum of R5 and R1
LDRSB R0,   [R5, R1, LSL #1]    ; Read byte value from an address equal to
                                ; sum of R5 and two times R1, sign extended it
                                ; to a word value and put it in R0
STR    R0,   [R1, R2, LSL #2]   ; Stores R0 to an address equal to sum of R1
                                ; and four times R2
```

A.3.4 LDR and STR, Unprivileged
Load and Store with unprivileged access.

Syntax
```
op{type}T{cond} Rt, [Rn {, #offset}]   ; immediate offset
```

where

op	is one of the following:	
	LDR	Load Register
	STR	Store Register
type	is one of the following:	
	B	Unsigned byte, zero extends to 32 bits on loads
	SB	Signed byte, sign extends to 32 bits (LDR only)
	H	Unsigned halfword, zero extends to 32 bits on loads
	SH	Signed halfword, sign extends to 32 bits (LDR only)
	-	Omit, for word
cond	is an optional condition code; see "Conditional Execution" section on page 358.	
Rt	is the register to load or store.	
Rn	is the register on which the memory address is based.	
Offset	is an offset from Rn and can be 0–255. If offset is omitted, the address is the value in Rn.	

Operation
These load and store instructions perform the same function as the memory access instructions with immediate offset; see "LDR and STR, Immediate Offset" section on page 362. The difference is that these instructions have only unprivileged access even when used in privileged software.

When used in unprivileged software, these instructions behave exactly the same way as normal memory access instructions with immediate offset.

Restrictions
In these instructions:

- Rn must not be PC.
- Rt must not be SP and must not be PC.

Condition Flags
These instructions do not change the flags.

Examples
```
STRBTEQ R4, [R7]      ; Conditionally store least significant byte in
                      ; R4 to an address in R7, with unprivileged access
LDRHT   R2, [R2, #8]  ; Load halfword value from an address equal to
                      ; sum of R2 and 8 into R2, with unprivileged access
```

A.3.5 LDR, PC-Relative
Load register from memory.

Syntax
```
LDR{type}{cond} Rt, label
LDRD{cond} Rt, Rt2, label          ; Load two words
```

where

type	is one of the following:

        B   Unsigned byte, zero extends to 32 bits
        SB   Signed byte, sign extends to 32 bits
        H   Unsigned halfword, zero extends to 32 bits
        SH   Signed halfword, sign extends to 32 bits
        -         Omit, for word

cond	is an optional condition code; see "Conditional Execution" section on page 358.
Rt	is the register to load or store.
Rt2	is the second register to load or store.
label	is a PC-relative expression; see "PC-Relative Expressions" section on page 358.

Operation
LDR loads a register with a value from a PC-relative memory address. The memory address is specified by a label or by an offset from the PC.

The value to load or store can be a byte, halfword, or word. For load instructions, bytes and halfwords can either be signed or be unsigned; see "Address Alignment" section on page 357.

label must be within a limited range of the current instruction. Table A.5 shows the possible offsets between label and PC.

Table A.5 Offset Ranges	
Instruction Type	**Offset Range**
Word, halfword, signed halfword, byte, signed byte	−4095 to 4095
Two words	−1020 to 1020

Note
You might have to use the .W suffix to get the maximum offset range; see "Instruction Width Selection" section on page 360.

Restrictions
In these instructions:

- Rt can be SP or PC only for word loads.
- $Rt2$ must not be SP and must not be PC.
- Rt must be different from $Rt2$.

When Rt is PC in a word load instruction:

- bit[0] of the loaded value must be 1 for correct execution, and a branch occurs to this halfword-aligned address.
- If the instruction is conditional, it must be the last instruction in the IT block.

Condition Flags
These instructions do not change the flags.

Examples
```
LDR     R0, LookUpTable    ; Load R0 with a word of data from an address
                           ; labelled as LookUpTable
LDRSB   R7, localdata      ; Load a byte value from an address labelled
                           ; as localdata, sign extend it to a word
                           ; value, and put it in R7
```

A.3.6 LDM and STM
Load and Store Multiple registers.

Syntax
```
op{addr_mode}{cond} Rn{!}, reglist
```

where

op	is one of the following:
	LDM Load Multiple registers
	STM Store Multiple registers
$addr\_mode$	is any one of the following:
	IA Increment address After each access; this is the default
	DB Decrement address Before each access
$Cond$	is an optional condition code; see "Conditional Execution" section on page 358.
Rn	is the register on which the memory addresses are based.
!	is an optional writeback suffix. If ! is present, the final address, that is loaded from or stored to, is written back into Rn.

reglist   is a list of one or more registers to be loaded or stored, enclosed in braces. It can contain register ranges. It must be comma separated if it contains more than one register or register range; see "Examples" section on page 370.

- LDM and LDMFD are synonyms for LDMIA. LDMFD refers to its use for popping data from Full Descending stacks.
- LDMEA is a synonym for LDMDB and refers to its use for popping data from Empty Ascending stacks.
- STM and STMEA are synonyms for STMIA. STMEA refers to its use for pushing data onto Empty Ascending stacks.
- STMFD is synonym for STMDB and refers to its use for pushing data onto Full Descending stacks

Operation

LDM instructions load the registers in *reglist* with word values from memory addresses based on *Rn*. STM instructions store the word values in the registers in *reglist* to memory addresses based on *Rn*.

For LDM, LDMIA, LDMFD, STM, STMIA, and STMEA, the memory addresses used for the accesses are at 4-byte intervals ranging from *Rn* to *Rn* + 4 * (*n*–1), where *n* is the number of registers in *reglist*. The accesses happen in order of increasing register numbers, with the lowest numbered register using the lowest memory address and the highest number register using the highest memory address. If the writeback suffix is specified, the value of *Rn* + 4 * (*n*–1) is written back to *Rn*.

For LDMDB, LDMEA, STMDB, and STMFD, the memory addresses used for the accesses are at 4-byte intervals ranging from *Rn* to *Rn* − 4 * (*n*–1), where *n* is the number of registers in *reglist*. The accesses happen in order of decreasing register numbers, with the highest numbered register using the highest memory address and the lowest number register using the lowest memory address. If the writeback suffix is specified, the value of *Rn* − 4 * (*n*–1) is written back to *Rn*.

The PUSH and POP instructions can be expressed in this form; see "PUSH and POP" section on page 370 for details.

Restrictions

In these instructions:

- *Rn* must not be PC.
- *reglist* must not contain SP.
- In any STM instruction, *reglist* must not contain PC.
- In any LDM instruction, *reglist* must not contain PC if it contains LR.
- *reglist* must not contain *Rn* if you specify the writeback suffix.

When PC is in *reglist* in an LDM instruction:

- bit[0] of the value loaded to the PC must be 1 for correct execution, and a branch occurs to this halfword-aligned address.
- If the instruction is conditional, it must be the last instruction in the IT block.

Condition Flags

These instructions do not change the flags.

Examples

```
LDM   R8,{R0,R2,R9}       ; LDMIA is a synonym for LDM
STMDB R1!,{R3-R6,R11,R12}
```

Incorrect Examples

```
STM R5!,{R5,R4,R9}   ; Value stored for R5 is unpredictable
LDM R2, {}           ; There must be at least one register in the list
```

A.3.7 PUSH and POP

Push registers onto and pop registers off a full-descending stack.

Syntax

```
PUSH{cond} reglist
POP{cond} reglist
```

where

cond	is an optional condition code; see "Conditional Execution" section on page 358.
Reglist	is a nonempty list of registers, enclosed in braces. It can contain register ranges. It must be comma separated if it contains more than one register or register range.

PUSH and POP are synonyms for STMDB and LDM (or LDMIA) with the memory addresses for the access based on SP and with the final address for the access written back to the SP. PUSH and POP are the preferred mnemonics in these cases.

Operation

PUSH stores registers on the stack, with the lowest numbered register using the lowest memory address and the highest numbered register using the highest memory address.

POP loads registers from the stack, with the lowest numbered register using the lowest memory address and the highest numbered register using the highest memory address.

PUSH uses the value in the SP register minus four as the highest memory address, POP uses the value in the SP register as the lowest memory address, implementing a full-descending stack. On completion, PUSH updates the SP register to point to the location of the lowest store value, POP updates the SP register to point to the location above the highest location loaded.

If a POP instruction includes PC in its *reglist*, a branch to this location is performed when the POP instruction has completed. Bit[0] of the value read for the PC is used to update the APSR T-bit. This bit must be 1 to ensure correct operation.

See *LDM and STM* on page 368 for more information.

Restrictions

In these instructions:

- *reglist* must not contain SP.

- For the PUSH instruction, *reglist* must not contain PC.
- For the POP instruction, *reglist* must not contain PC if it contains LR.

When PC is in *reglist* in a POP instruction:

- bit[0] of the value loaded to the PC must be 1 for correct execution, and a branch occurs to this halfword-aligned address.
- If the instruction is conditional, it must be the last instruction in the IT block.

Condition Flags
These instructions do not change the flags.

Examples
```
PUSH {R0,R4-R7}      ; Push R0, R4, R5, R6, R7 onto the stack
PUSH {R2,LR}         ; Push R2 and the link-register onto the stack
POP  {R0,R6,PC}      ; Pop R0, R6 and PC from the stack, then branch to the new PC.
```

A.3.8 LDREX and STREX

Load and Store Register Exclusive.

Syntax
```
LDREX{cond}   Rt, [Rn {, #offset}]
STREX{cond}   Rd, Rt, [Rn {, #offset}]
LDREXB{cond}  Rt, [Rn]
STREXB{cond}  Rd, Rt, [Rn]
LDREXH{cond}  Rt, [Rn]
STREXH{cond}  Rd, Rt, [Rn]
```

where

cond	is an optional condition code; see "Conditional Execution" section on page 358.
Rd	is the destination register for the returned status.
Rt	is the register to load or store.
Rn	is the register on which the memory address is based.
offset	is an optional offset applied to the value in *Rn*. If *offset* is omitted, the address is the value in *Rn*.

Operation
LDREX, LDREXB, and LDREXH load a word, byte, and halfword, respectively, from a memory address.

STREX, STREXB, and STREXH attempt to store a word, byte, and halfword, respectively, to a memory address. The address used in any store-exclusive instruction must be the same as the address in the most recently executed load-exclusive instruction. The values stored by the store-exclusive instruction must also have the same data size as the value loaded by the preceding load-exclusive instruction. This means software must always use a load-exclusive instruction and a matching store-exclusive instruction to perform a synchronization operation.

If a store-exclusive instruction performs the store, it writes 0 to its destination register. If it does not perform the store, it writes 1 to its destination register. If the store-exclusive instruction writes 0 to the destination register, it is guaranteed that no other process in the system has accessed the memory location between the load-exclusive and store-exclusive instructions.

For reasons of performance, keep the number of instructions between corresponding load-exclusive and store-exclusive instruction to a minimum.

Note

The result of executing a store-exclusive instruction to an address that is different from that used in the preceding load-exclusive instruction is unpredictable.

Restrictions

In these instructions:

- Do not use PC.
- Do not use SP for *Rd* and *Rt*.
- For STREX, *Rd* must be different from both *Rt* and *Rn*.
- The value of *offset* must be a multiple of 4 in the range 0–1020.

Condition Flags

These instructions do not change the flags.

Examples

```
MOV      R1, #0x1              ; Initialize the 'lock taken' value try
LDREX    R0, [LockAddr]        ; Load the lock value
CMP      R0, #0                ; Is the lock free?
ITT      EQ                    ; IT instruction for STREXEQ and CMPEQ
STREXEQ R0, R1, [LockAddr]    ; Try and claim the lock
CMPEQ    R0, #0                ; Did this succeed?
BNE      try                   ; No – try again
....                           ; Yes – we have the lock
```

A.3.9 CLREX

Clear Exclusive.

Syntax

```
CLREX{cond}
```

where

cond   is an optional condition code; see "Conditional Execution" section on page 358.

Operation

Use CLREX to make the next STREX, STREXB, or STREXH instructions write 1 to its destination register and fail to perform the store. It is useful in exception handler code to force the failure of the store exclusive if the exception occurs between a load-exclusive instruction and the matching store-exclusive instruction in a synchronization operation.

Condition Flags

These instructions do not change the flags.

Examples
```
CLREX
```

A.4 GENERAL DATA-PROCESSING INSTRUCTIONS

Table A.6 shows the data-processing instructions.

Table A.6 Data-Processing Instructions

Mnemonic	Brief Description	See
ADC	Add with Carry	*ADD, ADC, SUB, SBC, and RSB* on page 374
ADD	Add	*ADD, ADC, SUB, SBC, and RSB* on page 374
ADDW	Add	*ADD, ADC, SUB, SBC, and RSB* on page 374
AND	Logical AND	*AND, ORR, EOR, BIC, and ORN* on page 376
ASR	Arithmetic Shift Right	*ASR, LSL, LSR, ROR, and RRX* on page 377
BIC	Bit Clear	*AND, ORR, EOR, BIC, and ORN* on page 376
CLZ	Count leading zeros	*CLZ* on page 378
CMN	Compare Negative	*CMP and CMN* on page 378
CMP	Compare	*CMP and CMN* on page 378
EOR	Exclusive OR	*AND, ORR, EOR, BIC, and ORN* on page 376
LSL	Logical Shift Left	*ASR, LSL, LSR, ROR, and RRX* on page 377
LSR	Logical Shift Right	*ASR, LSL, LSR, ROR, and RRX* on page 377
MOV	Move	*MOV and MVN* on page 379
MOVT	Move Top	*MOVT* on page 381
MOVW	Move 16-bit constant	*MOV and MVN* on page 379
MVN	Move NOT	*MOV and MVN* on page 379
ORN	Logical OR NOT	*AND, ORR, EOR, BIC, and ORN* on page 376
ORR	Logical OR	*AND, ORR, EOR, BIC, and ORN* on page 376
RBIT	Reverse Bits	*REV, REV16, REVSH, and RBIT* on page 381
REV	Reverse byte order in a word	*REV, REV16, REVSH, and RBIT* on page 381
REV16	Reverse byte order in each halfword	*REV, REV16, REVSH, and RBIT* on page 381
REVSH	Reverse byte order in bottom halfword and sign extend	*REV, REV16, REVSH, and RBIT* on page 381
ROR	Rotate Right	*ASR, LSL, LSR, ROR, and RRX* on page 377
RRX	Rotate Right with Extend	*ASR, LSL, LSR, ROR, and RRX* on page 377
RSB	Reverse Subtract	*ADD, ADC, SUB, SBC, and RSB* on page 374
SBC	Subtract with Carry	*ADD, ADC, SUB, SBC, and RSB* on page 374
SUB	Subtract	*ADD, ADC, SUB, SBC, and RSB* on page 374
SUBW	Subtract	*ADD, ADC, SUB, SBC, and RSB* on page 374
TEQ	Test Equivalence	*TST and TEQ* on page 382
TST	Test	*TST and TEQ* on page 382

A.4.1 ADD, ADC, SUB, SBC, and RSB

Add, Add with Carry, Subtract, Subtract with Carry, and Reverse Subtract.

Syntax

```
op{S}{cond} {Rd,} Rn, Operand2
op{cond} {Rd,} Rn, #imm12      ; ADD and SUB only
```

where

op	is one of the following:

ADD	Add
ADC	Add with Carry
SUB	Subtract
SBC	Subtract with Carry
RSB	Reverse Subtract

S is an optional suffix. If *S* is specified, the condition code flags are updated on the result of the operation; see "Conditional Execution" section on page 358.

Cond is an optional condition code; see "Conditional Execution" section on page 358.

Rd is the destination register. If *Rd* is omitted, the destination register is *Rn*.

Rn is the register holding the first operand.

Operand2 is a flexible second operand; see "Flexible Second Operand" section on page 353 for details of the options.

imm12 is any value in the range 0–4095.

Operation

The ADD instruction adds the value of *Operand2* or *imm12* to the value in *Rn*. The ADC instruction adds the values in *Rn* and *Operand2*, together with the carry flag.

The SUB instruction subtracts the value of *Operand2* or *imm12* from the value in *Rn*. The SBC instruction subtracts the value of *Operand2* from the value in *Rn*. If the carry flag is clear, the result is reduced by one.

The RSB instruction subtracts the value in *Rn* from the value of *Operand2*. This is useful because of the wide range of options for *Operand2*.

Use ADC and SBC to synthesize multiword arithmetic; see "Multiword Arithmetic Examples" section on page 376; see also "ADR" section on page 362.

Note

ADDW is equivalent to the ADD syntax that uses the *imm12* operand. SUBW is equivalent to the SUB syntax that uses the *imm12* operand.

Restrictions

In these instructions:

- *Operand2* must not be SP and must not be PC.
- *Rd* can be SP only in ADD and SUB and only with the additional restrictions.
  - *Rn* must also be SP.
  - Any shift in *Operand2* must be limited to a maximum of 3 bits using LSL.

- *Rn* can be SP only in ADD and SUB.
- *Rd* can be PC only in the ADD{*cond*} PC, PC, Rm instruction where:
  - You must not specify the S suffix.
  - *Rm* must not be PC and must not be SP.
  - If the instruction is conditional, it must be the last instruction in the IT block.
- With the exception of the ADD{*cond*} PC, PC, Rm instruction, *Rn* can be PC only in ADD and SUB, and only with the additional restrictions:
  - You must not specify the S suffix.
  - The second operand must be a constant in the range 0–4095.

Note

- When using the PC for an addition or a subtraction, bits[1:0] of the PC are rounded to b00 before performing the calculation, making the base address for the calculation word-aligned.
- If you want to generate the address of an instruction, you have to adjust the constant based on the value of the PC. ARM recommends that you use the ADR instruction instead of ADD or SUB with *Rn* equal to the PC, because your assembler automatically calculates the correct constant for the ADR instruction.

When *Rd* is PC in the ADD{*cond*} PC, PC, Rm instruction:

- bit[0] of the value written to the PC is ignored.
- A branch occurs to the address created by forcing bit[0] of that value to 0.

Condition Flags

If S is specified, these instructions update the N, Z, C, and V flags according to the result.

Examples

```
ADD     R2, R1, R3
SUBS    R8, R6, #240      ; Sets the flags on the result
RSB     R4, R4, #1280     ; Subtracts contents of R4 from 1280
ADCHI   R11, R0, R3       ; Only executed if C flag set and Z
                          ; flag clear
```

Multiword Arithmetic Examples

Example A.4 shows two instructions that add a 64-bit integer contained in R2 and R3 to another 64-bit integer contained in R0 and R1 and place the result in R4 and R5.

Multiword values do not have to use consecutive registers. Example A.5 shows instructions that subtract a 96-bit integer contained in R9, R1, and R11 from another contained in R6, R2, and R8. The example stores the result in R6, R9, and R2.

Example A.4 64-Bit Addition

```
ADDS    R4, R0, R2        ; add the least significant words
ADC     R5, R1, R3        ; add the most significant words with carry
```

Example A.5 96-Bit Subtraction			
SUBS	R6, R6, R9	; subtract the least significant words
SBCS	R9, R2, R1	; subtract the middle words with carry
SBC	R2, R8, R11	; subtract the most significant words with carry

A.4.2 **AND, ORR, EOR, BIC, and ORN**

Logical AND, OR, Exclusive OR, Bit Clear, and OR NOT.

Syntax

```
op{S}{cond} {Rd,} Rn, Operand2
```

where

op	is one of the following:

AND	Logical AND
ORR	Logical OR or bit set
EOR	Logical Exclusive OR

BIC	Logical AND NOT or Bit Clear
ORN	Logical OR NOT

S	is an optional suffix. If S is specified, the condition code flags are updated on the result of the operation; see "Conditional Execution" section on page 358.
Cond	is an optional condition code; see "Conditional Execution" section on page 358.
Rd	is the destination register.
Rn	is the register holding the first operand.
Operand2	is a flexible second operand; see "Flexible Second Operand" section on page 353 for details of the options.

Operation

The AND, EOR, and ORR instructions perform bitwise AND, Exclusive OR, and OR operations on the values in *Rn* and *Operand2*.

The BIC instruction performs an AND operation on the bits in *Rn* with the complements of the corresponding bits in the value of *Operand2*.

The ORN instruction performs an OR operation on the bits in *Rn* with the complements of the corresponding bits in the value of *Operand2*.

Restrictions

Do not use SP and do not use PC.

Condition Flags

If S is specified, these instructions

- Update the N and Z flags according to the result
- Can update the C flag during the calculation of *Operand2*; see "Flexible Second Operand" section on page 353
- Do not affect the V flag.

Examples
```
AND     R9, R2, #0xFF00
ORREQ   R2, R0, R5
ANDS    R9, R8, #0x19
EORS    R7, R11, #0x18181818
BIC     R0, R1, #0xab
ORNS    R7, R11, R14, ROR #4
ORNS    R7, R11, R14, ASR #32
```

A.4.3 ASR, LSL, LSR, ROR, and RRX

Arithmetic Shift Right, Logical Shift Left, Logical Shift Right, Rotate Right, and Rotate Right with Extend.

Syntax
```
op{S}{cond} Rd, Rm, Rs
op{S}{cond} Rd, Rm, #n
RRX{S}{cond} Rd, Rm
```

where

op	is one of the following:

	ASR	Arithmetic Shift Right
	LSL	Logical Shift Left
	LSR	Logical Shift Right
	ROR	Rotate Right

S   is an optional suffix. If S is specified, the condition code flags are updated on the result of the operation; see "Conditional Execution" section on page 358.

Rd   is the destination register.

Rm   is the register holding the value to be shifted.

Rs   is the register holding the shift length to apply to the value in Rm. Only the least significant byte is used and can be in the range 0–255.

n   is the shift length. The range of shift length depends on the instruction:

	ASR	Shift length from 1 to 32
	LSL	Shift length from 0 to 31
	LSR	Shift length from 1 to 32
	ROR	Shift length from 1 to 31

Note
MOVS Rd, Rm is the preferred syntax for LSLS Rd, Rm, #0.

Operation
ASR, LSL, LSR, and ROR move the bits in the register Rm to the left or right by the number of places specified by constant n or register Rs. RRX moves the bits in register Rm to the right by 1.

In all these instructions, the result is written to Rd, but the value in register Rm remains unchanged. For details on what result is generated by the different instructions, see "Shift Operations" section on page 354.

Restrictions
Do not use SP and do not use PC.

Condition Flags
If S is specified:

- These instructions update the N and Z flags according to the result.
- The C flag is updated to the last bit shifted out, except when the shift length is 0; see "Shift Operations" section on page 354.

Examples
```
ASR    R7, R8, #9   ; Arithmetic shift right by 9 bits
LSLS   R1, R2, #3   ; Logical shift left by 3 bits with flag update
LSR    R4, R5, #6   ; Logical shift right by 6 bits
ROR    R4, R5, R6   ; Rotate right by the value in the bottom byte of R6
RRX    R4, R5       ; Rotate right with extend
```

A.4.4 CLZ
Count Leading Zeros.

Syntax
```
CLZ{cond} Rd, Rm
```

where

cond	is an optional condition code; see "Conditional Execution" section on page 358.
Rd	is the destination register.
Rm	is the operand register.

Operation
The CLZ instruction counts the number of leading zeros in the value in Rm and returns the result in Rd. The result value is 32 if no bits are set in the source register and 0 if bit[31] is set.

Restrictions
Do not use SP and do not use PC.

Condition Flags
This instruction does not change the flags.

Examples
```
CLZ     R4,R9
CLZNE   R2,R3
```

A.4.5 CMP and CMN
Compare and Compare Negative.

Syntax
```
CMP{cond} Rn, Operand2
CMN{cond} Rn, Operand2
```

where

cond	is an optional condition code; see "Conditional Execution" section on page 358.
Rn	is the register holding the first operand.
Operand2	is a flexible second operand; see "Flexible Second Operand" section on page 353 for details of the options.

Operation

These instructions compare the value in a register with *Operand2*. They update the condition flags on the result but do not write the result to a register.

The CMP instruction subtracts the value of *Operand2* from the value in *Rn*. This is the same as a SUBS instruction, except that the result is discarded.

The CMN instruction adds the value of *Operand2* to the value in *Rn*. This is the same as an ADDS instruction, except that the result is discarded.

Restrictions

In these instructions:

- Do not use PC.
- *Operand2* must not be SP.

Condition Flags

These instructions update the N, Z, C, and V flags according to the result.

Examples

```
CMP    R2, R9
CMN    R0, #6400
CMPGT  SP, R7, LSL #2
```

A.4.6 MOV and MVN

Move and Move NOT.

Syntax

```
MOV{S}{cond} Rd, Operand02
MOV{cond} Rd, #imm16
MVN{S}{cond} Rd, Operand2
```

where

S	is an optional suffix. If S is specified, the condition code flags are updated on the result of the operation; see "Conditional Execution" section on page 358.
cond	is an optional condition code; see "Conditional Execution" section on page 358.
Rd	is the destination register.
Operand2	is a flexible second operand; see "Flexible Second Operand" section on page 353 for details of the options.
imm16	is any value in the range 0–65535.

Operation

The MOV instruction copies the value of *Operand2* into *Rd*. When *Operand2* in a MOV instruction is a register with a shift other than LSL #0, the preferred syntax is the corresponding shift instruction:

- ASR{S}{cond} Rd, Rm, #n is the preferred syntax for MOV{S}{cond} Rd, Rm, ASR #n.
- LSL{S}{cond} Rd, Rm, #n is the preferred syntax for MOV{S}{cond} Rd, Rm, LSL #n if *n* != 0.
- LSR{S}{cond} Rd, Rm, #n is the preferred syntax for MOV{S}{cond} Rd, Rm, LSR #n.
- ROR{S}{cond} Rd, Rm, #n is the preferred syntax for MOV{S}{cond} Rd, Rm, ROR #n.
- RRX{S}{cond} Rd, Rm is the preferred syntax for MOV{S}{cond} Rd, Rm, RRX.

Also, the MOV instruction permits additional forms of *Operand2* as synonyms for shift instructions:

- MOV{S}{cond} Rd, Rm, ASR Rs is a synonym for ASR{S}{cond} Rd, Rm, Rs.
- MOV{S}{cond} Rd, Rm, LSL Rs is a synonym for LSL{S}{cond} Rd, Rm, Rs.
- MOV{S}{cond} Rd, Rm, LSR Rs is a synonym for LSR{S}{cond} Rd, Rm, Rs.
- MOV{S}{cond} Rd, Rm, ROR Rs is a synonym for ROR{S}{cond} Rd, Rm, Rs.

See ASR, LSL, LSR, ROR, and RRX on page 377.

The MVN instruction takes the value of *Operand2*, performs a bitwise logical NOT operation on the value, and places the result into *Rd*.

Note

The MOVW instruction provides the same function as MOV but is restricted to using the *imm16* operand.

Restrictions

You can use SP and PC only in the MOV instruction, with the following restrictions:

- The second operand must be a register without shift.
- You must not specify the S suffix.

When *Rd* is PC in a MOV instruction:

- bit[0] of the value written to the PC is ignored.
- A branch occurs to the address created by forcing bit[0] of that value to 0.

Note

Though it is possible to use MOV as a branch instruction, ARM strongly recommends the use of a BX or BLX instruction to branch for software portability to the ARM instruction set.

Condition Flags

If S is specified, these instructions:

- Update the N and Z flags according to the result
- Can update the C flag during the calculation of *Operand2*; see "Flexible Second Operand" section on page 353
- Do not affect the V flag.

Example

```
MOVS    R11, #0x000B  ; Write value of 0x000B to R11, flags get updated
MOV     R1, #0xFA05   ; Write value of 0xFA05 to R1, flags are not updated
MOVS    R10, R12      ; Write value in R12 to R10, flags get updated
MOV     R3, #23       ; Write value of 23 to R3
MOV     R8, SP        ; Write value of stack pointer to R8
MVNS    R2, #0xF      ; Write value of 0xFFFFFFF0 (bitwise inverse of0xF)
                      ; to the R2 and update flags
```

A.4.7 **MOVT**

Move Top.

Syntax
```
MOVT{cond} Rd, #imm16
```

where

    cond   is an optional condition code; see "Conditional Execution" section on page 358.
    Rd   is the destination register.
    imm16   is a 16-bit immediate constant.

Operation
MOVT writes a 16-bit immediate value, *imm16*, to the top halfword, *Rd*[31:16], of its destination register. The write does not affect *Rd*[15:0].

    The MOV, MOVT instruction pair enables you to generate any 32-bit constant.

Restrictions
Rd must not be SP and must not be PC.

Condition Flags
This instruction does not change the flags.

Examples
```
MOVT  R3, #0xF123 ; Write 0xF123 to upper halfword of R3, lower halfword
                  ; and APSR are unchanged
```

A.4.8 **REV, REV16, REVSH, and RBIT**

Reverse bytes and Reverse bits.

Syntax
```
op{cond} Rd, Rn
```

where

    op   is any of the following:
             REV   Reverse byte order in a word

REV16	Reverse byte order in each halfword independently
REVSH	Reverse byte order in the bottom halfword, and sign extends to 32 bits
RBIT	Reverse the bit order in a 32-bit word

cond	is an optional condition code; see "Conditional Execution" section on page 358.
Rd	is the destination register.
Rn	is the register holding the operand.

Operation

Use these instructions to change endianness of data:

REV	converts 32-bit big-endian data into little-endian data or 32-bit little-endian data into big-endian data.
REV16	converts 16-bit big-endian data into little-endian data or 16-bit little-endian data into big-endian data.
REVSH	converts either: 16-bit signed big-endian data into 32-bit signed little-endian data 16-bit signed little-endian data into 32-bit signed big-endian data.

Restrictions

Do not use SP and do not use PC.

Condition Flags

These instructions do not change the flags.

Examples

```
REV     R3, R7   ; Reverse byte order of value in R7 and write it to R3
REV16   R0, R0   ; Reverse byte order of each 16-bit halfword in R0
REVSH   R0, R5   ; Reverse Signed Halfword
REVHS   R3, R7   ; Reverse with Higher or Same condition
RBIT    R7, R8   ; Reverse bit order of value in R8 and write the result to R7
```

A.4.9 TST and TEQ

Test bits and Test Equivalence.

Syntax

```
TST{cond} Rn, Operand2
TEQ{cond} Rn, Operand2
```

where

cond	is an optional condition code; see "Conditional Execution" section on page 358.
Rn	is the register holding the first operand.
Operand2	is a flexible second operand; see "Flexible Second Operand" section on page 353 for details of the options.

Operation

These instructions test the value in a register against *Operand2*. They update the condition flags based on the result but do not write the result to a register.

The TST instruction performs a bitwise AND operation on the value in *Rn* and the value of *Operand2*. This is the same as the ANDS instruction, except that it discards the result.

To test whether a bit of *Rn* is 0 or 1, use the TST instruction with an *Operand2* constant that has bit set to 1 and all other bits cleared to 0.

The TEQ instruction performs a bitwise Exclusive OR operation on the value in *Rn* and the value of *Operand2*. This is same as the EORS instruction, except that it discards the result.

Use the TEQ instruction to test if two values are equal without affecting the V or C flags.

TEQ is also useful for testing the sign of a value. After the comparison, the N flag is the logical Exclusive OR of the sign bits of the two operands.

Restrictions

Do not use SP and do not use PC.

Condition Flags

These instructions:

- Update the N and Z flags according to the result
- Can update the C flag during the calculation of *Operand2*; see "Flexible Second Operand" section on page 353.
- Do not affect the V flag.

Examples

```
TST    R0, #0x3F8 ; Perform bitwise AND of R0 value to 0x3F8,
                   ; APSR is updated but result is discarded
TEQEQ R10, R9      ; Conditionally test if value in R10 is equal to
                   ; value in R9, APSR is updated but result is discarded
```

A.5 MULTIPLY AND DIVIDE INSTRUCTIONS

Table A.7 shows the multiply and divide instructions.

A.5.1 MUL, MLA, and MLS

Multiply, Multiply with Accumulate, and Multiply with Subtract, using 32-bit operands and producing a 32-bit result.

Syntax

```
MUL{S}{cond}  Rd, Rn, Rm ; Multiply
MLA{cond} Rd, Rn, Rm, Ra  ; Multiply with accumulate
MLS{cond} Rd, Rn, Rm, Ra  ; Multiply with subtract
```

where

cond   is an optional condition code; see "Conditional Execution" section on page 358.
S   is an optional suffix. If S is specified, the condition code flags are updated on the result of the operation; see "Conditional Execution" section on page 358.

Table A.7 Multiply and Divide Instructions

Mnemonic	Brief Description	See
MLA	Multiply with Accumulate, 32-bit result	*MUL, MLA, and MLS* on page 383
MLS	Multiply and Subtract, 32-bit result	*MUL, MLA, and MLS* on page 383
MUL	Multiply, 32-bit result	*MUL, MLA, and MLS* on page 383
SDIV	Signed Divide	*SDIV and UDIV* on page 386
SMLAL	Signed Multiply with Accumulate (32 × 32 + 64), 64-bit result	*UMULL, UMLAL, SMULL, and SMLAL* on page 385
SMULL	Signed Multiply (32 × 32), 64-bit result	*UMULL, UMLAL, SMULL, and SMLAL* on page 385
UDIV	Unsigned Divide	*SDIV and UDIV* on page 386
UMLAL	Unsigned Multiply with Accumulate (32 × 32 + 64), 64-bit result	*UMULL, UMLAL, SMULL, and SMLAL* on page 385
UMULL	Unsigned Multiply (32 × 32), 64-bit result	*UMULL, UMLAL, SMULL, and SMLAL* on page 385

Rd	is the destination register. If Rd is omitted, the destination register is Rn.
Rn, Rm	are registers holding the values to be multiplied.
Ra	is a register holding the value to be added or subtracted from.

Operation

The MUL instruction multiplies the values from Rn and Rm and places the least significant 32 bits of the result in Rd.

The MLA instruction multiplies the values from Rn and Rm, adds the value from Ra, and places the least significant 32 bits of the result in Rd.

The MLS instruction multiplies the values from Rn and Rm, subtracts the product from the value from Ra, and places the least significant 32 bits of the result in Rd.

The results of these instructions do not depend on whether the operands are signed or unsigned.

Restrictions

In these instructions, do not use SP and do not use PC.

If you use the S suffix with the MUL instruction:

- Rd, Rn, and Rm must all be in the range R0–R7.
- Rd must be the same as Rm.
- You must not use the *cond* suffix.

Condition Flags

If S is specified, the MUL instruction:

- Updates the N and Z flags according to the result
- Does not affect the C and V flags.

Examples

```
MUL      R10, R2, R5      ; Multiply, R10 = R2 × R5
MLA      R10, R2, R1, R5  ; Multiply with accumulate, R10 = (R2 × R1) + R5
MULS     R0, R2, R2       ; Multiply with flag update, R0 = R2 × R2
MULLT    R2, R3, R2       ; Conditionally multiply, R2 = R3 × R2
MLS      R4, R5, R6, R7   ; Multiply with subtract, R4 = R7 - (R5 × R6)
```

A.5.2 UMULL, UMLAL, SMULL, and SMLAL

Signed and Unsigned Long Multiply, with optional Accumulate, using 32-bit operands and producing a 64-bit result.

Syntax

> *op{cond} RdLo, RdHi, Rn, Rm*

where

op	is one of the following:
	UMULL Unsigned Long Multiply
	UMLAL Unsigned Long Multiply, with Accumulate
	SMULL Signed Long Multiply
	SMLAL Signed Long Multiply, with Accumulate
Cond	is an optional condition code; see "Conditional Execution" section on page 358.
RdHi, RdLo	are the destination registers. For UMLAL and SMLAL, they also hold the accumulating value.
Rn, Rm	are registers holding the operands.

Operation

The UMULL instruction interprets the values from *Rn* and *Rm* as unsigned integers. It multiplies these integers and places the least significant 32 bits of the result in *RdLo* and the most significant 32 bits of the result in *RdHi*.

The UMLAL instruction interprets the values from *Rn* and *Rm* as unsigned integers. It multiplies these integers, adds the 64-bit result to the 64-bit unsigned integer contained in *RdHi* and *RdLo*, and writes the result back to *RdHi* and *RdLo*.

The SMULL instruction interprets the values from *Rn* and *Rm* as two's complement signed integers. It multiplies these integers and places the least significant 32 bits of the result in *RdLo* and the most significant 32 bits of the result in *RdHi*.

The SMLAL instruction interprets the values from *Rn* and *Rm* as two's complement signed integers. It multiplies these integers, adds the 64-bit result to the 64-bit signed integer contained in *RdHi* and *RdLo*, and writes the result back to *RdHi* and *RdLo*.

Restrictions

In these instructions:

- Do not use SP and do not use PC.
- *RdHi* and *RdLo* must be different registers.

Condition Flags
These instructions do not affect the condition code flags.

Examples
```
UMULL    R0, R4, R5, R6    ; Unsigned (R4,R0) = R5 × R6
SMLAL    R4, R5, R3, R8    ; Signed (R5,R4) = (R5,R4) + R3 × R8
```

A.5.3 SDIV and UDIV
Signed Divide and Unsigned Divide.

Syntax
```
SDIV{cond} {Rd,} Rn, Rm
UDIV{cond} {Rd,} Rn, Rm
```

where

cond	is an optional condition code; see "Conditional Execution" section on page 358.
Rd	is the destination register. If *Rd* is omitted, the destination register is *Rn*.
Rn	is the register holding the value to be divided.
Rm	is a register holding the divisor.

Operation
SDIV performs a signed integer division of the value in *Rn* by the value in *Rm*. UDIV performs an unsigned integer division of the value in *Rn* by the value in *Rm*.

For both instructions, if the value in *Rn* is not divisible by the value in *Rm*, the result is rounded toward zero.

Restrictions
Do not use SP and do not use PC.

Condition Flags
These instructions do not change the flags.

Examples
```
SDIV R0, R2, R4 ; Signed divide, R0 = R2/R4
UDIV R8, R8, R1 ; Unsigned divide, R8 = R8/R1
```

A.6 SATURATING INSTRUCTIONS
This section describes the saturating instructions: SSAT and USAT.

A.6.1 SSAT and USAT
Signed Saturate and Unsigned Saturate to any bit position, with optional shift before saturating.

Syntax

```
op{cond} Rd, #n, Rm {, shift #s}
```

where

op	is one of the following:
	SSAT Saturates a signed value to a signed range
	USAT Saturates a signed value to an unsigned range
cond	is an optional condition code; see "Conditional Execution" section on page 358.
Rd	is the destination register.
n	specifies the bit position to saturate to:
	n ranges from 1 to 32 for SSAT
	n ranges from 0 to 31 for USAT
Rm	is the register containing the value to saturate.
shift #s	is an optional shift applied to Rm before saturating. It must be one of the following:
	ASR #s where s is in the range 1–31
	LSL #s where s is in the range 0–31

Operation

These instructions saturate to a signed or unsigned n-bit value.

The SSAT instruction applies the specified shift and then saturates to the signed range $-2^{n-1} \leq x \leq 2^{n-1} - 1$. The USAT instruction applies the specified shift and then saturates to the unsigned range $0 \leq x \leq 2^n - 1$.

For signed n-bit saturation using SSAT, this means that

- If the value to be saturated is less than -2^{n-1}, the result returned is -2^{n-1}.
- If the value to be saturated is greater than $2^{n-1} - 1$, the result returned is $2^{n-1} - 1$.
- Otherwise, the result returned is the same as the value to be saturated.

For unsigned n-bit saturation using USAT, this means that

- If the value to be saturated is less than 0, the result returned is 0.
- If the value to be saturated is greater than $2^n - 1$, the result returned is $2^n - 1$.
- Otherwise, the result returned is the same as the value to be saturated.

If the returned result is different from the value to be saturated, it is called *saturation*. If saturation occurs, the instruction sets the Q flag to 1 in the APSR. Otherwise, it leaves the Q flag unchanged. To clear the Q flag to 0, you must use the MSR instruction; see *MSR* on page 400.

To read the state of the Q flag, use the MRS instruction; see "MRS" section on page 400.

Restrictions

Do not use SP and do not use PC.

Condition Flags

These instructions do not affect the condition code flags.

If saturation occurs, these instructions set the Q flag to 1.

Examples

```
SSAT    R7, #16, R7, LSL #4 ; Logical shift left value in R7 by 4, then
                            ; saturate it as a signed 16-bit value and
                            ; write it back to R7
USATNE  R0, #7,  R5         ; Conditionally saturate value in R5 as an
                            ; unsigned 7 bit value and write it to R0
```

A.7 BITFIELD INSTRUCTIONS

Table A.8 shows the instructions that operate on adjacent sets of bits in registers or bitfields.

A.7.1 BFC and BFI

Bit Field Clear and Bit Field Insert.

Syntax

```
BFC{cond} Rd, #lsb, #width
BFI{cond} Rd, Rn, #lsb, #width
```

where

cond	is an optional condition code; see "Conditional Execution" section on page 358.
Rd	is the destination register.
Rn	is the source register.
lsb	is the position of the least significant bit of the bitfield. lsb must be in the range 0–31.
width	is the width of the bitfield and must be in the range 1–32−lsb.

Operation

BFC clears a bitfield in a register. It clears width bits in Rd, starting at the low bit position lsb. Other bits in Rd are unchanged.

Table A.8 Packing and Unpacking Instructions

Mnemonic	Brief Description	See
BFC	Bit Field Clear	BFC and BFI on page 388
BFI	Bit Field Insert	BFC and BFI on page 388
SBFX	Signed Bit Field Extract	SBFX and UBFX on page 389
SXTB	Sign extend a byte	SXT and UXT on page 390
SXTH	Sign extend a halfword	SXT and UXT on page 390
UBFX	Unsigned Bit Field Extract	SBFX and UBFX on page 389
UXTB	Zero extend a byte	SXT and UXT on page 390
UXTH	Zero extend a halfword	SXT and UXT on page 390

BFI copies a bitfield into one register from another register. It replaces *width* bits in *Rd* starting at the low bit position *lsb*, with *width* bits from *Rn* starting at bit[0]. Other bits in *Rd* are unchanged.

Restrictions
Do not use SP and do not use PC.

Condition Flags
These instructions do not affect the flags.

Examples
```
BFC R4, #8, #12      ; Clear bit 8 to bit 19 (12 bits) of R4 to 0
BFI R9, R2, #8, #12  ; Replace bit 8 to bit 19 (12 bits) of R9 with
                     ; bit 0 to bit 11 from R2
```

A.7.2 SBFX and UBFX
Signed Bit Field Extract and Unsigned Bit Field Extract.

Syntax
```
SBFX{cond} Rd, Rn, #lsb, #width
UBFX{cond} Rd, Rn, #lsb, #width
```

where

cond	is an optional condition code; see "Conditional Execution" section on page 358.
Rd	is the destination register.
Rn	is the source register.
lsb	is the position of the least significant bit of the bitfield. *lsb* must be in the range 0–31.
width	is the width of the bitfield and must be in the range 1–32–*lsb*.

Operation
SBFX extracts a bitfield from one register; sign extends it to 32 bits and writes the result to the destination register.

UBFX extracts a bitfield from one register; zero extends it to 32 bits and writes the result to the destination register.

Restrictions
Do not use SP and do not use PC.

Condition Flags
These instructions do not affect the flags.

Examples
```
SBFX R0, R1,  #20, #4   ; Extract bit 20 to bit 23 (4 bits) from R1 and sign
                        ; extend to 32 bits and then write the result to R0.
UBFX R8, R11, #9,  #10  ; Extract bit 9 to bit 18 (10 bits) from R11 and zero
                        ; extend to 32 bits and then write the result to R8
```

A.7.3 SXT and UXT

Sign extend and zero extend.

Syntax

```
SXT extend{cond} Rd, Rm {, ROR #n}
UXT extend{cond} Rd, Rm {, ROR #n}
```

where

extend	is one of the following:
	B extends an 8-bit value to a 32-bit value.
	H extends a 16-bit value to a 32-bit value.
cond	is an optional condition code; see "Conditional Execution" section on page 358.
Rd	is the destination register.
Rm	is the register holding the value to extend.
ROR #n	is one of the following:
	ROR #8 value from Rm is rotated right 8 bits.
	ROR #16 value from Rm is rotated right 16 bits.
	ROR #24 value from Rm is rotated right 24 bits.
	If ROR #n is omitted, no rotation is performed.

Operation

These instructions do the following:

1. Rotate the value from Rm right by 0, 8, 16, or 24 bits.
2. Extract bits from the resulting value:
   a. SXTB extracts bits[7:0], and sign extends to 32 bits.
   b. UXTB extracts bits[7:0], and zero extends to 32 bits.
   c. SXTH extracts bits[15:0], and sign extends to 32 bits.
   d. UXTH extracts bits[15:0], and zero extends to 32 bits.

Restrictions

Do not use SP and do not use PC.

Condition Flags

These instructions do not affect the flags.

Examples

```
SXTH R4, R6, ROR #16 ; Rotate R6 right by 16 bits, then obtain the lower
                     ; halfword of the result and then sign extend to
                     ; 32 bits and write the result to R4.
UXTB R3, R10         ; Extract lowest byte of the value in R10 and zero
                     ; extend it, and write the result to R3
```

A.8 BRANCH AND CONTROL INSTRUCTIONS

Table A.9 shows the branch and control instructions.

A.8.1 B, BL, BX, and BLX

Branch instructions.

Syntax
```
B{cond} label
BL{cond} label
BX{cond} Rm
BLX{cond} Rm
```

where

B	is branch (immediate).
BL	is branch with link (immediate).
BX	is branch indirect (register).
BLX	is branch indirect with link (register).
cond	is an optional condition code; see "Conditional Execution" section on page 358.
label	is a PC-relative expression; see "PC-Relative Expressions" section on page 358.
Rm	is a register that indicates an address to branch to. Bit[0] of the value in Rm must be 1, but the address to branch to is created by changing bit[0] to 0.

Operation

All these instructions cause a branch to label or to the address indicated in Rm. In addition:

- The BL and BLX instructions write the address of the next instruction to LR (the link register, R14).
- The BX and BLX instructions cause a usage fault exception if bit[0] of Rm is 0.

Bcond label is the only conditional instruction that can be either inside or outside an IT block. All other branch instructions must be conditional inside the IT block and must be unconditional outside the IT block; see "IT" section on page 393.

Table A.10 shows the ranges for the various branch instructions.

Table A.9 Branch and Control Instructions

Mnemonic	Brief Description	See
B	Branch	B, BL, BX, and BLX on page 391
BL	Branch with Link	B, BL, BX, and BLX on page 391
BLX	Branch indirect with Link	B, BL, BX, and BLX on page 391
BX	Branch indirect	B, BL, BX, and BLX on page 391
CBNZ	Compare and Branch if Nonzero	CBZ and CBNZ on page 393
CBZ	Compare and Branch if zero	CBZ and CBNZ on page 393
IT	If-Then	IT on page 393
TBB	Table Branch Byte	TBB and TBH on page 395
TBH	Table Branch Halfword	TBB and TBH on page 395

Table A.10 Branch Ranges

Instruction	Branch Range
B label	–16 to +16 MB
Bcond label (outside IT block)	–1 to +1 MB
Bcond label (inside IT block)	–16 to +16 MB
BL{cond} label	–16 to +16 MB
BX{cond} Rm	Any value in register
BLX{cond} Rm	Any value in register

Note

You might have to use the .W suffix to get the maximum branch range; see "Instruction Width Selection" section on page 360.

Restrictions

The restrictions are as follows:

- Do not use PC in the BLX instruction.
- For BX and BLX, bit[0] of Rm must be 1 for correct execution, but a branch occurs to the target address created by changing bit[0] to 0.
- When any of these instructions is inside an IT block, it must be the last instruction of the IT block.

Note

Bcond is the only conditional instruction that is not required to be inside an IT block. However, it has a longer branch range when it is inside an IT block.

Condition Flags

These instructions do not change the flags.

Examples

```
B        loopA    ; Branch to loopA
BLE      ng       ; Conditionally branch to label ng
B.W      target   ; Branch to target within 16MB range
BEQ      target   ; Conditionally branch to target
BEQ.W    target   ; Conditionally branch to target within 1MB
BL       funC     ; Branch with link (Call) to function funC, return
                  address
                  ; stored in LR
BX       LR       ; Return from function call
BXNE     R0       ; Conditionally branch to address stored in R0
BLX      R0       ; Branch with link and exchange (Call) to a address
                  stored
                  ; in R0
```

A.8.2 **CBZ and CBNZ**

Compare and Branch on Zero and Compare and Branch on Nonzero.

Syntax

```
CBZ  Rn, label
CBNZ Rn, label
```

where

Rn	is the register holding the operand.
label	is the branch destination.

Operation

Use the CBZ or CBNZ instructions to avoid changing the condition code flags and to reduce the number of instructions.

CBZ Rn, label does not change condition flags but is otherwise equivalent to

```
CMP  Rn, #0
BEQ  label
```

CBNZ Rn, label does not change condition flags but is otherwise equivalent to

```
CMP  Rn, #0
BNE  label
```

Restrictions

The restrictions are as follows:

- Rn must be in the range of R0–R7.
- The branch destination must be within 4–130 bytes after the instruction.
- These instructions must not be used inside an IT block.

Condition Flags

These instructions do not change the flags.

Examples

```
CBZ  R5, target ;Forward branch if R5 is zero
CBNZ R0, target ;Forward branch if R0 is not zero
```

A.8.3 **IT**

If-Then condition instruction.

Syntax

```
IT{x{y{z}}} cond
```

where

x	specifies the condition switch for the second instruction in the IT block.
y	specifies the condition switch for the third instruction in the IT block.

z   specifies the condition switch for the fourth instruction in the IT block.

cond   specifies the condition for the first instruction in the IT block.

The condition switch for the second, third, and fourth instruction in the IT block can be either

T   Then, applies the condition *cond* to the instruction

E   Else, applies the inverse condition of *cond* to the instruction.

Note

It is possible to use AL (the *always* condition) for *cond* in an IT instruction. If this is done, all the instructions in the IT block must be unconditional, and each of *x*, *y*, and *z* must be T or omitted but not E.

Operation

The IT instruction makes up to four following instructions conditional. The conditions can be all the same, or some of them can be the logical inverse of the others. The conditional instructions following the IT instruction form the *IT block*.

The instructions in the IT block, including any branches, must specify the condition in the {cond} part of their syntax.

Note

Your assembler might be able to generate the required IT instructions for conditional instructions automatically so that you do not need to write them yourself. See your assembler documentation for details.

A BKPT instruction in an IT block is always executed, even if its condition fails.

Exceptions can be taken between an IT instruction and the corresponding IT block or within an IT block. Such an exception results in entry to the appropriate exception handler, with suitable return information in LR and stacked PSR.

Instructions designed for use for exception returns can be used as normal to return from the exception, and execution of the IT block resumes correctly. This is the only way that a PC-modifying instruction is permitted to branch to an instruction in an IT block.

Restrictions

The following instructions are not permitted in an IT block:

- IT
- CBZ and CBNZ
- CPSID and CPSIE

Other restrictions when using an IT block are as follows:

- a branch or any instruction that modifies the PC must either be outside an IT block or must be the last instruction inside the IT block. These are as follows:
  - ADD PC, PC, Rm
  - MOV PC, Rm
  - B, BL, BX, BLX
  - Any LDM, LDR, or POP instruction that writes to the PC
  - TBB and TBH

- Do not branch to any instruction inside an IT block, except when returning from an exception handler.
- All conditional instructions except B*cond* must be inside an IT block. B*cond* can be either outside or inside an IT block but has a larger branch range if it is inside one.
- Each instruction inside the IT block must specify a condition code suffix that is either the same or logical inverse as for the other instructions in the block.

Note
Your assembler might place extra restrictions on the use of IT blocks, such as prohibiting the use of assembler directives within them.

Condition Flags
This instruction does not change the flags.

Example

```
ITTE    NE              ; Next 3 instructions are conditional
ANDNE   R0, R0, R1      ; ANDNE does not update condition flags
ADDSNE  R2, R2, #1      ; ADDSNE updates condition flags
MOVEQ   R2, R3          ; Conditional move
CMP     R0, #9          ; Convert R0 hex value (0 to 15) into ASCII
                        ; ('0'-'9', 'A'-'F')
ITE     GT              ; Next 2 instructions are conditional
ADDGT   R1, R0, #55     ; Convert 0xA -> 'A'
ADDLE   R1, R0, #48     ; Convert 0x0 -> '0'
IT      GT              ; IT block with only one conditional instruction
ADDGT   R1, R1, #1      ; Increment R1 conditionally
ITTEE   EQ              ; Next 4 instructions are conditional
MOVEQ   R0, R1          ; Conditional move
ADDEQ   R2, R2, #10     ; Conditional add
ANDNE   R3, R3, #1      ; Conditional AND
BNE.W   dloop           ; Branch instruction can only be used in the last
                        ; instruction of an IT block
IT      NE              ; Next instruction is conditional
ADD     R0, R0, R1      ; Syntax error: no condition code used in IT block
```

A.8.4 TBB and TBH
Table Branch Byte and Table Branch Halfword.

Syntax
```
TBB [Rn, Rm]
TBH [Rn, Rm, LSL #1]
```

where

Rn   is the register containing the address of the table of branch lengths. If *Rn* is PC, then the address of the table is the address of the byte immediately following the TBB or TBH instruction.

Rm   is the index register. This contains an index into the table. For halfword tables, LSL #1 doubles the value in *Rm* to form the right offset into the table.

Operation

These instructions cause a PC-relative forward branch using a table of single byte offsets for TBB or halfword offsets for TBH. *Rn* provides a pointer to the table, and *Rm* supplies an index into the table. For TBB, the branch offset is twice the unsigned value of the byte returned from the table, and for TBH, the branch offset is twice the unsigned value of the halfword returned from the table. The branch occurs to the address at that offset from the address of the byte immediately after the TBB or TBH instruction.

Restrictions

The restrictions are as follows:

- *Rn* must not be SP.
- *Rm* must not be SP and must not be PC.
- When any of these instructions is used inside an IT block, it must be the last instruction of the IT block.

Condition Flags

These instructions do not change the flags.

Examples

```
ADR.W R0, BranchTable_Byte
TBB    [R0, R1]                       ; R1 is the index, R0 is the base address of the
                                      ; branch table
Case1
                                      ; an instruction sequence follows
Case2
                                      ; an instruction sequence follows
Case3
                                      ; an instruction sequence follows
BranchTable_Byte
DCB 0                                 ; Case1 offset calculation
DCB ((Case2-Case1)/2)                 ; Case2 offset calculation
DCB ((Case3-Case1)/2)                 ; Case3 offset calculation

TBH [PC, R1, LSL #1]                  ; R1 is the index, PC is used as base of the
                                      ; branch table
BranchTable_H
DCI ((CaseA - BranchTable_H)/2)       ; CaseA offset calculation
DCI ((CaseB - BranchTable_H)/2)       ; CaseB offset calculation
DCI ((CaseC - BranchTable_H)/2)       ; CaseC offset calculation
CaseA
                                      ; an instruction sequence follows
CaseB
                                      ; an instruction sequence follows
CaseC
                                      ; an instruction sequence follows
```

A.9 MISCELLANEOUS INSTRUCTIONS

Table A.11 shows the remaining Cortex-M3 instructions.

A.9.1 BKPT

Breakpoint.

Syntax

```
BKPT #imm
```

where

   imm   is an expression evaluating to an integer in the range 0–255 (8-bit value).

Operation

The BKPT instruction causes the processor to enter Debug state. Debug tools can use this to investigate system state when the instruction at a particular address is reached.

imm is ignored by the processor. If required, a debugger can use it to store additional information about the breakpoint. ARM does not recommend the use of the BKPT instruction with an immediate value set to 0xAB for any purpose other than Semi-hosting.

The BKPT instruction can be placed inside an IT block, but it executes unconditionally, unaffected by the condition specified by the IT instruction.

Condition Flags

This instruction does not change the flags.

Table A.11 Miscellaneous Instructions

Mnemonic	Brief Description	See
BKPT	Breakpoint	*BKPT* on page 397
CPSID	Change Processor State, Disable Interrupts	*CPS* on page 398
CPSIE	Change Processor State, Enable Interrupts	*CPS* on page 398
DMB	Data Memory Barrier	*DMB* on page 398
DSB	Data Synchronization Barrier	*DSB* on page 399
ISB	Instruction Synchronization Barrier	*ISB* on page 399
MRS	Move from special register to register	*MRS* on page 400
MSR	Move from register to special register	*MSR* on page 400
NOP	No Operation	*NOP* on page 401
SEV	Send Event	*SEV* on page 402
SVC	Supervisor Call	*SVC* on page 402
WFE	Wait For Event	*WFE* on page 403
WFI	Wait For Interrupt	*WFI* on page 403

Examples

```
BKPT #0x3 ; Breakpoint with immediate value set to 0x3 (debugger can
          ; extract the immediate value by locating it using the PC)
```

Note

ARM does not recommend the use of the BKPT instruction with an immediate value set to 0xAB for any purpose other than Semi-hosting.

A.9.2 CPS

Change Processor State.

Syntax

```
CPSeffect iflags
```

where

effect	is one of the following:	
	IE	clears the special purpose register.
	ID	sets the special purpose register.
iflags	is a sequence of one or more flags:	
	i	sets or clears PRIMASK.
	f	sets or clears FAULTMASK.

Operation

CPS changes the PRIMASK and FAULTMASK special register values.

Restrictions

The restrictions are as follows:

- Use CPS only from privileged software; it has no effect if used in unprivileged software.
- CPS cannot be conditional and so must not be used inside an IT block.

Condition Flags

This instruction does not change the condition flags.

Examples

```
CPSID i ; Disable interrupts and configurable fault handlers (set PRIMASK)
CPSID f ; Disable interrupts and all fault handlers (set FAULTMASK)
CPSIE i ; Enable interrupts and configurable fault handlers (clear PRIMASK)
CPSIE f ; Enable interrupts and fault handlers (clear FAULTMASK)
```

A.9.3 DMB

Data Memory Barrier.

Syntax

```
DMB{cond}
```

where

cond   is an optional condition code; see "Conditional Execution" section on page 358.

Operation
DMB acts as a Data Memory Barrier. It ensures that all explicit memory accesses that appear, in program order, before the DMB instruction are completed before any explicit memory accesses that appear, in program order, after the DMB instruction. DMB does not affect the ordering or execution of instructions that do not access memory.

Condition Flags
This instruction does not change the flags.

Examples
```
DMB  ; Data Memory Barrier
```

A.9.4 DSB
Data Synchronization Barrier.

Syntax
```
DSB{cond}
```

where

cond   is an optional condition code; see "Conditional Execution" section on page 358.

Operation
DSB acts as a special data synchronization memory barrier. Instructions that come after the DSB, in program order, do not execute until the DSB instruction completes. The DSB instruction completes when all explicit memory accesses before it complete.

Condition Flags
This instruction does not change the flags.

Examples
```
DSB  ; Data Synchronization Barrier
```

A.9.5 ISB
Instruction Synchronization Barrier.

Syntax
```
ISB{cond}
```

where

cond   is an optional condition code; see "Conditional Execution" section on page 358.

Operation
ISB acts as an Instruction Synchronization Barrier. It flushes the pipeline of the processor so that all instructions following the ISB are fetched from cache or memory again, after the ISB instruction has been completed.

Condition Flags
This instruction does not change the flags.

Examples
```
ISB ; Instruction Synchronization Barrier
```

A.9.6 MRS
Move the contents of a special register to a general-purpose register.

Syntax
```
MRS{cond} Rd, spec_reg
```

where

cond	is an optional condition code; see "Conditional Execution" section on page 358.
Rd	is the destination register.
spec_reg	can be any of APSR, IPSR, EPSR, IEPSR, IAPSR, EAPSR, PSR, MSP, PSP, PRIMASK, BASEPRI, BASEPRI_MAX, FAULTMASK, or CONTROL.

Operation
Use MRS in combination with MSR as part of a read-modify-write sequence for updating a PSR, for example, to clear the Q flag.

In process swap code, the programmer model state of the process being swapped out must be saved, including relevant PSR contents. Similarly, the state of the process being swapped in must also be restored. These operations use MRS in the state-saving instruction sequence and MSR in the state-restoring instruction sequence.

Note
BASEPRI_MAX is an alias of BASEPRI when used with the MRS instruction; see *MSR* on page 400.

Restrictions
Rd must not be SP and must not be PC.

Condition Flags
This instruction does not change the flags.

Examples
```
MRS R0, PRIMASK ; Read PRIMASK value and write it to R0
```

A.9.7 MSR
Move the contents of a general-purpose register into the specified special register.

Syntax
```
MSR{cond} spec_reg, Rn
```

where

cond   is an optional condition code; see "Conditional Execution" section on page 358.
Rn   is the source register.
spec_reg   can be any of APSR, IPSR, EPSR, IEPSR, IAPSR, EAPSR, PSR, MSP, PSP,
           PRIMASK, BASEPRI, BASEPRI_MAX, FAULTMASK, or CONTROL.

Operation

The register access operation in MSR depends on the privilege level. Unprivileged software can only access the APSR. Privileged software can access all special registers.

In unprivileged software, writes to unallocated or execution state bits in the PSR are ignored.

Note

When you write to BASEPRI_MAX, the instruction writes to BASEPRI only if either

* *Rn* is nonzero and the current BASEPRI value is 0.
* *Rn* is nonzero and less than the current BASEPRI value.

See "MRS" section on page 400.

Restrictions

Rn must not be SP and must not be PC.

Condition Flags

This instruction updates the flags explicitly based on the value in *Rn*.

Examples

```
MSR CONTROL, R1 ; Read R1 value and write it to the CONTROL register
```

A.9.8 NOP

No Operation.

Syntax

```
NOP{cond}
```

where

cond   is an optional condition code; see "Conditional Execution" section on page 358.

Operation

NOP does nothing. NOP is not necessarily a time-consuming NOP. The processor might remove it from the pipeline before it reaches the execution stage.

Use NOP for padding, for example, to place the subsequence instruction on a 64-bit boundary.

Condition Flags

This instruction does not change the flags.

Examples
```
NOP ; No operation
```

A.9.9 SEV

Send Event.

Syntax
```
SEV{cond}
```

where

    cond   is an optional condition code; see "Conditional Execution" section on page 358.

Operation

SEV is a hint instruction that causes an event to be signaled to all processors within a multiprocessor system. It also sets the local event register to 1. More details can be found in Section 14.3, Multiprocessor communication.

Condition Flags

This instruction does not change the flags.

Examples
```
SEV ; Send Event
```

A.9.10 SVC

Supervisor Call.

Syntax
```
SVC{cond} #imm
```

where

    cond   is an optional condition code; see "Conditional Execution" section on page 358.
    imm   is an expression evaluating to an integer in the range 0–255 (8-bit value).

Operation

The SVC instruction causes the SVC exception.

imm is ignored by the processor. If required, it can be retrieved by the exception handler to determine what service is being requested.

Condition Flags

This instruction does not change the flags.

Examples
```
SVC  #0x32 ; Supervisor Call (SVC handler can extract the immediate value
           ; by locating it via the stacked PC)
```

A.9.11 **WFE**

Wait For Event.

Syntax
```
WFE{cond}
```

where

cond   is an optional condition code; see "Conditional Execution" section on page 358.

Operation

WFE is a hint instruction.

If the event register is 0, WFE suspends execution until one of the following events occurs:

- An exception, unless masked by the exception mask registers or the current priority level.
- An exception enters the pending state, if SEVONPEND in the System Control register is set.
- A Debug Entry request, if Debug is enabled.
- An event signaled by a peripheral or another processor in a multiprocessor system using the SEV instruction.

If the event register is 1, WFE clears it to 0 and returns immediately. Please refer to Section 14.2.1, Sleep Mode, for details of WFE and the event register.

Condition Flags

This instruction does not change the flags.

Examples
```
WFE ; Wait for event
```

A.9.12 **WFI**

Wait for Interrupt.

Syntax
```
WFI{cond}
```

where

cond   is an optional condition code; see "Conditional Execution" section on page 358.

Operation

WFI is a hint instruction that suspends execution until one of the following events occurs:

- An exception.
- A Debug Entry request, regardless of whether Debug is enabled.

Condition Flags

This instruction does not change the flags.

Examples
```
WFI ; Wait for interrupt
```

The 16-Bit Thumb Instructions and Architecture Versions

B

Most of the 16-bit Thumb® instructions are available in architecture v4T (ARM7TDMI). However, a number of them are added in architecture v5, v6, and v7. Table B.1 lists these instructions.

Table B.1 Change of 16-bit Instruction Support in Various Recent ARM Architecture Versions

Instruction	V4T	v5	v6	Cortex™-M3 (v7-M)
BKPT	N	Y	Y	Y
BLX	N	Y	Y	BLX <reg> only
CBZ, CBNZ	N	N	N	Y
CPS	N	N	Y	CPSIE <i/f>, CPSID <i/f>
CPY	N	N	Y	Y
NOP	N	N	N	Y
IT	N	N	N	Y
REV (various forms)	N	N	Y	REV, REV16, REVSH
SEV	N	N	N	Y
SETEND	N	N	Y	N
SWI	Y	Y	Y	Changed to SVC
SXTB, SXTH	N	N	Y	Y
UXTB, UXTH	N	N	Y	Y
WFE, WFI	N	N	N	Y

Cortex-M3 Exceptions
Quick Reference

C.1 EXCEPTION TYPES AND ENABLES

Table C.1 Quick Summary of Cortex™-M3 Exception Types and Their Priority Configurations

Exception Type	Name	Priority (Level Address)	Enable
1	Reset	−3	Always
2	NMI	−2	Always
3	Hard fault	−1	Always
4	MemManage fault	Programmable (0xE000ED18)	NVIC SHCSR (0xE000ED24) bit[16]
5	Bus fault	Programmable (0xE000ED19)	NVIC SHCSR (0xE000ED24) bit[17]
6	Usage fault	Programmable (0xE000ED1A)	NVIC SHCSR (0xE000ED24) bit[18]
7–10	—	—	—
11	SVC	Programmable (0xE000ED1F)	Always
12	Debug monitor	Programmable (0xE000ED20)	NVIC DEMCR (0xE000EDFC) bit[16]
13	—	—	—
14	PendSV	Programmable (0xE000ED22)	Always
15	SYSTICK	Programmable (0xE000ED23)	SYSTICK CTRLSTAT (0xE000E010) bit[1]
16–255	IRQ	Programmable (0xE000E400)	NVIC SETEN (0xE000E100)

C.2 STACK CONTENTS AFTER EXCEPTION STACKING

Table C.2 Exception Stack Frame

Address	Data	Push Order
Old SP (N+32)→	(Previously pushed data)	—
(N+28)	PSR	2
(N+24)	PC	1
(N+20)	LR	8
(N+16)	R12	7
(N+12)	R3	6
(N+8)	R2	5
(N+4)	R1	4
New SP (N)→	R0	3

Note: If double word stack alignment feature is used and the SP was not double word aligned when the exception occurred, the stack frame top might begin at ([OLD_SP-4] AND 0xFFFFFFF8), and the rest of the table moves one word down.

Nested Vectored Interrupt Controller and System Control Block Registers Quick Reference

Table D.1 Interrupt Controller Type Register (0xE000E004)

Bits	Name	Type	Reset Value	Description
4:0	INTLINESNUM	R	—	Number of interrupt inputs in steps of 32 0 = 1–32 1 = 33–64 …

Table D.2 Auxiliary Control Register (0xE000E008)

Bits	Name	Type	Reset Value	Description
2	DISFOLD	R/W	0	When this bit is set, it disables the overlapping of the IT execution cycle with another instruction. The overlapping (called *IT folding*) is an optimization to allow faster execution of conditional execution.
1	DISDEFWBUF	R/W	0	When this bit is set, it disables the use of write buffers within the processor so that an instruction following a store instruction must not start until the store operation is completed. This bit does not affect write buffer outside the processor (e.g., in bus bridge).
0	DISMCYCINT	R/W	0	When this bit is set, it disables interruption of multicycle instructions.

Note: The Auxiliary Control register is available from Cortex™-M3 revision 2.

Table D.3 SYSTICK Control and Status Register (0xE000E010)

Bits	Name	Type	Reset Value	Description
16	COUNTFLAG	R	0	Read as 1 if counter reaches 0 since this is the last time this register is read. Clear to 0 automatically when read or when current counter value is cleared.
2	CLKSOURCE	R/W	0	0 = External reference clock (STCLK) 1 = Use core clock
1	TICKINT	R/W	0	1 = Enable SYSTICK interrupt generation when SYSTICK timer reaches 0 0 = Do not generate interrupt
0	ENABLE	R/W	0	SYSTICK timer enable

Table D.4 SYSTICK Reload Value Register (0xE000E014)

Bits	Name	Type	Reset Value	Description
23:0	RELOAD	R/W	0	Reload value when timer reaches 0.

Table D.5 SYSTICK Current Value Register (0xE000E018)

Bits	Name	Type	Reset Value	Description
23:0	CURRENT	R/Wc	0	Read to return current value of the timer. Write to clear counter to 0; clearing of current value should also clear COUNTFLAG in SYSTICK Control and Status register.

Table D.6 SYSTICK Calibration Value Register (0xE000E01C)

Bits	Name	Type	Reset Value	Description
31	NOREF	R	—	1 = No external reference clock (STCLK not available) 0 = External reference clock available
30	SKEW	R	—	1 = Calibration value is not exactly 10 ms 0 = Calibration value is accurate
23:0	TENMS	R/W	0	Calibration value for 10 ms. SoC designer should provide this value through Cortex-M3 input signals. If this value is read as 0, it means calibration value is not available.

Table D.7 External Interrupt SETEN Registers (0xE000E100-0xE000E11C)

Address	Name	Type	Reset Value	Description
0xE000E100	SETENA0	R/W	0	Enable for external Interrupt #0–#31 bit[0] for Interrupt #0 bit[1] for Interrupt #1 ... bit[31] for Interrupt #31
0xE000E104	SETENA1	R/W	0	Enable for external Interrupt #32–#63
...	...	...	...	...

Table D.8 External Interrupt CLREN Registers (0xE000E180-0xE000E19C)

Address	Name	Type	Reset Value	Description
0xE000E180	CLRENA0	R/W	0	Clear Enable for external Interrupt #0–#31 bit[0] for Interrupt #0 bit[1] for Interrupt #1 ... bit[31] for Interrupt #31
0xE000E184	CLRENA1	R/W	0	Clear Enable for external Interrupt #32–#63
...	...	...	...	...

Table D.9 External Interrupt SETPEND Registers (0xE000E200-0xE000E21C)

Address	Name	Type	Reset Value	Description
0xE000E200	SETPEND0	R/W	0	Pending for external Interrupt #0–#31 bit[0] for Interrupt #0 bit[1] for Interrupt #1 … bit[31] for Interrupt #31
0xE000E204	SETPEND1	R/W	0	Pending for external Interrupt #32–#63
…	…	…	…	…

Table D.10 External Interrupt CLRPEND Registers (0xE000E280-0xE000E29C)

Address	Name	Type	Reset Value	Description
0xE000E280	CLRPEND0	R/W	0	Clear Pending for external Interrupt #0–#31 bit[0] for Interrupt #0 bit[1] for Interrupt #1 … bit[31] for Interrupt #31
0xE000E284	CLRPEND1	R/W	0	Clear Pending for external Interrupt #32–#63
…	…	…	…	…

Table D.11 External Interrupt ACTIVE Registers (0xE000E300-0xE000E31C)

Address	Name	Type	Reset Value	Description
0xE000E300	ACTIVE0	R	0	Active status for external Interrupt #0–#31 bit[0] for Interrupt #0 bit[1] for Interrupt #1 … bit[31] for Interrupt #31
0xE000E304	ACTIVE1	R	0	Active status for external Interrupt #32–#63
…	…	…	…	…

Table D.12 External Interrupt Priority Level Register (0xE000E400-0xE000E4EF; listed as byte addresses)

Address	Name	Type	Reset Value	Description
0xE000E400	PRI_0	R/W	0	Priority level external Interrupt #0
0xE000E401	PRI_1	R/W	0	Priority level external Interrupt #1
…	…	…	…	…
0xE000E41F	PRI_31	R/W	0	Priority level external Interrupt #31
…	…	…	…	…

Table D.13 CPU ID Base Register (address 0xE000ED00)

Bits	Name	Type	Reset Value	Description
31:24	IMPLEMENTER	R	0x41	Implementer code; ARM is 0x41
23:20	VARIANT	R	0x0/0x1/0x2	Implementation defined variant number
19:16	Constant	R	0xF	Constant
15:4	PARTNO	R	0xC23	Part number
3:0	REVISION	R	0x0/0x1	Revision code

Table D.14 Interrupt Control and State Register (0xE000ED04)

Bits	Name	Type	Reset Value	Description
31	NMIPENDSET	R/W	0	NMI Pended
28	PENDSVSET	R/W	0	Write 1 to pend system call; Read value indicates pending status
27	PENDSVCLR	W	0	Write 1 to clear PendSV pending status
26	PENDSTSET	R/W	0	Write 1 to pend SYSTICK exception; Read value indicates pending status
25	PENDSTCLR	W	0	Write 1 to clear SYSTICK pending status
23	ISRPREEMPT	R	0	Indicates that a pending interrupt is going to be active in next step (for debug).
22	ISRPENDING	R	0	External Interrupt Pending (excluding system exceptions like NMI for fault)
21:12	VECTPENDING	R	0	Pending ISR number
11	RETTOBASE	R	0	Set to 1 when the processor is running an exception handler and will return to thread level if interrupt return and no other exceptions pending
8:0	VECTACTIVE	R	0	Current running interrupt service routine

Table D.15 Vector Table Offset Register (address 0xE000ED08)

Bits	Name	Type	Reset Value	Description
29	TBLBASE	R/W	0	Table base in Code (0) or RAM (1) memory region
28:7	TBLOFF	R/W	0	Table offset value from Code region or RAM region

Table D.16 Application Interrupt and Reset Control Register (address 0xE000ED0C)

Bits	Name	Type	Reset Value	Description
31:16	VECTKEY	R/W	—	Access key; 0x05FA must be written to this field to write to this register, otherwise the write will be ignored. The read back value is 0xFA05.
15	ENDIANESS	R	—	Indicates endianness for data: 1 for big endian (BE8) and 0 for little endian. This can only change after a reset.

Table D.16 Application Interrupt and Reset Control Register (address 0xE000ED0C)   *Continued*

Bits	Name	Type	Reset Value	Description
10:8	PRIGROUP	R/W	0	Priority group
2	SYSRESETREQ	W	—	Request chip control logic to generate a reset
1	VECTCLRACTIVE	W	—	Clear all active state information for exceptions; typically used in debug or OS to allow system to recover from system error (Reset is safer).
0	VECTRESET	W	—	Reset Cortex-M3 (except debug logic); but this will not reset circuits outside the processor.

Table D.17 System Control Register (0xE000ED10)

Bits	Name	Type	Reset Value	Description
4	SEVONPEND	R/W	0	Send Event on Pending. Wake up from WFE if a new interrupt is pended regardless of whether the interrupt has priority higher than current level.
3	Reserved	—	—	—
2	SLEEPDEEP	R/W	0	Enable SLEEPDEEP output signal when entering sleep mode.
1	SLEEPONEXIT	R/W	0	Enable Sleep on Exit feature
0	Reserved	—	—	—

Table D.18 Configuration Control Register (0xE000ED14)

Bits	Name	Type	Reset Value	Description
9	STKALIGN	R/W	0 or 1	Force exception stacking start in double word aligned address. This bit is reset as zero on Cortex-M3 revision 1, and is reset as one on revision 2. Revision 0 does not have this feature.
8	BFHFNMIGN	R/W	0	Ignore data bus fault during hard fault and NMI handlers.
7:5	Reserved	—	—	Reserved
4	DIV_0_TRP	R/W	0	Trap on divide by 0
3	UNALIGN_TRP	R/W	0	Trap on unaligned accesses
2	Reserved	—	—	Reserved
1	USERSETMPEND	R/W	0	If set to 1, allow user code to write to Software Trigger Interrupt register
0	NONBASETHRDENA	R/W	0	Nonbase Thread Enable. If set to 1, allows exception handler to return to thread state at any level by controlling return value.

Table D.19 System Exceptions Priority Level Register (0xE000ED18–0xE000ED23; listed as byte addresses)

Address	Name	Type	Reset Value	Description
0xE000ED18	PRI_4	R/W	0	Priority level for memory management fault
0xE000ED19	PRI_5	R/W	0	Priority level for bus fault
0xE000ED1A	PRI_6	R/W	0	Priority level for usage fault
0xE000ED1B	—	—	—	—
0xE000ED1C	—	—	—	—
0xE000ED1D	—	—	—	—
0xE000ED1E	—	—	—	—
0xE000ED1F	PRI_11	R/W	0	Priority level for SVC
0xE000ED20	PRI_12	R/W	0	Priority level for debug monitor
0xE000ED21	—	—	—	—
0xE000ED22	PRI_14	R/W	0	Priority level for PendSV
0xE000ED23	PRI_15	R/W	0	Priority level for SYSTICK

Table D.20 System Handler Control and State Register (0xE000ED24)

Bits	Name	Type	Reset Value	Description
18	USGFAULTENA	R/W	0	Usage Fault Handler Enable
17	BUSFAULTENA	R/W	0	Bus Fault Handler Enable
16	MEMFAULTENA	R/W	0	Memory Management Fault Enable
15	SVCALLPENDED	R/W	0	SVC pended; SVC is started but was replaced by a higher priority exception
14	BUSFAULTPENDED	R/W	0	Bus fault pended; bus fault is started, but was replaced by a higher priority exception
13	MEMFAULTPENDED	R/W	0	Memory management fault pended; memory management fault started but was replaced by a higher priority exception
12	USGFAULTPENDED	R/W	0	Usage fault pended; usage fault started but was replaced by a higher-priority exception[a]
11	SYSTICKACT	R/W	0	Read as 1 if SYSTICK exception is active
10	PENDSVACT	R/W	0	Read as 1 if PendSV exception is active
8	MONITORACT	R/W	0	Read as 1 if debug monitor exception is active
7	SVCALLACT	R/W	0	Read as 1 if SVC exception is active
3	USGFAULTACT	R/W	0	Read as 1 if usage fault exception is active
1	BUSFAULTACT	R/W	0	Read as 1 if bus fault exception is active
0	MEMFAULTACT	R/W	0	Read as 1 if memory management fault is active

[a]Bit 12 (USGFAULTPENDED) is not available on revision 0 of Cortex-M3 processor.

Table D.21 Memory Management Fault Status Register (0xE000ED28; byte size)

Bits	Name	Type	Reset Value	Description
7	MMARVALID	—	0	Indicates MMAR is valid
6:5	—	—	—	—
4	MSTKERR	R/Wc	0	Stacking error
3	MUNSTKERR	R/Wc	0	Unstacking error
2	—	—	—	—
1	DACCVIOL	R/Wc	0	Data access violation
0	IACCVIOL	R/Wc	0	Instruction access violation

Table D.22 Bus Fault Status Register (0xE000ED29; byte size)

Bits	Name	Type	Reset Value	Description
7	BFARVALID	—	0	Indicates BFAR is valid
6:5	—	—	—	—
4	STKERR	R/Wc	0	Stacking error
3	UNSTKERR	R/Wc	0	Unstacking error
2	IMPREISERR	R/Wc	0	Imprecise data access violation
1	PRECISERR	R/Wc	0	Precise data access violation
0	IBUSERR	R/Wc	0	Instruction access violation

Table D.23 Usage Fault Status Register (0xE000ED2A; half word size)

Bits	Name	Type	Reset Value	Description
9	DIVBYZERO	R/Wc	0	Indicates divide by zero takes place (can only be set if DIV_0_TRP is set)
8	UNALIGNED	R/Wc	0	Indicates unaligned access takes place (can only be set if UNALIGN_TRP is set)
7:4	—	—	—	—
3	NOCP	R/Wc	0	Attempts to execute a coprocessor instruction
2	INVPC	R/Wc	0	Attempts to do exception with bad value in EXC_RETURN number
1	INVSTATE	R/Wc	0	Attempts to switch to invalid state (e.g., ARM)
0	UNDEFINSTR	R/Wc	0	Attempts to execute an undefined instruction

Table D.24 Hard Fault Status Register (0xE000ED2C)

Bits	Name	Type	Reset Value	Description
31	DEBUGEVT	R/Wc	0	Indicates hard fault is triggered by debug event
30	FORCED	R/Wc	0	Indicates hard fault is taken because of bus fault/ memory management fault/usage fault
29:2	—	—	—	—
1	VECTBL	R/Wc	0	Indicates hard fault is caused by failed vector fetch
0	—	—	—	—

Table D.25 Debug Fault Status Register (0xE000ED30)

Bits	Name	Type	Reset Value	Description
4	EXTERNAL	R/Wc	0	EDBGRQ signal asserted
3	VCATCH	R/Wc	0	Vector fetch occurred
2	DWTTRAP	R/Wc	0	DWT match occurred
1	BKPT	R/Wc	0	BKPT instruction executed
0	HALTED	R/Wc	0	Halt requested in NVIC

Table D.26 Memory Manage Address Register MMAR (0xE000ED34)

Bits	Name	Type	Reset Value	Description
31:0	MMAR	R	—	Address that caused memory manage fault

Table D.27 Bus Fault Manage Address Register BFAR (0xE000ED38)

Bits	Name	Type	Reset Value	Description
31:0	BFAR	R	—	Address that caused bus fault

Table D.28 Auxiliary Fault Status Register (0xE000ED3C)

Bits	Name	Type	Reset Value	Description
31:0	Vendor controlled	R/Wc	0	Vendor controlled (optional)

Table D.29 MPU Type Register (0xE000ED90)

Bits	Name	Type	Reset Value	Description
23:16	IREGION	R	—	Number Instruction region. Because ARM v7-M architecture uses a unified MPU, this is always 0.
15:8	DREGION	R	—	Number of regions supported by this MPU
0	SEPARATE	R	—	This is always 0 as the MPU is always unified.

Table D.30 MPU Control Register (0xE000ED94)

Bits	Name	Type	Reset Value	Description
2	PRIVDEFENA	R/W	0	Privileged Default memory map enable
1	HFNMIENA	R/W	0	If set to 1, it enables MPU during hard fault handler and NMI handler. Otherwise, the MPU is not enabled for hard fault handler and NMI.
0	ENABLE	R/W	0	Enable the MPU if set to 1.

Table D.31 MPU Region Number Register (0xE000ED98)

Bits	Name	Type	Reset Value	Description
7:0	REGION	R/W	0	Select which region is being programmed

Table D.32 MPU Region Base Address Register (0xE000ED9C)

Bits	Name	Type	Reset Value	Description
31:N	ADDR	R/W	0	Base address of the region. N is dependent on the region size.
4	VALID	R/W	0	If this is 1, the region defined in bit[3:0] will be used in this programming step, otherwise, the region selected by MPU Region Number register is used.
3:0	REGION	R/W	0	This field overrides MPU Region Number register if VALID is 1, otherwise, this is ignored.

Table D.33 MPU Region Base Attribute and Size Register (0xE000EDA0)

Bits	Name	Type	Reset Value	Description
31:29	Reserved	—	—	
28	XN	R/W	0	Instruction access disable (1 = Disable)
27	Reserved	—	—	
26:24	AP	R/W	000	Data access permission field
23:22	Reserved	—	—	
21:19	TEX	R/W	000	Type extension field
18	S	R/W	—	Shareable
17	C	R/W	—	Cacheable
16	B	R/W	—	Bufferable
15:8	SRD	R/W	0x00	Sub region disable
7:6	Reserved	—	—	
5:1	REGION SIZE	R/W	—	MPU protection region size
0	ENABLE	R/W	0	Region enable

Table D.34 MPU Alias Registers (0xE000EDA4–0xE000EDB8)

Address	Name	Description
0xE000EDA4	Alias of D9C	MPU Alias 1 Region Base Address register
0xE000EDA8	Alias of DA0	MPU Alias 1 Region Attribute and Size register
0xE000EDAC	Alias of D9C	MPU Alias 2 Region Base Address register
0xE000EDB0	Alias of DA0	MPU Alias 2 Region Attribute and Size register
0xE000EDB4	Alias of D9C	MPU Alias 3 Region Base Address register
0xE000EDB8	Alias of DA0	MPU Alias 3 Region Attribute and Size register

Table D.35 Debug Halting Control and Status Register (0xE000EDF0)

Bits	Name	Type	Reset Value	Description
31:16	KEY	W	—	Debug key; value of 0xA05F must be written to this field to write to this register, otherwise, the write will be ignored.
25	S_RESET_ST	R	—	Core has been reset or is being reset. This bit is cleared on read.
24	S_RETIRE_ST	R	—	Instruction is completed since last read. This bit is cleared on read.
19	S_LOCKUP	R	—	When this bit is 1, the core is in locked-up state.
18	S_SLEEP	R	—	When this bit is 1, the core is in sleep mode.
17	S_HALT	R	—	When this bit is 1, the core is halted.
16	S_REGRDY	R	—	Register read/write operation is completed.
15:6	Reserved	—	—	Reserved
5	C_SNAPSTALL	R/W	—	Used to break a stalled memory access
4	Reserved	—	—	Reserved
3	C_MASKINTS	R/W	—	Mask interrupts while stepping; can only be modified when the processor is halted.
2	C_STEP	R/W	—	Single step the processor, valid only if C_DEBUGEN is set.
1	C_HALT	R/W	—	Halt the processor core, valid only if C_DEBUGEN is set.
0	C_DEBUGEN	R/W	—	Enable halt mode debug

Table D.36 Debug Core Register Selector Register (0xE000EDF4)

Bits	Name	Type	Reset Value	Description
16	REGWnR	W	—	Direction of data transfer Write = 1, Read = 0
15:5	Reserved	—	—	—
4:0	REGSEL	W	—	
				Register to be accessed 00000 = R0 00001 = R1 ... 01111 = R15 10000 = xPSR/Flags 10001 = MSP (Main Stack Pointer) 10010 = PSP (Process Stack Pointer) 10100 = Special registers: [31:24] CONTROL, [23:16] FAULTMASK, [15:8] BASEPRI, [7:0] PRIMASK. Others values are reserved

Table D.37 Debug Core Register Data Register (0xE000EDF8)

Bits	Name	Type	Reset Value	Description
31:0	Data	R/W	—	Data register to hold register read result or to write data into the selected register.

Table D.38 Debug Exception and Monitor Control Register (0xE000EDFC)

Bits	Name	Type	Reset Value	Description
24	TRCENA	R/W	0	Trace system enable; to use DWT, ETM, ITM, and TPIU, this bit must be set to 1.
23:20	Reserved	—	—	Reserved
19	MON_REQ	R/W	0	Indication that the debug monitor is caused by a manual pending request rather than hardware debug events.
18	MON_STEP	R/W	0	Single step the processor; valid only if MON_EN is set.
17	MON_PEND	R/W	0	Pend the monitor exception request; the core will enter monitor exception when priority is allowed.
16	MON_EN	R/W	0	Enable the debug monitor exception
15:11	Reserved	—	—	Reserved

Continued

Table D.38 Debug Exception and Monitor Control Register (0xE000EDFC) *Continued*

Bits	Name	Type	Reset Value	Description
10	VC_HARDERR	R/W	0	Debug trap on hard faults
9	VC_INTERR	R/W	0	Debug trap on interrupt/exception service errors
8	VC_BUSERR	R/W	0	Debug trap on bus faults
7	VC_STATERR	R/W	0	Debug trap on usage fault state errors
6	VC_CHKERR	R/W	0	Debug trap on usage fault enabled checking errors (e.g., unaligned, divide by zero)
5	VC_NOCPERR	R/W	0	Debug trap on usage fault; no coprocessor errors
4	VC_MMERR	R/W	0	Debug trap on memory management fault
3:1	Reserved	—	—	Reserved
0	VC_CORERESET	R/W	0	Debug trap on core reset

Table D.39 Software Trigger Interrupt Register (0xE000EF00)

Bits	Name	Type	Reset Value	Description
8:0	INTID	W	—	Writing the interrupt number sets the pending bit of the interrupt.

Table D.40 NVIC Peripheral ID Registers (0xE000EFD0-0xE000EFFC)

Address	Name	Type	Reset Value	Description
0xE000EFD0	PERIPHID4	R	0x04	Peripheral ID register
0xE000EFD4	PERIPHID5	R	0x00	Peripheral ID register
0xE000EFD8	PERIPHID6	R	0x00	Peripheral ID register
0xE000EFDC	PERIPHID7	R	0x00	Peripheral ID register
0xE000EFE0	PERIPHID0	R	0x00	Peripheral ID register
0xE000EFE4	PERIPHID1	R	0xB0	Peripheral ID register
0xE000EFE8	PERIPHID2	R	0x0B/0x1B/0x2B	Peripheral ID register
0xE000EFEC	PERIPHID3	R	0x00	Peripheral ID register
0xE000EFF0	PCELLID0	R	0x0D	Component ID register
0xE000EFF4	PCELLID1	R	0xE0	Component ID register
0xE000EFF8	PCELLID2	R	0x05	Component ID register
0xE000EFFC	PCELLID0	R	0xB1	Component ID register

Note: PERIPHID2 value is 0x0B for Cortex-M3 revision 0, 0x1B for revision 1, and 0x2B for revision 2.

Cortex-M3 Troubleshooting Guide

E.1 OVERVIEW

One of the challenges of using the Cortex™-M3 is to locate problems when the program goes wrong. The Cortex-M3 processor provides a number of Fault Status registers to assist in troubleshooting (see Table E.1).

The MMSR, BFSR, and UFSR registers can be accessed in one go using a word transfer instruction. In this situation, the combined fault status register is called the *Configurable Fault Status register* (CFSR) (see Figure E.1).

For users of CMSIS compliant device drivers, these Fault Status registers can be accessed using the following symbols:

- `SCB->CFSR`: Configurable Fault Status register
- `SCB->HFSR`: Hard Fault Status register
- `SCB->DFSR`: Debug Fault Status register
- `SCB->AFSR`: Auxiliary Fault Status register

Another important piece of information is the stacked program counter (PC). This is located in memory address [SP + 24] when a fault exception handler is entered. Because there are two stack pointers in the Cortex-M3, the fault handler might need to determine which stack pointer was used before obtaining the stacked PC.

In addition, for bus faults and memory management faults, you might also be able to determine the address that caused the fault. This is done by accessing the MemManage (Memory Management) Fault Address register (MMAR) and the Bus Fault Address register (BFAR). The contents of these two registers are only valid when the MMAVALID bit (in MMSR) or BFARVALID bit (in BFSR) is set. The MMAR and BFAR are physically the same register, so only one of them can be valid at a time (see Table E.2).

For users of CMSIS compliant device drivers, these Fault Address registers can be accessed using the following symbols:

- `SCB->MMAR`: MemManage Fault Address register
- `SCB->BFAR`: Bus Fault Address register

Finally, the link register (LR) value when entering the fault handler might also provide hints about the cause of the fault. In the case of faults caused by invalid EXC_RETURN values, the value of LR when the fault handler is entered shows the previous LR value when the fault occurred. Fault handler can report the faulty LR value, and software programmers can then use this information to check why the LR ends up with an illegal return value.

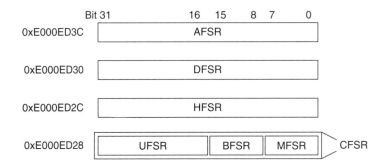

FIGURE E.1

Accessing Fault Status Registers.

Table E.1 Fault Status Registers on Cortex-M3

Address	Register	Full Name	Size
0xE000ED28	MMSR	MemManage Fault Status register	Byte
0xE000ED29	BFSR	Bus Fault Status register	Byte
0xE000ED2A	UFSR	Usage Fault Status register	Half word
0xE000ED2C	HFSR	Hard Fault Status register	Word
0xE000ED30	DFSR	Debug Fault Status register	Word
0xE000ED3C	AFSR	Auxiliary Fault Status register	Word

Table E.2 Fault Address Registers on Cortex-M3

Address	Register	Full Name	Size
0xE000ED34	MMAR	MemManage Fault Address register	Word
0xE000ED38	BFAR	Bus Fault Address register	Word

E.2 DEVELOPING FAULT HANDLERS

In most cases, fault handlers for development and for real running systems differ from one another. For software development, the fault handler should focus on reporting the type of error, whereas the fault handler for running systems will likely focus on system recovery actions. Here, we cover only the fault reporting because system recovery actions highly depend on design type and requirements.

In complex software, instead of outputting the results inside the fault handler, the contents of these registers can be copied to a memory block and then PendSV can be used to report the fault details later. This avoids potential faults in display or outputting routines causing lockup. For simple applications this might not matter, and the fault details can be output directly within the fault handler routine.

E.2.1 **Report Fault Status Registers**

The most basic step of a fault handler is to report the Fault Status register values. These include the following:

- UFSR
- BFSR
- MMSR
- HFSR
- DFSR
- AFSR (optional)

E.2.2 **Report Stacked PC and Other Stacked Registers**

In a fault handler, the step for getting the stacked PC is similar to the SVC example in this book.
This process can be carried out in assembly language as follows:

```
TST    LR, #0x4      ; Test EXC_RETURN number in LR bit 2
ITTEE EQ             ; if zero (equal) then
MRSEQ R0, MSP        ; Main Stack was used, put MSP in R0
LDREQ R0,[R0,#24]    ; Get stacked PC from stack.
MRSNE R0, PSP        ; else, Process Stack was used, put PSP in R0
LDRNE R0,[R0,#24]    ; Get stacked PC from stack.
```

Most Cortex-M3 developers use C for their projects. However, in C, it is difficult to locate and directly access the stack frame (stacked register values) as you cannot obtain the stack point value in C. To report the stack frame contents in your fault handler in C, you need to use a short assembly code to obtain stack point value, and then pass it to the fault reporting function in C as a parameter (see Figure E.2). The mechanism is identical to the SVC example in Chapter 12. The following example

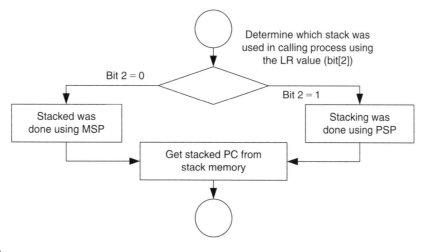

FIGURE E.2

Getting the Value of a Stacked PC from Stack Memory.

uses embedded assembler, which can work with RealView Development Suite (RVDS) and Keil Real-View Microcontroller Development Kit (MDK-ARM).

The first part of the program is an assembly wrapper. The vector table should have the starting address of this wrapper code in the hard fault entry. This wrapper code copies the correct stack pointer value into R0, and passes it to the C function as a parameter.

```
// hard fault handler wrapper in assembly
// it extract the location of stack frame and pass it
// to handler in C as pointer.
__asm void hard_fault_handler_asm(void)
{
TST       LR, #4
ITE       EQ
MRSEQ     R0, MSP
MRSNE     R0, PSP
B         __cpp(hard_fault_handler_c )
}
```

The second part of the handler is in C. Here, we demonstrate how the stacked register contents and the Fault Status registers can be accessed.

```
// hard fault handler in C,
// with stack frame location as input parameter
void hard_fault_handler_c(unsigned int * hardfault_args)
{
unsigned int stacked_r0;
unsigned int stacked_r1;
unsigned int stacked_r2;
unsigned int stacked_r3;
unsigned int stacked_r12;
unsigned int stacked_lr;
unsigned int stacked_pc;
unsigned int stacked_psr;

stacked_r0  = ((unsigned long) hardfault_args[0]);
stacked_r1  = ((unsigned long) hardfault_args[1]);
stacked_r2  = ((unsigned long) hardfault_args[2]);
stacked_r3  = ((unsigned long) hardfault_args[3]);

stacked_r12 = ((unsigned long) hardfault_args[4]);
stacked_lr  = ((unsigned long) hardfault_args[5]);
stacked_pc  = ((unsigned long) hardfault_args[6]);
stacked_psr = ((unsigned long) hardfault_args[7]);

printf ("[Hard fault handler]\n");
printf ("R0  = %x\n", stacked_r0);
printf ("R1  = %x\n", stacked_r1);
printf ("R2  = %x\n", stacked_r2);
printf ("R3  = %x\n", stacked_r3);
printf ("R12 = %x\n", stacked_r12);
printf ("LR  = %x\n", stacked_lr);
printf ("PC  = %x\n", stacked_pc);
```

```
printf ("PSR  = %x\n", stacked_psr);
printf ("BFAR = %x\n", (*((volatile unsigned long *)(0xE000ED38))));
printf ("CFSR = %x\n", (*((volatile unsigned long *)(0xE000ED28))));
printf ("HFSR = %x\n", (*((volatile unsigned long *)(0xE000ED2C))));
printf ("DFSR = %x\n", (*((volatile unsigned long *)(0xE000ED30))));
printf ("AFSR = %x\n", (*((volatile unsigned long *)(0xE000ED3C))));

exit(0); // terminate

return;
}
```

Please note that this handler will not work correctly if the stack pointer is pointing to an invalid memory region (e.g., because of stack overflow). This affects all C code as stack is required in C functions in most cases.

To help with debugging, we should also create a disassembled code list file so that we can locate the problem easily.

E.2.3 Read Fault Address Register

The Fault Address register can be erased after the MMARVALID or BFARVALID is cleared. To correctly access the Fault Address register, the following procedure should be used:

1. Read BFAR/MMAR.
2. Read BFARVALID/MMARVALID. If it is 0, the BFAR/MMAR read should be discarded.
3. Clear BFARVALID/MMARVALID.

The reason for this procedure instead of reading valid bits first is to prevent a fault handler being preempted by another higher-priority fault handler after the valid bit is read, which could lead to the following erroneous fault-reporting sequence:

1. Read BFARVALID/MMARVALID.
2. Valid bit is set, going to read BFAR/MMAR.
3. Higher-priority exception preempts existing fault handler, which generates another fault, causing another fault handler to be executed.
4. The higher-priority fault handler clears the BFARVALID/MMARVALID bit, causing the BFAR/MMAR to be erased.
5. After returning to the original fault handler, the BFAR/MMAR is read, but now the content is invalid and leads to incorrect reporting of the fault address.

Therefore, it is important to read the BFARVALID/MMARVALID after reading the Fault Address register to ensure that the address register content is valid.

E.2.4 Clear Fault Status Bits

After the fault reporting is done, the fault status bit in the FSR should be cleared so that next time the fault handler is executed, the previous faults will not confuse the fault handler. In addition, if the fault address valid bit is not clear, the Fault Address register will not get an update for the next fault.

E.2.5 Others

It is often necessary to save the contents of LR in the beginning of a fault handler. However, if the fault is caused by a stack error, pushing the LR to stack might just make things worst. As we know, R0–R3 and R12 should already been saved, so that we could copy LR to one of these registers before doing any function calls.

E.3 UNDERSTANDING THE CAUSE OF THE FAULT

After obtaining the information we need, we can establish the cause of the problem. Tables E.3 through E.7 list some of the common reasons that faults occur.

Table E.3 MemManage Fault Status Register

Bit	Possible Causes
MMARVALID (bit 7)	Indicates the Memory Manage Address register (MMAR) contains a valid fault addressing value.
MSTKERR (bit 4)	Error occurred during stacking (starting of exception). **1.** Stack pointer is corrupted. **2.** Stack size is too large, reaching a region not defined by the MPU or disallowed in the MPU configuration.
MUNSTKERR (bit 3)	Error occurred during unstacking (ending of exception). If there was no error stacking but error occurred during unstacking, it might be because of the following reasons: **1.** Stack pointer was corrupted during exception. **2.** MPU configuration was changed by exception handler.
DACCVIOL (bit 1)	Violation to memory access protection, which is defined by MPU setup. For example, user application trying to access privileged-only region.
IACCVIOL (bit 0)	**1.** Violation to memory access protection, which is defined by MPU setup. For example, user application trying to access privileged-only region. Stacked PC might be able to locate the code that has caused the problem. **2.** Branch to nonexecutable regions. **3.** Invalid exception return code. **4.** Invalid entry in exception vector table. For example, loading of an executable image for traditional ARM core into the memory, or exception happened before vector table was set. **5.** Stacked PC corrupted during exception handling.

Table E.4 Bus Fault Status Register

Bit	Possible Causes
BFARVALID (bit 7)	Indicates the Bus Fault Address register contains a valid bus fault address.
STKERR (bit 4)	Error occurred during stacking (starting of exception). **1.** Stack pointer is corrupted. **2.** Stack size became too large, reaching an undefined memory region. **3.** PSP is used but not initialized.

Table E.4 Bus Fault Status Register *Continued*

Bit	Possible Causes
UNSTKERR(bit 3)	Error occurred during unstacking (ending of exception). If there was no error stacking but error occurred during unstacking, it might be that the stack pointer was corrupted during exception.
IMPRECISERR (bit 2)	Bus error during data access. Bus error could be caused by the device not being initialized, access of privileged-only device in user mode, or the transfer size is incorrect for the specific device.
PRECISERR (bit 1)	Bus error during data access. The fault address may be indicated by BFAR. A bus error could be caused by the device not being initialized, access of privileged-only device in user mode, or the transfer size is incorrect for the specific device.
IBUSERR (bit 0)	1. Branch to invalid memory regions; for example, caused by incorrect function pointers in program code. 2. Invalid exception return code; for example, a stacked EXC_RETURN code is corrupted, and as a result, an exception return incorrectly treated as a branch. 3. Invalid entry in exception vector table. For example, loading of an executable image for traditional ARM core into the memory, or exception, happens before the vector table is set. 4. Stacked PC corrupted during function calls. 5. Access to NVIC or SCB in user mode (nonprivileged).

Table E.5 Usage Fault Status Register

Bit	Possible Causes
DIVBYZERO (bit 9)	Divide by 0 takes place and DIV_0_TRP is set. The code causing the fault can be located using stacked PC.
UNALIGNED (bit 8)	Unaligned access attempted with UNALIGN_TRP is set. The code causing the fault can be located using stacked PC.
NOCP (bit 3)	Attempt to execute a coprocessor instruction. The code causing the fault can be located using stacked PC.
INVPC (bit 2)	1. Invalid value in EXC_RETURN number during exception return. For example, a. Return to thread with EXC_RETURN = 0xFFFFFFF1 b. Return to handler with EXC_RETURN = 0xFFFFFFF9 To investigate the problem, the current LR value provides the value of LR at the failing exception return. 2. Invalid exception active status. For example, a. Exception return with exception active bit for the current exception already cleared. Possibly caused by use of VECTCLRACTIVE, or clearing of exception active status in NVIC SHCSR. b. Exception return to thread with one or more exception active bits still active. 3. Stack corruption causing the stacked IPSR to be incorrect. For INVPC fault, the stacked PC shows the point where the faulting exception interrupted the main/preempted program. To investigate the cause of the problem, it is best to use exception trace feature in the DWT. 4. ICI/IT bit invalid for current instruction. This can happen when a multiple-load/store instruction gets interrupted and, during the interrupt handler, the stacked PC is modified. When the interrupt return takes place, the nonzero ICI bit is applied to an instruction that does not use ICI bits. The same problem can also happen because of corruption of stacked PSR.

Continued

Table E.5 Usage Fault Status Register *Continued*

Bit	Possible Causes
INVSTATE (bit 1)	1. Loading branch target address to PC with LSB equals 0. Stacked PC should show the branch target. 2. LSB of vector address in vector table is 0. Stacked PC should show the starting of exception handler. 3. Stacked PSR corrupted during exception handling, so after the exception the core tries to return to the interrupted code in ARM state.
UNDEFINSTR (bit 0)	1. Use of instructions not supported in Cortex-M3. 2. Bad/corrupted memory contents. 3. Loading of ARM object code during link stage. Checks compile steps. 4. Instruction align problem. For example, if GNU Tool chain is used, omitting of .align after .ascii might cause next instruction to be unaligned (start in odd memory address instead of halfword addresses).

Table E.6 Hard Fault Status Register

Bit	Possible Causes
DEBUGEVF (bit 31)	Fault is caused by debug event: 1. Breakpoint/watchpoint events. 2. If the hard fault handler is executing, it might be caused by execution of BKPT without enable monitor handler (MON_EN = 0) and halt debug is not enabled (C_DEBUGEN = 0). By default, some C compilers might include semihosting code that use BKPT.
FORCED (bit 30)	1. Trying to run SVC/BKPT within SVC/monitor or another handler with same or higher priority. 2. A fault occurred if the corresponding handler is disabled or cannot be started because another exception with the same or higher priority is running, or because exception mask is set.
VECTBL (bit 1)	Vector fetch failed. It could be caused by 1. Bus fault at vector fetch. 2. Incorrect vector table offset setup.

Table E.7 Debug Fault Status Register

Bit	Possible Causes
EXTERNAL (bit 4)	EDBGRQ signal has been asserted.
VCATCH (bit 3)	Vector catch event has occurred.
DWTTRAP (bit 2)	DWT watchpoint event has occurred.
BKPT (bit 1)	1. Breakpoint instruction is executed. 2. FPB unit generated a breakpoint event. In some cases, BKPT instructions are inserted by C startup code as part of the semihosting debugging setup. This should be removed for a real application code. Please refer to your compiler document for details.
HALTED (bit 0)	Halt request in NVIC.

E.4 OTHER POSSIBLE PROBLEMS

A number of other common problems are in Table E.8.

Table E.8 Other Possible Problems

Situations	Possible Causes
No program execution	Vector table could be set up incorrectly. **1.** Located in incorrect memory location. **2.** LSB of vectors (including hard fault handler) is not set to 1. **3.** Use of branch instruction (as in vector table in traditional ARM processor) in the vector table. Please generate a disassembly code listing to check if the vector table is set up correctly.
Program crashes after a few numbers of instructions	Possibly caused by incorrect endian setting, or incorrect stack pointer setup (check vector table), or use of C object library for traditional ARM processor (ARM code instead of Thumb® code). The offending C object library code could be part of the C startup routine. Please check compiler and linker options to ensure that Thumb or Thumb-2 library files are used.
Processor does not enter sleep mode when WFE is executed	A WFE instruction does not always result in sleep. If the internal event register was set before the WFE instruction, it will clear the event register and act as NOP. Therefore, in normal coding WFE should be used with a loop.
Processor stops executing unexpectedly	When sleep-on-exit feature is enabled, the processor enters sleep mode when returning from exception handler to thread mode, even if no WFI or WFE instructions are used.
Unexpected SEVONPEND behavior	The SEVONPEND in the NVIC System Control register enables a disabled interrupt to wake up the processor from WFE, but not WFI. The wake up event is generated only at a new pending of an interrupt. If the interrupt pending status was already set before execution of WFE, arrival of a new interrupt request will not generate the wake up event and hence will not wake up the processor.

Interrupt priority level not working as expected	Unlike many other processors, the Cortex-M3 processor uses value 0 for highest programmable exception priority level. The larger the priority level value, the lower the priority is.
	    When programming the priority level registers for interrupt, make sure the priority values are written to the implemented bits of the registers. The least significant bits of the priority level registers are not implemented.
	    Most Cortex-M3 microcontrollers are either 3 bits (8 levels) or 4 bits (16 levels). When there is less than 8 bits of priority level, the LSBs are not implemented. So, if you write priority level values like 0x03, 0x07, and so forth to these registers, the value will become 0x00.

Continued

Table E.8 Other Possible Problems *Continued*

Situations	Possible Causes
SVC instruction result in fault exception	The Cortex-M3 processor does not support recursive exception— an exception cannot preempt unless it is higher priority than the current level. As a result, you cannot use SVC within an SVC, hard fault or NMI handler, or any other exception handlers which have the same or higher priority than the SVC exception.
Parameters passing to SVC corrupted	When passing parameters to exception handlers like SVC, the extraction of parameters (R0–R3) should be carried by getting the parameters from the stack frame instead of using values on the register bank. This is because there could be a chance that another exception was processed just before entering the SVC (new arrival exception case). Because the other exception handler can change R0–R3, R12 (AAPCS does not require a C function to keep R0–R3, R12 unchanged), the values of R0–R3 and R12 can be undefined when entering the SVC handler. To obtain the parameters correctly, the stacked data should be used. This involved using a simple SVC wrapper in assembly to extract the correct stack pointer and pass it on to the C handler as a C pointer. Example code of this can be found in chapter 11 of this book and ARM application note AN179. A similar arrangement can be found in returning data from the exception handler to the interrupted program. The handler should store the return data into the stack frame. Otherwise, the value in the register bank will be overwritten during unstacking.
SYSTICK exception occur after clearing TICKINT	The TICKINT bit in the SYSTICK Control and Status register enables and disables the generation of SYSTICK exception. However, if the SYSTICK exception is already in pending state, clearing of TICKINT will not stop the SYSTICK exception from getting fired. To ensure the SYSTICK exception will not be generated, you need to clear TICKINT and the SYSTICK pending status in the Interrupt Control and Status register in the NVIC.
JTAG locked out	In many Cortex-M3-based microcontrollers, the JTAG and I/O pins are shared. If the I/O functions for these pins are enabled right in the start of the program, you could find that you will not be able to debug or erase the Flash again.
Unexpected extra interrupt	Some microcontrollers have a write buffer in the bus bridge for the peripheral bus. This makes it possible for an exception handler to clear a peripheral interrupt by writing to the peripherals, exiting the handler, and then entering the interrupt again as the peripheral cannot deassert the interrupt request fast enough. There are several work-arounds for this problem: **1.** The interrupt service routine (ISR) could carry out a dummy access to the peripheral before exception exit. However, this can increase the duration of the ISR. **2.** The clearing of the interrupt can be moved to the beginning of the ISR so that the interrupt is cleared before the ISR end. This might not work if the ISR duration is shorter than the buffered write delay, so extensive testing should be carried out for various peripheral clock ratios.

Table E.8 Other Possible Problems *Continued*

Situations	Possible Causes
Problem with using the normal interrupt as software interrupt	Some users might intend to use unassigned exception types available in NVIC for software interrupt functions. However, the external interrupt and the PendSV exceptions are imprecise. That means the exception handler might not happen immediately after pending it. To handle this, a polling loop can be used to ensure that the exception is carried out.
Unexpected BLX or BX instructions which switched to ARM state	For users of GNU assembler, if a function symbol is from a different file, you need to ensure the function name is declared as a function:

	`.text`
	`.global my_function_name`
	`.type my_function_name, %function`

	Otherwise, a branch or call to this function might result in accidental switching to ARM state. In addition, during linking stage, you might need "-mcortex-m3 -mthumb" options for the GNU linker. Otherwise, the GNU linker might pull in the wrong version of C library code.
Unexpected disabling of interrupt	The behavior of the Cortex-M3 and ARM7TDMI processors are different regarding exception return. In ARM7TDMI, if the interrupt is disabled inside an interrupt handler, it will be automatically reenabled at exception return due to restore of CPSR (I bit). For the Cortex-M3 processor, if you disable interrupts manually using PRIMASK (e.g., "CPSID I" instruction or __disable_irq()), you will need to reenable it at a later stage inside the interrupt handler. Otherwise, the PRIMASK register will remain set after exception return and will block all interrupts from occuring.
Unexpected unaligned accesses	The Cortex-M3 processor supports unaligned transfers on single load/store. Normally C compilers do not generate unaligned transfer except for packed structures and manual manipulation of pointers. For programming in assembly language, it could be a bigger issue as a user might accidentally use unaligned transfers and not know it. For example, when a program reads a peripheral with word transfer of an unaligned address, the lowest two bits of the addresses might be ignored when using other ARM processors (as the AHB to APB bridge might force these two bits to 0). In the case of the Cortex-M3 processor, if the same software code is used, it will divide the unaligned transfer into multiple aligned transfers. This could end up with different results. Equally, issues can be found when porting software for the Cortex-M3 processor to other ARM processors that do not support unaligned transfers. It is easy to detect if the software generates unaligned transfers. This can be done by setting the UNALIGN_TRP bit in the NVIC Configuration Control register in the System Control Block. By doing this, a usage fault will be triggered when an unaligned transfer happens, and you can then eliminate all unexpected unaligned transfers.

Example Linker Script
for CodeSourcery G++

F.1 EXAMPLE LINKER SCRIPT FOR CORTEX-M3

The following linker script is modified from a generic linker script included in CodeSourcery G++ Lite (generic-m.ld), which is the target for Cortex™-M3 processors. This linker script assumes that the CS3 start up sequence and vector table is used, and therefore, it is toolchain specific. For other GNU C compilers, please refer to the documentation and examples provided in the package.

[cortexm3.ld – Used in Example 5 and Example 6 of Chapter 19]

```
/* Linker script for generic-m
 *
 * Version:Sourcery G++ Lite 2009q1-161
 * BugURL:https://support.codesourcery.com/GNUToolchain/
 *
 * Copyright 2007, 2008 CodeSourcery, Inc.
 *
 * The authors hereby grant permission to use, copy, modify, distribute,
 * and license this software and its documentation for any purpose, provided
 * that existing copyright notices are retained in all copies and that this
 * notice is included verbatim in any distributions. No written agreement,
 * license, or royalty fee is required for any of the authorized uses.
 * Modifications to this software may be copyrighted by their authors
 * and need not follow the licensing terms described here, provided that
 * the new terms are clearly indicated on the first page of each file where
 * they apply.
 * */
OUTPUT_FORMAT ("elf32-littlearm", "elf32-bigarm", "elf32-littlearm")
ENTRY(_start)
SEARCH_DIR(.)
GROUP(-lgcc -lc -lcs3 -lcs3unhosted -lcs3micro)
MEMORY
{
  /* ROM is a readable (r), executable region (x)        */
  rom (rx) : ORIGIN = 0, LENGTH = 32k

  /* RAM is a readable (r), writable (w) and              */
  /* executable region (x)                                */
  ram (rwx) : ORIGIN = 0x20000000, LENGTH = 16k
}
```

433

```
/*   These force the linker to search for particular symbols from
 *   the start of the link process and thus ensure the user's
 *   overrides are picked up
 */
EXTERN(__cs3_reset_generic_m)
INCLUDE micro-names.inc
EXTERN(__cs3_interrupt_vector_micro)
EXTERN(__cs3_start_c main __cs3_stack __cs3_heap_end)

PROVIDE(__cs3_heap_start = _end);
PROVIDE(__cs3_heap_end = __cs3_region_start_ram + __cs3_region_size_ram);
PROVIDE(__cs3_region_num = (__cs3_regions_end - __cs3_regions) / 20);
PROVIDE(__cs3_stack = __cs3_region_start_ram + __cs3_region_size_ram);

SECTIONS
{
  .text :
  {
    CREATE_OBJECT_SYMBOLS
    __cs3_region_start_rom = .;
    *(.cs3.region-head.rom)
    ASSERT (. == __cs3_region_start_rom, ".cs3.region-head.rom not permitted");

    /* Vector table */
    __cs3_interrupt_vector = __cs3_interrupt_vector_micro;
    *(.cs3.interrupt_vector) /* vector table */
    /* Make sure we pulled in an interrupt vector. */
    ASSERT (. != __cs3_interrupt_vector_micro, "No interrupt vector");

    /* Map CS3 vector symbols to handler names in C */
    __cs3_reset         = Reset_Handler;
    __cs3_isr_nmi       = NMI_Handler;
    __cs3_isr_hard_fault = HardFault_Handler;

    *(.text .text.* .gnu.linkonce.t.*)
    *(.plt)
    *(.gnu.warning)
    *(.glue_7t) *(.glue_7) *(.vfp11_veneer)

    *(.ARM.extab* .gnu.linkonce.armextab.*)
    *(.gcc_except_table)
  } >rom
  .eh_frame_hdr : ALIGN (4)
  {
    KEEP (*(.eh_frame_hdr))
  } >rom
  .eh_frame : ALIGN (4)
  {
    KEEP (*(.eh_frame))
  } >rom
  /* .ARM.exidx is sorted, so has to go in its own output section. */
  __exidx_start = .;
  .ARM.exidx :
```

```
{
  *(.ARM.exidx* .gnu.linkonce.armexidx.*)
} >rom
__exidx_end = .;
.rodata : ALIGN (4)
{
  *(.rodata .rodata.* .gnu.linkonce.r.*)

  . = ALIGN(4);
  KEEP(*(.init))

  . = ALIGN(4);
  __preinit_array_start = .;
  KEEP (*(.preinit_array))
  __preinit_array_end = .;

  . = ALIGN(4);
  __init_array_start = .;
  KEEP (*(SORT(.init_array.*)))
  KEEP (*(.init_array))
  __init_array_end = .;

  . = ALIGN(4);
  KEEP(*(.fini))

  . = ALIGN(4);
  __fini_array_start = .;
  KEEP (*(.fini_array))
  KEEP (*(SORT(.fini_array.*)))
  __fini_array_end = .;

  . = ALIGN(0x4);
  KEEP (*crtbegin.o(.ctors))
  KEEP (*(EXCLUDE_FILE (*crtend.o) .ctors))
  KEEP (*(SORT(.ctors.*)))
  KEEP (*crtend.o(.ctors))

  . = ALIGN(0x4);
  KEEP (*crtbegin.o(.dtors))
  KEEP (*(EXCLUDE_FILE (*crtend.o) .dtors))
  KEEP (*(SORT(.dtors.*)))
  KEEP (*crtend.o(.dtors))

  /* Add debug information
  . = ALIGN(4);
  __my_debug_regions = .;
  LONG (__cs3_heap_start)
  LONG (__cs3_heap_end)
  LONG (__cs3_stack) */

  . = ALIGN(4);
  __cs3_regions = .;
  LONG (0)
  LONG (__cs3_region_init_ram)
  LONG (__cs3_region_start_ram)
```

```
    LONG (__cs3_region_init_size_ram)
    LONG (__cs3_region_zero_size_ram)
    __cs3_regions_end = .;

    . = ALIGN (8);
    *(.rom)
    *(.rom.b)
    _etext = .;
} >rom
.data : ALIGN (8)
{
    __cs3_region_start_ram = .;
    _data = .;
    *(.cs3.region-head.ram)
    KEEP(*(.jcr))
    *(.got.plt) *(.got)
    *(.shdata)
    *(.data .data.* .gnu.linkonce.d.*)
    . = ALIGN (8);
    *(.ram)
    _edata = .;
} >ram AT>rom
.bss :
{
    _bss = .;
    *(.shbss)
    *(.bss .bss.* .gnu.linkonce.b.*)
    *(COMMON)
    . = ALIGN (8);
    *(.ram.b)
    _ebss = .;
    _end = .;
    __end = .;
} >ram AT>rom
__cs3_region_init_ram = LOADADDR (.data);
__cs3_region_init_size_ram = _edata - ADDR (.data);
__cs3_region_zero_size_ram = _end - _edata;
__cs3_region_size_ram = LENGTH(ram);


.stab           0 (NOLOAD) : { *(.stab) }
.stabstr        0 (NOLOAD) : { *(.stabstr) }
/* DWARF debug sections.
 * Symbols in the DWARF debugging sections are relative to the beginning
 * of the section so we begin them at 0. */
/* DWARF 1 */
.debug          0 : { *(.debug) }
.line           0 : { *(.line) }
/* GNU DWARF 1 extensions */
.debug_srcinfo  0 : { *(.debug_srcinfo) }
.debug_sfnames  0 : { *(.debug_sfnames) }
/* DWARF 1.1 and DWARF 2 */
.debug_aranges  0 : { *(.debug_aranges) }
```

```
    .debug_pubnames  0 : { *(.debug_pubnames) }
    /* DWARF 2 */
    .debug_info      0 : { *(.debug_info .gnu.linkonce.wi.*) }
    .debug_abbrev    0 : { *(.debug_abbrev) }
    .debug_line      0 : { *(.debug_line) }
    .debug_frame     0 : { *(.debug_frame) }
    .debug_str       0 : { *(.debug_str) }
    .debug_loc       0 : { *(.debug_loc) }
    .debug_macinfo   0 : { *(.debug_macinfo) }
    /* DWARF 2.1 */
    .debug_ranges    0 : { *(.debug_ranges) }
    /* SGI/MIPS DWARF 2 extensions */
    .debug_weaknames 0 : { *(.debug_weaknames) }
    .debug_funcnames    0 : { *(.debug_funcnames) }
    .debug_typenames    0 : { *(.debug_typenames) }
    .debug_varnames     0 : { *(.debug_varnames) }
    .note.gnu.arm.ident 0 : { KEEP (*(.note.gnu.arm.ident)) }
    .ARM.attributes    0 : { KEEP (*(.ARM.attributes)) }
    /DISCARD/ : { *(.note.GNU-stack) }
}
```

CMSIS Core Access Functions Reference

The Cortex™ Microcontroller Software Interface Standard (CMSIS) contains the following standardized functions:

- Core peripheral access functions
- Intrinsic functions

In this appendix, the basic information about these standardized functions is covered. Some of the functions in CMSIS use the standard data types defined in "*stdint.h*". For example,

Table G.1 Standard Data Types Used in CMSIS

Type	Description
uint32_t	Unsigned 32-bit integer
uint16_t	Unsigned 16-bit integer
uint8_t	Unsigned 8-bit integer

G.1 EXCEPTION AND INTERRUPT NUMBERS

A number of functions in CMSIS use interrupt numbers to access interrupt features. The interrupt number definition is different from the processor Interrupt Status register (IPSR) definition. In CMSIS, peripheral interrupts start from value of 0, and negative numbers are used to indicate system exceptions.

Table G.2 Exception and Interrupt Number

CMSIS Interrupt Number	Exception Number In Processor (IPSR)	Exception	Exception Type Name (enum) – "IRQn_Type"	Exception Handler Name
—	—	Reset	—	Reset_Handler
−14	2	NMI	NonMaskableInt_IRQn	NMI_Handler

Continued

Table G.2 Exception and Interrupt Number *Continued*

CMSIS Interrupt Number	Exception Number In Processor (IPSR)	Exception	Exception Type Name (enum) – "IRQn_Type"	Exception Handler Name
–13	3	Hard fault	—	HardFault_Handler
–12	4	Memory Management fault	MemoryManagement_IRQn	MemManage_Handler
–11	5	Bus fault	BusFault_IRQn	BusFault_Handler
–10	6	Usage fault	UsageFault_IRQn	UsageFault_Handler
–5	11	SVC	SVCall_IRQn	SVC_Handler
–4	12	Debug monitor	DebugMonitor_IRQn	DebugMon_Handler
–2	14	PendSV	PendSV_IRQn	PendSV_Handler
–1	15	SysTick	SysTick_IRQn	SysTick_Handler
0	16	Peripheral interrupt 0	<MCU specific>	<MCU specific>
1	17	Peripheral interrupt 1	<MCU specific>	<MCU specific>
2	18	Peripheral interrupt 2	<MCU specific>	<MCU specific>
…	…	…	<MCU specific>	<MCU specific>

G.2 NVIC ACCESS FUNCTIONS

The following functions are available for NVIC feature accesses.

Function Name	void NVIC_SetPriorityGrouping(uint32_t PriorityGroup)
Description	Set the priority grouping in NVIC Interrupt Controller. (This function is not available on the Cortex-M0/M1.)
Parameter	PriorityGroup is priority grouping field.
Return	None

Function Name	uint32_t NVIC_GetPriorityGrouping(void)
Description	Get the priority grouping from NVIC Interrupt Controller. (This function is not available on Cortex-M0/M1.)
Parameter	None
Return	Priority grouping field.

Function Name	void NVIC_EnableIRQ(IRQn_Type IRQn)
Description	Enable Interrupt in NVIC Interrupt Controller.
Parameter	IRQn_Type IRQn specifies the positive interrupt number. It cannot be system exception.
Return	None

Function Name	void NVIC_DisableIRQ(IRQn_Type IRQn)
Description	Disable Interrupt in NVIC Interrupt Controller.
Parameter	IRQn_Type IRQn is the positive number of the external interrupt. It cannot be system exception.
Return	None

Function Name	uint32_t NVIC_GetPendingIRQ(IRQn_Type IRQn)
Description	Read the interrupt pending bit for a device-specific interrupt source.
Parameter	IRQn_Type IRQn is the number of the device-specific interrupt. This function does not support system exception.
Return	1 if pending interrupt, else 0.

Function Name	void NVIC_SetPendingIRQ(IRQn_Type IRQn)
Description	Set the pending bit for an external interrupt.
Parameter	IRQn_Type IRQn is the number of the interrupt. This function does not support system exception.
Return	None

Function Name	void NVIC_ClearPendingIRQ(IRQn_Type IRQn)
Description	Clear the pending bit for an external interrupt.
Parameter	IRQn_Type IRQn is the number of the interrupt. This function does not support system exception.
Return	None

Function Name	uint32_t NVIC_GetActive(IRQn_Type IRQn)
Description	Read the active bit for an external interrupt. (This function is not available on Cortex-M0/M1.)
Parameter	IRQn_Type IRQn is the number of the interrupt. This function does not support system exception.
Return	1 if active, else 0.

Function Name	void NVIC_SetPriority(IRQn_Type IRQn, uint32_t priority)
Description	Set the priority for an interrupt or system exception with programmable priority level.
Parameter	IRQn_Type IRQn is the number of the interrupt. unint32_t priority is the priority for the interrupt. This function automatically shifts the input priority value left to put priority value in implemented bits.
Return	None

Function Name	uint32_t NVIC_GetPriority(IRQn_Type IRQn)
Description	Read the priority for an interrupt or system exception with programmable priority level.
Parameter	IRQn_Type IRQn is the number of the interrupt.
Return	Return value (type uint32_t) is the priority for the interrupt. This function automatically shifts the input priority value right to remove unimplemented bits in the priority value register.

Function Name	uint32_t NVIC_EncodePriority (uint32_t PriorityGroup, uint32_t PreemptPriority, uint32_t SubPriority)
Description	Encode the priority for an interrupt: Encode the priority for an interrupt with the given priority group, preemptive priority value, and subpriority value. In case of a conflict between priority grouping and available priority bits (__NVIC_PRIO_BITS), the smallest possible priority group is set. (This function is not available on Cortex-M0/M1.)
Parameter	PriorityGroup is the used priority group. PreemptPriority is the preemptive priority value (starting from 0). SubPriority is the subpriority value (starting from 0).
Return	The priority for the interrupt.

Function Name	void NVIC_DecodePriority (uint32_t Priority, uint32_t PriorityGroup, uint32_t* pPreemptPriority, uint32_t* pSubPriority)
Description	Decode the priority of an interrupt: Decode an interrupt priority value with the given priority group to preemptive priority value and subpriority value. In case of a conflict between priority grouping and available priority bits (__NVIC_PRIO_BITS), the smallest possible priority group is set. (This function is not available on Cortex-M0/M1.)
Parameter	Priority is the priority for the interrupt. PriorityGroup is the used priority group. pPreemptPriority is the preemptive priority value (starting from 0). pSubPriority is the subpriority value (starting from 0).
Return	None

G.3 SYSTEM AND SYSTICK FUNCTIONS

The following functions are for system setup.

Function Name	void SystemInit (void)
Description	Initialize the system
Parameter	None
Return	None

Function Name	void SystemCoreClockUpdate(void)
Description	Update the *SystemCoreClock* variable. This function should be used each time after the processor clock frequency is changed. This function is introduced from CMSIS version 1.30. Earlier version of CMSIS do not have this function and use a different variable called *SystemFrequency* for timing information.
Parameter	None
Return	None

Function Name	void NVIC_SystemReset(void)
Description	Initiate a system reset request
Parameter	None
Return	None

Function Name	uint32_t SysTick_Config(uint32_t ticks)
Description	Initialize and start the SysTick counter and its interrupt. This function program the SysTick to generate SysTick exception for every "ticks" number of core clock cycles.
Parameter	*ticks* is the number of clock ticks between two interrupts.
Return	None

G.4 CORE REGISTERS ACCESS FUNCTIONS

The following functions are for accessing special registers in the processor core.

Function Name	Description
uint32_t __get_MSP(void)	Get MSP value
void __set_MSP(uint32_t topOfMainStack)	Change MSP value

Continued

Function Name	Description
uint32_t __get_PSP(void)	Get PSP value
void __set_PSP(uint32_t topOfProcStack)	Change PSP value
uint32_t __get_BASEPRI(void)	Get BASEPRI value
void __set_BASEPRI(uint32_t basePri)	Change BASEPRI value
uint32_t __get_PRIMASK(void)	Get PRIMASK value
void __set_PRIMASK(uint32_t priMask)	Change PRIMASK value
uint32_t __get_FAULTMASK(void)	Get FAULTMASK value
void __set_FAULTMASK(uint32_t faultMask)	Change FAULTMASK value
uint32_t __get_CONTROL(void)	Get CONTROL value
void __set_CONTROL(uint32_t control)	Change CONTROL value

G.5 CMSIS INTRINSIC FUNCTIONS

The CMSIS provides a number of intrinsic functions for access to instructions that cannot be generated by ISO/IEC C. The function "__enable_fault_irq" and "__disable_fault_irq" are not available for Cortex-M0/M1.

Functions for system features.

Function Name	Instruction	Description
void __WFI(void)	WFI	Wait for interrupt (sleep)
void __WFE(void)	WFE	Wait for event (sleep)
void __SEV(void)	SEV	Send event
void __enable_irq(void)	CPSIE i	Enable interrupt (clear PRIMASK)
void __disable_irq(void)	CPSID i	Disable interrupt (set PRIMASK)
void __enable_fault_irq(void)	CPSIE f	Enable interrupt (clear FAULTMASK)
void __disable_fault_irq(void)	CPSID f	Disable interrupt (set FAULTMASK)
void __NOP(void)	NOP	No operation
void __ISB(void)	ISB	Instruction synchronisation barrier
void __DSB(void)	DSB	Data synchronisation barrier
void __DMB(void)	DMB	Data memory barrier

Functions for exclusive memory accesses shown in the next table – these functions are not available on Cortex-M0/M1. Functions for data processing – The "__RBIT" function in the table on the next page is not available for Cortex-M0/M1.

Function Name	Instruction	Description
uint8_t __LDREXB(uint8_t *addr)	LDREXB	Exclusive load byte
uint16_t __LDREXH(uint16_t *addr)	LDREXH	Exclusive load half word
uint32_t __LDREXW(uint32_t *addr)	LDREX	Exclusive load word
uint32_t __STREXB(uint8_t value, uint8_t *addr)	STREXB	Exclusive store byte. Return value is the access status (success = 0, failed = 1).
uint32_t __STREXH(uint16_t value, uint8_t *addr)	STREXH	Exclusive store half word. Return value is the access status (success = 0, failed = 1).
uint32_t __STREXW(uint32_t value, uint8_t *addr)	STREX	Exclusive store word. Return value is the access status (success = 0, failed = 1).
void __CLREX(void)	CLREX	Reset exclusive lock created by exclusive read.

Function Name	Instruction	Description
uint32_t __REV(uint32_t value)	REV	Reverse byte order inside a word.
uint32_t __REV16(uint32_t value)	REV16	Reverse byte order inside each of the two half words.
uint32_t __REVSH(uint32_t value)	REVSH	Reverse byte order in the lower half word and then extend the result to 32-bit.
uint32_t __RBIT(uint32_t value)	RBIT	Reverse bit order in the word.

G.6   DEBUG MESSAGE OUTPUT FUNCTION

A debug message output function is defined to use ITM for message output.

Function Name	uint32_t ITM_putchar(uint32_t chr)
Description	Output a character through the ITM output channel 0. When no debugger is connected, the function returns immediately. If debugger is connected and instrumentation trace is enabled, the function outputs the character to ITM and stalls if the ITM is still busy on the last transfer.
Parameter	"chr" is the character to be output.
Return	The output character "chr".

Connectors for Debug and Tracers

H.1 OVERVIEW

A number of commonly used debug connectors are shown here. Most of the ARM development tools use one of these pins out. When developing your ARM circuit board, it is recommended to use a standard debug signal arrangement to make connection to the debugger easier.

H.2 THE 20-PIN CORTEX DEBUG + ETM CONNECTOR

Newer ARM microcontroller boards use a 0.05" 20 pin header (Samtec FTSH-120) for both debug and trace. (The signals greyed out in the following figures are not available on the Cortex™-M3.)

The 20-pin Cortex Debug + ETM connector supports both JTAG and Serial-Wire debug protocols (see Figures H.1 and H.2). When the Serial debug protocol is used, the TDO signal can be used for Serial-Wire Viewer (SWV) output for trace capture. The connector also provides a 4-bit wide trace port for capturing of trace that requires a higher trace bandwidth (e.g., when ETM trace is enabled).

The FTSH-120 connector is smaller than the traditional IDC connector and is recommended for new designs. An example development board that uses this new connector is the Keil MCBSTM32E evaluation board.

FIGURE H.1

The 20-Pin Cortex Debug + ETM Connector.

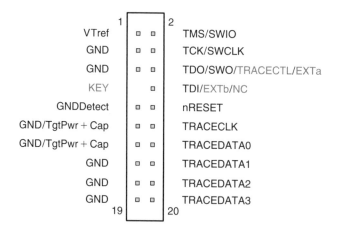

FIGURE H.2

The 20-Pin Cortex Debug + ETM Connector Pin Layout.

FIGURE H.3

The 10-Pin Cortex Debug Connector.

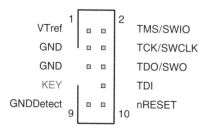

FIGURE H.4

The 10-Pin Cortex Debug Connector Pin Layout.

H.3 THE 10-PIN CORTEX DEBUG CONNECTOR

For devices without ETM, you can use an even smaller 0.05" 10-pin connector for debug. Similar to the 20-pin Cortex Debug + ETM connector, both JTAG and Serial-Wire debug protocols are supported in the 10-pin version (see Figures H.3 and H.4).

H.4 LEGACY 20-PIN IDC CONNECTOR

A common debug connector used in ARM development boards is the 20-pin IDC connector (see Figure H.5). The 20 pin IDC connector arrangement support JTAG debug, Serial-Wire debug (SWIO and SWCLK), and SWV. The nICEDETECT pin allows the target system to detect if a debugger is connected. When no debugger is attached, this pin is pulled high. A debugger connection connects this pin to the ground. This is used in some development boards that support multiple JTAG configurations. The nSRST connection is optional; debugger can reset a Cortex-M3 system through the NVIC so this connection is often omitted from the top level of microcontroller designs.

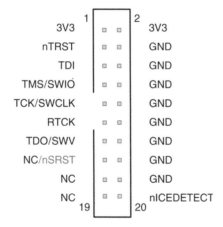

FIGURE H.5

The 20-Pin IDC Connector.

H.5 LEGACY 38-PIN MICTOR CONNECTOR

In some ARM system designs, a Mictor connector is used when trace port is required (e.g., for instruction trace with ETM; see Figure H.6). It can also be used for JTAG/SWD connection. The 20-pin IDC connector can be connected in parallel with the Mictor connector (only one is used at a time).

1	NC
3	NC
5	GND
7	Pulldown
9	NC/nSRST
11	TDO/SWV
13	RTCK
15	TCK/SWCLK
17	TMS/SWIO
19	TDI
21	nTRST
23	0/TRACEDATA[15]
25	0/TRACEDATA[14]
27	0/TRACEDATA[13]
29	0/TRACEDATA[12]
31	0/TRACEDATA[11]
33	0/TRACEDATA[10]
35	0/TRACEDATA[9]
37	0/TRACEDATA[8]

NC	2
NC	4
TRACECLK	6
Pulldown	8
Pulldown	10
Pullup (Vref)	12
VSupply	14
0/TRACEDATA[7]	16
0/TRACEDATA[6]	18
0/TRACEDATA[5]	20
0/TRACEDATA[4]	22
TRACEDATA[3]	24
TRACEDATA[2]	26
TRACEDATA[1]	28
0	30
0	32
1	34
0/TRACECTRL	36
TRACEDATA[0]	38

FIGURE H.6

The 38-Pin Mictor Connector.

Typically, a Cortex-M3 microcontroller only has 4 bits of trace data signals, so most of the trace data pins on the Mictor connectors are not used. The Mictor connector is used mostly in other ARM Cortex processors (Cortex-A8/A9, Cortex-R4); in some multiprocessor systems the trace system might require a wider trace port. In such cases, some of the other unused pins on the connector will also be used. For Cortex-M3 systems, the Cortex Debug + ETM connector is recommended.

References

1. *Cortex-M3 Technical Reference Manual (TRM)*: downloadable from the ARM documentation web site at *http://infocenter.arm.com/help/topic/com.arm.doc.ddi0337g/index.html*

2. *ARMv7-M Architecture Application Level Reference Manual*: downloadable from the ARM documentation web site at *www.arm.com/products/CPUs/ARM_Cortex-M3_v7.html*

3. *CoreSight Technology System Design Guide*: downloadable from the ARM documentation web site at *http://infocenter.arm.com/help/topic/com.arm.doc.dgi0012b/index.html*

4. *AMBA Specification*: downloadable from the ARM documentation web site at *www.arm.com/products/solutions/AMBA_Spec.html*

5. *AAPCS Procedure Call Standard for the ARM Architecture*: downloadable from the ARM documentation web site at *http://infocenter.arm.com/help/topic/com.arm.doc.ihi0042c/index.html*

6. *RVCT 4.0 Compilation Tools Compiler User Guide*: downloadable from the ARM documentation web site at *http://infocenter.arm.com/help/topic/com.arm.doc.dui0205i/index.html*

7. *ARM Application Note 179: Cortex-M3 Embedded Software Development*: downloadable from the ARM documentation web site at *http://infocenter.arm.com/help/topic/com.arm.doc.dai0179b/index.html*

8. *RVCT 4.0 Compilation Tools Compiler Reference Guide*: downloadable from the ARM documentation web site at *http://infocenter.arm.com/help/topic/com.arm.doc.dui0348b/index.html*

Index

A

AAPCS (Procedure Call Standard for ARM Architecture),
    159, 204
    assembly code and C program interactions, 170
    double-word stack alignment, 204
Access port (AP), 245
AFSR (Auxiliary Fault Status Register), 126, 416, 421
AHB (Advanced High-performance Bus), 80, 101, 146, 207
    AHB-AP, 102, 245, 264–265
    AHB-to-APB, 102, 107
    in BE-8 Big Endian mode, 95, 96
    BusMatrix, 102, 105–107
    error responses, causes, 121
    in word-invariant big endian, 95, 96
AIRCR (NVIC Application Interrupt and Reset Control
    Register), 113, 125, 241, 254, 412
AMBA (Advanced Microcontroller Bus Architecture), 101, 244
APB (Advanced Peripheral Bus), 80, 101, 104, 244
    APB-AP, 246
API (Application Programming Interface), 126, 193
APSR (Application Program Status Register), 29, 279, 358,
    359
    flag bits for conditional branches, 62
    and MSR instruction, 29, 55
    signed saturation results, 69
    updating instructions, 58
    with traditional Thumb instruction syntax, 45
ARM Architecture Reference Manual, The, 8
ATB (Advanced Trace Bus), 103, 246, 255, 256
ATB funnel, 246, 256
Auxiliary Control Register, 275, 277, 281

B

Background region (MPU), 212, 225
BASEPRI, 16, 30
    special register, 136–137
    use, 31
BFAR (Bus Fault Address Register), 122, 421
BFSR (Bus Fault Status Register), 121, 122, 152, 153,
    415, 426
Big Endian
    in ARM7, 95, 96, 284
    in Cortex-M3, 95, 96
    memory views, 95
Bit band
    alias, 79, 80, 289
    vs. bit bang, 87
    operations, 84–91
    semaphore operation, 179–180

Breakpoint, 21, 262
    in cortex-M3, 251–253
    and Flash Patch, 21, 103, 253, 255, 262–264
    Insert/Remove breakpoint, 323
Bus Fault, 121–122
    precise and imprecise, 122
    stacking error, 152
    status register, 121, 122, 152, 153, 415, 426
    unstacking error, 153
BusMatrix, 102, 105–107
Byte-invariant big endian, 95, 96

C

CFSR (Configurable Fault Status Register), 421
CMSIS (Cortex Microcontroller Software Interface
    Standard), 67, 95, 164–165, 185
    areas of standardization, 165–166
    benefit of, 168–169
    core access functions, 186, 439–445
    example, 168
    intrinsic functions, 167, 444–445
    MPU register names in, 218
    organization of, 166
    port existing applications using, 334
    stopwatch example with interrupts, 327–333
Context Switching, 127
    example, 128
    in simple OS, 203
CONTROL (one of the special registers), 14, 31–32
CoreSight architecture, 21, 255
    debugging hardware, 21
    overview, 244–248
Cortex-A8, 5, 7
Cortex-M0, 278–281, 444
Cortex-M3
    advantages, 1–2, 18, 22–24, 277–278
    applications, 9, 334
    barrier instructions, 67
    bit-band operation
        advantages, 87–90
        in C programs, 90
        of different data sizes, 90
    breakpoint instruction in, 251–253
    bus faults, 121
    bus interfaces on, 17–18, 104–105
    connection of AHB-AP in, 265
    data transfers, 53
    debugging components, 11
    debugging features, 243

Cortex-M3 *Continued*
    debugging functions in, 244
    debugging support, 21–22
    debug modes in, 248–250
    debug systems in, 247
    default configuration, 83
    default ROM table values, 266
    differences between Cortex-M0, 278–281
    differences among other versions, 272–277
    ETM in, 260–261
    exception types and enables, 407–408
    instructions, 19, 57, 58, 60, 70, 349
    interrupt and exceptions, 19–21, 35–36
    linker script for, 433–437
    link register (LR), 28
    memory attributes 82–83
    memory map, 16–17, 79–82
    MPU, 18
      registers, 212–217
    multiprocessor communication, 236–241
    nested interrupt support in, 148
    NVIC in, 15–16
    operation modes, 14–15, 32–34, 285
    priority levels, 111–116
    privilege levels in, 15
    processor-based microcontrollers, 2
    program counter, 28
    registers, 12–14, 25–26, 29–32
    reset types and signals on, 107–108
    simple timer, 141
    sleep modes, 232–234
    stack memory operations, 36–40
    stack pointer (SP) in, 26–28
    supporting endian modes, 95–97
    tail chaining interrupt, 148–149
    three-stage pipeline in, 99–101
    trace interfaces in, 246
    trace system in, 255–256
    troubleshooting guide, 421
    unaligned transfers in, 92–93
    vector table definition in CS3, 302
    vs. Cortex-M3-based MCUs, 3
Cortex-R4, 5, 7
CPI (Cycle Per Instruction), 257
CS3, 301, 302
CYCCNT (Cycle Counter in DWT), 256, 257

D

DAP (Debug Access Port), 21, 102, 104, 244, 245
D-Code bus, 17, 103, 273
Data abort, 121
Debug registers
    DCRDR (Debug Core Register Data Register), 253,
      254, 419

DCRSR (Debug Core Register Selector Register), 253,
    254, 419
DEMCR (Debug Exception and Monitor Control Register),
    249, 250, 419–420
DFSR (Debug Fault Status Register), 252, 254, 416, 428
DHCSR (Debug Halting Control and Status Register), 248,
    249, 418
DP (Debug Port), 21, 244, 245
DWT (Data Watchpoint and Trace unit), 21, 80, 102, 256–258
    and ETM, 260
    and ITM, 260

E

Embedded Assembler, 163–164, 197, 288, 423
EPSR (Execution Program Status Register), 29, 152
ETM (Embedded Trace Macrocell), 21, 80, 102, 246, 256,
    260–261, 267
Exception exit, 119, 147–148
Exception Return, 148, 149–151
Exceptions
    ARM7TDMI mapping, 285
    configuration registers, 137–138
    exception handler, 14, 33, 88, 117, 121, 147, 149, 189,
      327
    exits, 147–148
    fault exceptions, 120–126
    handling, 19, 36, 125, 148, 149, 152, 204
    and interrupts, 19–21, 35
    PendSV, 126–129
    PRIMASK register, 135–136
    priority levels, 111–117
    priority setup, 185
    register updates, 147
    return value, 149–151
    stacking, 145–147, 408
    SVC, 126–129
    SYSTICK, 141, 229, 232, 328
    types, 35, 109–111, 407
    vector, 117, 147
    vector table, 36, 117–118
Exclusive accesses, 93–95
    for semaphores, 177–179
EXC_RETURN, 147, 149–151, 153, 202

F

FAULTMASK, 14, 16, 30, 31, 135–136, 210
FPB (Flash Patch and Breakpoint Unit), 21, 103, 253,
    255, 262–264

H

Halt mode debug, 250, 251, 254
Hard fault
    avoiding lockup, 210

priority level, 111
    status register, 125, 416, 428
HFSR (Hard Fault Status Register), 125, 416, 428
High registers, 25

I

I-Code interface bus, 17, 103
ICI (Interrupt-Continuable Instructions)
    bit field in PSR, 30
Inline assembler, 163–164, 198–199, 288, 305
Instruction Barrier (ISB), 67
Instruction trace, 12, 21
    ETM, 102, 260
Instrumentation Trace, 172
Intellectual property (IP) licensing, 3
Interrupt latency, 16, 22, 23, 152, 207
Interrupt return, 147–148, 284, 287
Intrinsic functions, 135, 163, 165, 167,
    444–445
IPSR (Interrupt Program Status Register), 29, 168, 206
IRQ (Interrupt Request), 20, 131, 189
IT (IF-THEN), 65, 152, 393–394
    assembler language, 65–66
    Thumb-2 instructions, 70–72
ITM (Instrumentation Trace Macrocell)
    ATB interface, 105
    debugging component, 22, 258–260
    functionalities, 258–259
    hardware trace, 260
    software trace, 259
    timestamp feature, 260

L

LabVIEW, 335–336
    for ARM porting, 345–347
    application areas, 337
    development of, 337–339
    features in, 344–345
    project, example of, 339–343
    working, 343–344
Literal pool, 263
Load/store operations, 84, 152, 287, 427
Lockup, 422
    situations, 208–210
Low registers, 25
LR (link register), 149
    branch and link instructions, 60
    R14, 13, 28
    saving, 62
    stacking, 145, 146
    update, 147, 149
    value, 421
LSU (Load Store Unit), 257

M

Memory Barrier Instructions, 67
Memory Management fault, 122–123, 137
    MMAR, 416, 421
    and MPU violation, 152, 218
    status register, 415
Memory Map, 16–17, 67, 79–82, 83, 103, 161–163, 211,
    284, 325
MFSR (Memory-management Fault Status Register), 123,
    152, 426
MMAR (Memory-management Fault Address Register),
    416, 421
Monitor exception, 21, 35, 110, 248, 251–253
MPU (Memory Protection Unit), 6, 9, 11, 18, 83, 102,
    122, 211
    registers, 212–217
    setup, 218–224
    system characteristics, 285
MSP (Main Stack Pointer), 12, 26, 28, 39, 40, 145, 183
MSTKERR (Memory Management Stacking Error), 152,
    426
MUNSTKERR (Memory Management Unstacking Error),
    153, 426

N

NMI (nonmaskable interrupt), 2, 23, 35
    double fault situations, 209
    and FIQ, 286
Nonbase Thread Enable, 205–206, 413
NVIC (Nested Vectored Interrupt Controller), 131
    accessing, 186
    and CPU core, 101–103
    DCRDR, 253, 254
    DCRSR, 253, 254
    debugging features, 254
    enabling and disabling interrupts, 187
    fault status register, 121–122, 123, 124
    features, 15–16
    registers, 409
    ROM table, 265–266
    SCS, 81, 131
    System Control register, 232
    SYSTICK registers, 141–143, 229

P

PC (Program Counter)
    R15, 13, 28
    register updation, 147
    stacked PC, 421
    value, 288
PendSV
    context switching, 128
    and SVC, 126–129

Pipeline, 99–100, 288
PPB (Private Peripheral Bus), 18
    AHB, 80
    APB, 80
    external PPB, 104–105
Preempt Priority, 113, 114, 115, 116
Prefetch abort, 121
PRIMASK, 29, 135–136, 178
    function, 14
    interrupt masking, 16, 30, 31
Priority Group, 113, 114, 115, 116, 132, 193
Privileged mode, 70, 131, 178, 205
Profiling (Data Watchpoint and Trace unit),
    256–258
PSP (Process Stack Pointer)
    ARM documentation, 26, 28
    MRS and MSR instructions, 40
    stacking, 145
    two-stack model, 39–41
PSR (Program Status register), 29, 145
    APSR, 29
    bit fields, 30
    EPSR, 29
    flags, 62
    IPSR, 29, 146, 147

Q

Q flag, 62, 69, 387

R

R13/SP, 28
Real time, 4
Reset
    control, 254
    fault handling method, 125
    self-reset control, 241–242
    signals, 107–108
    vector, 41, 46, 295
Reset sequence, 41–42
Retargeting, 302–304, 315, 317
ROM Table, 103, 265–267
RXEV (Receive Event), 232, 237

S

Saturation
    instructions, 68, 69
    operation, 68–70
Semaphores
    bit band, usage, 179–180
    exclusive access, usage, 93, 177–179, 287
Serial-Wire Viewer, 172, 257
Serial-Wire, 102, 244, 245
Sleep modes, 20, 23, 232–234, 276
Sleep-On-Exit, 234

Software Trace (Instrumentation Trace
    Macrocell), 259
Special registers, 14, 29, 70
    accessing, 304
    BASEPRI, 14, 30–31, 136–137
    control register, 31–32
    FAULTMASK, 14, 30–31, 135–136
    for MRS and MSR instructions, 71
    PRIMASK, 14, 30–31, 135–136
    PSRs, 29–30
Stack alignment, 204, 275, 277
Stack Pointer (SP), 204, 206
    R13, 12, 26–28
    stack memory operations, 36
    types, 26, 39
    updating, 147
Stacking
    error, 152
    exception sequence, 145–147
STIR (Software Trigger Interrupt register), 131, 141, 420
STKERR (stacking error), 152
Subpriority, 113, 114
Subregion, 215, 225
SVC (Supervisor Call), 126–129, 193, 206, 210
    handler, 205
    for output functions, 194–197
    and SWI, 127
    user applications, 193–194
    using with C, 197–199
SWI (Software Interrupt Instruction), 127, 287
SWJ-DP, Serial Wire JTAG – Debug Port, 21, 102, 274
System Control register, 233, 413
System Control Space (SCS), 32, 81, 131
SYSTICK
    context switching, 127
    registers, 141–143
    stopwatch, example, 328
    Timer, 102, 141–143, 229–232, 277

T

Table Branch, 75–77, 181
    and SVC, 194
Timestamp, 260
TPIU (Trace Port Interface Unit), 21, 103, 246,
    255, 261
Trace Enable (TRCENA), 256
    debug, 250
    in DEMCR, 259, 262
TXEV (Transmit Event), 105, 236

U

UFSR (Usage Fault Status Register), 124, 415, 427–428
Unaligned transfers, 92–93
    and D-Code bus, 103

Unified Assembler Language (UAL), 49–50
Unstacking
    and bus fault, 121
    error, 153
    interrupt return instruction, 147–148
UNSTKERR (Unstacking error), 153
Usage fault, 123–124, 137, 153
User mode, 131, 205

V

Vector catch (Debug event), 249
Vector fetch, 121, 147, 153, 207
Vector Table Offset register, 117, 132, 279, 412
Vector table relocation, 190–193
Vector table, 36, 190

and exceptions, 117–118
difference in traditional ARM cores, 286
modification, 326–327
remapping, 284
setup and enabling interrupt, 184–188
Virtual instrument (VI), 336, 337, 339, 340, 346

W

WIC (Wakeup Interrupt Controller), 21, 102, 234–236, 276, 277
Word-invariant big endian, 95, 96

X

xPSR – combined Program Status Register (PSR), 14, 29, 204, 287